PRINCIPLES OF QUANTUM MECHANICS
as Applied to Chemistry and Chemical Physics

This text presents a rigorous mathematical account of the principles of quantum mechanics, in particular as applied to chemistry and chemical physics. Applications are used as illustrations of the basic theory.

The first two chapters serve as an introduction to quantum theory, although it is assumed that the reader has been exposed to elementary quantum mechanics as part of an undergraduate physical chemistry or atomic physics course. Following a discussion of wave motion leading to Schrödinger's wave mechanics, the postulates of quantum mechanics are presented along with the essential mathematical concepts and techniques. The postulates are rigorously applied to the harmonic oscillator, angular momentum, the hydrogen atom, the variation method, perturbation theory, and nuclear motion. Modern theoretical concepts such as hermitian operators, Hilbert space, Dirac notation, and ladder operators are introduced and used throughout.

This advanced text is appropriate for beginning graduate students in chemistry, chemical physics, molecular physics, and materials science.

A native of the state of New Hampshire, Donald Fitts developed an interest in chemistry at the age of eleven. He was awarded an A.B. degree, *magna cum laude* with highest honors in chemistry, in 1954 from Harvard University and a Ph.D. degree in chemistry in 1957 from Yale University for his theoretical work with John G. Kirkwood. After one-year appointments as a National Science Foundation Postdoctoral Fellow at the Institute for Theoretical Physics, University of Amsterdam, and as a Research Fellow at Yale's Chemistry Department, he joined the faculty of the University of Pennsylvania, rising to the rank of Professor of Chemistry.

In Penn's School of Arts and Sciences, Professor Fitts also served as Acting Dean for one year and as Associate Dean and Director of the Graduate Division for fifteen years. His sabbatical leaves were spent in Britain as a NATO Senior Science Fellow at Imperial College, London, as an Academic Visitor in Physical Chemistry, University of Oxford, and as a Visiting Fellow at Corpus Christi College, Cambridge.

He is the author of two other books, *Nonequilibrium Thermodynamics* (1962) and *Vector Analysis in Chemistry* (1974), and has published research articles on the theory of optical rotation, statis' transport processes, nonequilibrium thermodyn mechanics, theory of liquids, intermolecular forces,

D0813510

UNH LIBRARY

3 4600 01087 8517

PRINCIPLES OF
QUANTUM MECHANICS

as Applied to Chemistry and Chemical Physics

DONALD D. FITTS

University of Pennsylvania

CAMBRIDGE
UNIVERSITY PRESS

PUBLISHED BY THE PRESS SYNDICATE OF THE UNIVERSITY OF CAMBRIDGE
The Pitt Building, Trumpington Street, Cambridge, United Kingdom

CAMBRIDGE UNIVERSITY PRESS
The Edinburgh Building, Cambridge CB2 2RU, UK   www.cup.cam.ac.uk
40 West 20th Street, New York, NY 10011-4211, USA   www.cup.org
10 Stamford Road, Oakleigh, Melbourne 3166, Australia

© D. D. Fitts 1999

This book is in copyright. Subject to statutory exception
and to the provisions of relevant collective licensing agreements,
no reproduction of any part may take place without
the written permission of Cambridge University Press.

First published 1999

Printed in the United Kingdom at the University Press, Cambridge

Typeset in Times 11/14pt, in 3B2 [KT]

A catalogue record for this book is available from the British Library

Library of Congress Cataloguing in Publication data
Fitts, Donald D., 1932–
Principles of quantum mechanics: as applied to chemistry and
chemical physics / Donald D. Fitts.
p.   cm.
Includes bibliographical references and index
ISBN 0 521 65124 7 (hc.). – ISBN 0 521 65841 1 (pbk.)
1. Quantum chemistry. I. Title.
QD462.F55   1999
541.2'8–dc21   98-39486 CIP

ISBN 0 521 65124 7 hardback
ISBN 0 521 65841 1 paperback

Chem
QD
462
. F55
1999

Contents

Physical constants

Preface

This book is intended as a text for a first-year physical-chemistry or chemical-physics graduate course in quantum mechanics. Emphasis is placed on a rigorous mathematical presentation of the principles of quantum mechanics with applications serving as illustrations of the basic theory. The material is normally covered in the first semester of a two-term sequence and is based on the graduate course that I have taught from time to time at the University of Pennsylvania. The book may also be used for independent study and as a reference throughout and beyond the student's academic program.

The first two chapters serve as an introduction to quantum theory. It is assumed that the student has already been exposed to elementary quantum mechanics and to the historical events that led to its development in an undergraduate physical chemistry course or in a course on atomic physics. Accordingly, the historical development of quantum theory is not covered. To serve as a rationale for the postulates of quantum theory, Chapter 1 discusses wave motion and wave packets and then relates particle motion to wave motion. In Chapter 2 the time-dependent and time-independent Schrödinger equations are introduced along with a discussion of wave functions for particles in a potential field. Some instructors may wish to omit the first or both of these chapters or to present abbreviated versions.

Chapter 3 is the heart of the book. It presents the postulates of quantum mechanics and the mathematics required for understanding and applying the postulates. This chapter stands on its own and does not require the student to have read Chapters 1 and 2, although some previous knowledge of quantum mechanics from an undergraduate course is highly desirable.

Chapters 4, 5, and 6 discuss basic applications of importance to chemists. In all cases the eigenfunctions and eigenvalues are obtained by means of raising and lowering operators. There are several advantages to using this ladder operator technique over the older procedure of solving a second-order differential equation by the series solution method. Ladder operators provide practice for the student in operations that are used in more advanced quantum theory and in advanced statistical mechanics. Moreover, they yield the eigenvalues and eigenfunctions more simply and more directly without the need to introduce generating functions and recursion relations and to consider asymptotic behavior and convergence. Although there is no need to invoke Hermite, Legendre, and Laguerre polynomials when using ladder operators, these functions are nevertheless introduced in the body of the chapters and their properties are discussed in the appendices. For traditionalists, the series-solution method is presented in an appendix.

Chapters 7 and 8 discuss spin and identical particles, respectively, and each chapter introduces an additional postulate. The treatment in Chapter 7 is limited to spin one-half particles, since these are the particles of interest to chemists. Chapter 8 provides the link between quantum mechanics and statistical mechanics. To emphasize that link, the free-electron gas and Bose–Einstein condensation are discussed. Chapter 9 presents two approximation procedures, the variation method and perturbation theory, while Chapter 10 treats molecular structure and nuclear motion.

The first-year graduate course in quantum mechanics is used in many chemistry graduate programs as a vehicle for teaching mathematical analysis. For this reason, this book treats mathematical topics in considerable detail, both in the main text and especially in the appendices. The appendices on Fourier series and the Fourier integral, the Dirac delta function, and matrices discuss these topics independently of their application to quantum mechanics. Moreover, the discussions of Hermite, Legendre, associated Legendre, Laguerre, and associated Laguerre polynomials in Appendices D, E, and F are more comprehensive than the minimum needed for understanding the main text. The intent is to make the book useful as a reference as well as a text.

I should like to thank Corpus Christi College, Cambridge for a Visiting Fellowship, during which part of this book was written. I also thank Simon Capelin, Jo Clegg, Miranda Fyte, and Peter Waterhouse of the Cambridge University Press for their efforts in producing this book.

Donald D. Fitts

1

The wave function

Quantum mechanics is a theory to explain and predict the behavior of particles such as electrons, protons, neutrons, atomic nuclei, atoms, and molecules, as well as the photon–the particle associated with electromagnetic radiation or light. From quantum theory we obtain the laws of chemistry as well as explanations for the properties of materials, such as crystals, semiconductors, superconductors, and superfluids. Applications of quantum behavior give us transistors, computer chips, lasers, and masers. The relatively new field of molecular biology, which leads to our better understanding of biological structures and life processes, derives from quantum considerations. Thus, quantum behavior encompasses a large fraction of modern science and technology.

Quantum theory was developed during the first half of the twentieth century through the efforts of many scientists. In 1926, E. Schrödinger interjected wave mechanics into the array of ideas, equations, explanations, and theories that were prevalent at the time to explain the growing accumulation of observations of quantum phenomena. His theory introduced the wave function and the differential wave equation that it obeys. Schrödinger's wave mechanics is now the backbone of our current conceptional understanding and our mathematical procedures for the study of quantum phenomena.

Our presentation of the basic principles of quantum mechanics is contained in the first three chapters. Chapter 1 begins with a treatment of plane waves and wave packets, which serves as background material for the subsequent discussion of the wave function for a free particle. Several experiments, which lead to a physical interpretation of the wave function, are also described. In Chapter 2, the Schrödinger differential wave equation is introduced and the wave function concept is extended to include particles in an external potential field. The formal mathematical postulates of quantum theory are presented in Chapter 3.

1.1 Wave motion

Plane wave

A simple stationary harmonic wave can be represented by the equation

$$\psi(x) = \cos\frac{2\pi x}{\lambda}$$

and is illustrated by the solid curve in Figure 1.1. The distance λ between peaks (or between troughs) is called the *wavelength* of the harmonic wave. The value of $\psi(x)$ for any given value of x is called the *amplitude* of the wave at that point. In this case the amplitude ranges from $+1$ to -1. If the harmonic wave is $A\cos(2\pi x/\lambda)$, where A is a constant, then the amplitude ranges from $+A$ to $-A$. The values of x where the wave crosses the x-axis, i.e., where $\psi(x)$ equals zero, are the *nodes* of $\psi(x)$.

If the wave moves without distortion in the positive x-direction by an amount x_0, it becomes the dashed curve in Figure 1.1. Since the value of $\psi(x)$ at any point x on the new (dashed) curve corresponds to the value of $\psi(x)$ at point $x - x_0$ on the original (solid) curve, the equation for the new curve is

$$\psi(x) = \cos\frac{2\pi}{\lambda}(x - x_0)$$

If the harmonic wave moves in time at a constant velocity v, then we have the relation $x_0 = vt$, where t is the elapsed time (in seconds), and $\psi(x)$ becomes

$$\psi(x,\ t) = \cos\frac{2\pi}{\lambda}(x - vt)$$

Suppose that in one second, ν cycles of the harmonic wave pass a fixed point on the x-axis. The quantity ν is called the *frequency* of the wave. The velocity

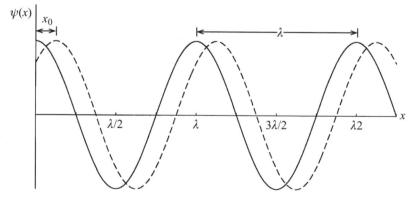

Figure 1.1 A stationary harmonic wave. The dashed curve shows the displacement of the harmonic wave by x_0.

v of the wave is then the product of v cycles per second and λ, the length of each cycle

$$v = v\lambda$$

and $\psi(x, t)$ may be written as

$$\psi(x, t) = \cos 2\pi \left(\frac{x}{\lambda} - vt \right)$$

It is convenient to introduce the *wave number k*, defined as

$$k \equiv \frac{2\pi}{\lambda} \tag{1.1}$$

and the *angular frequency* ω, defined as

$$\omega \equiv 2\pi v \tag{1.2}$$

Thus, the velocity v becomes $v = \omega/k$ and the wave $\psi(x, t)$ takes the form

$$\psi(x, t) = \cos(kx - \omega t)$$

The harmonic wave may also be described by the sine function

$$\psi(x, t) = \sin(kx - \omega t)$$

The representation of $\psi(x, t)$ by the sine function is completely equivalent to the cosine-function representation; the only difference is a shift by $\lambda/4$ in the value of x when $t = 0$. Moreover, *any* linear combination of sine and cosine representations is also an equivalent description of the simple harmonic wave. The most general representation of the harmonic wave is the complex function

$$\psi(x, t) = \cos(kx - \omega t) + \mathrm{i} \sin(kx - \omega t) = e^{\mathrm{i}(kx - \omega t)} \tag{1.3}$$

where i equals $\sqrt{-1}$ and equation (A.31) from Appendix A has been introduced. The real part, $\cos(kx - \omega t)$, and the imaginary part, $\sin(kx - \omega t)$, of the complex wave, (1.3), may be readily obtained by the relations

$$\mathrm{Re}\,[e^{\mathrm{i}(kx - \omega t)}] = \cos(kx - \omega t) = \frac{1}{2}[\psi(x, t) + \psi^*(x, t)]$$

$$\mathrm{Im}\,[e^{\mathrm{i}(kx - \omega t)}] = \sin(kx - \omega t) = \frac{1}{2\mathrm{i}}[\psi(x, t) - \psi^*(x, t)]$$

where $\psi^*(x, t)$ is the complex conjugate of $\psi(x, t)$

$$\psi^*(x, t) = \cos(kx - \omega t) - \mathrm{i} \sin(kx - \omega t) = e^{-\mathrm{i}(kx - \omega t)}$$

The function $\psi^*(x, t)$ also represents a harmonic wave moving in the positive x-direction.

The functions $\exp[\mathrm{i}(kx + \omega t)]$ and $\exp[-\mathrm{i}(kx + \omega t)]$ represent harmonic waves moving in the negative x-direction. The quantity $(kx + \omega t)$ is equal to $k(x + vt)$ or $k(x + x_0)$. After an elapsed time t, the value of the shifted harmonic wave at any point x corresponds to the value at the point $x + x_0$ at time $t = 0$. Thus, the harmonic wave has moved in the negative x-direction.

The moving harmonic wave $\psi(x, t)$ in equation (1.3) is also known as a *plane wave*. The quantity $(kx - \omega t)$ is called the *phase*. The velocity ω/k is known as the *phase velocity* and henceforth is designated by v_{ph}, so that

$$v_{ph} = \frac{\omega}{k} \tag{1.4}$$

Composite wave

A composite wave is obtained by the addition or superposition of any number of plane waves

$$\Psi(x, t) = \sum_{j=1}^{n} A_j e^{i(k_j x - \omega_j t)} \tag{1.5}$$

where A_j are constants. Equation (1.5) is a Fourier series representation of $\Psi(x, t)$. Fourier series are discussed in Appendix B. The composite wave $\Psi(x, t)$ is not a moving harmonic wave, but rather a superposition of n plane waves with different wavelengths and frequencies and with different amplitudes A_j. Each plane wave travels with its own phase velocity $v_{ph,j}$, such that

$$v_{ph,j} = \frac{\omega_j}{k_j}$$

As a consequence, the profile of this composite wave changes with time. The wave numbers k_j may be positive or negative, but we will restrict the angular frequencies ω_j to positive values. A plane wave with a negative value of k has a negative value for its phase velocity and corresponds to a harmonic wave moving in the negative x-direction. In general, the angular frequency ω depends on the wave number k. The dependence of $\omega(k)$ is known as the *law of dispersion* for the composite wave.

In the special case where the ratio $\omega(k)/k$ is the same for each of the component plane waves, so that

$$\frac{\omega_1}{k_1} = \frac{\omega_2}{k_2} = \cdots = \frac{\omega_n}{k_n}$$

then each plane wave moves with the same velocity. Thus, the profile of the composite wave does not change with time even though the angular frequencies and the wave numbers differ. For this *undispersed wave motion*, the angular frequency $\omega(k)$ is proportional to $|k|$

$$\omega(k) = c|k| \tag{1.6}$$

where c is a constant and, according to equation (1.4), is the phase velocity of each plane wave in the composite wave. Examples of undispersed wave motion are a beam of light of mixed frequencies traveling in a vacuum and the undamped vibrations of a stretched string.

For *dispersive wave motion*, the angular frequency $\omega(k)$ is not proportional to $|k|$, so that the phase velocity v_{ph} varies from one component plane wave to another. Since the phase velocity in this situation depends on k, the shape of the composite wave changes with time. An example of dispersive wave motion is a beam of light of mixed frequencies traveling in a dense medium such as glass. Because the phase velocity of each monochromatic plane wave depends on its wavelength, the beam of light is dispersed, or separated onto its component waves, when passed through a glass prism. The wave on the surface of water caused by dropping a stone into the water is another example of dispersive wave motion.

Addition of two plane waves

As a specific and yet simple example of composite-wave construction and behavior, we now consider in detail the properties of the composite wave $\Psi(x, t)$ obtained by the addition or superposition of the two plane waves $\exp[\text{i}(k_1 x - \omega_1 t)]$ and $\exp[\text{i}(k_2 x - \omega_2 t)]$

$$\Psi(x,\ t) = e^{\text{i}(k_1 x - \omega_1 t)} + e^{\text{i}(k_2 x - \omega_2 t)} \tag{1.7}$$

We define the average values $\overline{k}$ and $\overline{\omega}$ and the differences Δk and $\Delta \omega$ for the two plane waves in equation (1.7) by the relations

$$\overline{k} = \frac{k_1 + k_2}{2} \qquad \overline{\omega} = \frac{\omega_1 + \omega_2}{2}$$

$$\Delta k = k_1 - k_2 \qquad \Delta \omega = \omega_1 - \omega_2$$

so that

$$k_1 = \overline{k} + \frac{\Delta k}{2}, \qquad k_2 = \overline{k} - \frac{\Delta k}{2}$$

$$\omega_1 = \overline{\omega} + \frac{\Delta \omega}{2}, \qquad \omega_2 = \overline{\omega} - \frac{\Delta \omega}{2}$$

Using equation (A.32) from Appendix A, we may now write equation (1.7) in the form

$$\Psi(x,\ t) = e^{\text{i}(\overline{k}x - \overline{\omega}t)}[e^{\text{i}(\Delta kx - \Delta \omega t)/2} + e^{-\text{i}(\Delta kx - \Delta \omega t)/2}]$$

$$= 2\cos\left(\frac{\Delta kx - \Delta \omega t}{2}\right) e^{\text{i}(\overline{k}x - \overline{\omega}t)} \tag{1.8}$$

Equation (1.8) represents a plane wave $\exp[\text{i}(\overline{k}x - \overline{\omega}t)]$ with wave number $\overline{k}$, angular frequency $\overline{\omega}$, and phase velocity $\overline{\omega}/\overline{k}$, but with its amplitude modulated by the function $2\cos[(\Delta kx - \Delta \omega t)/2]$. The real part of the wave (1.8) at some fixed time t_0 is shown in Figure 1.2(*a*). The solid curve is the plane wave with wavelength $\lambda = 2\pi/\overline{k}$ and the dashed curve shows the profile of the amplitude of the plane wave. The profile is also a harmonic wave with wavelength

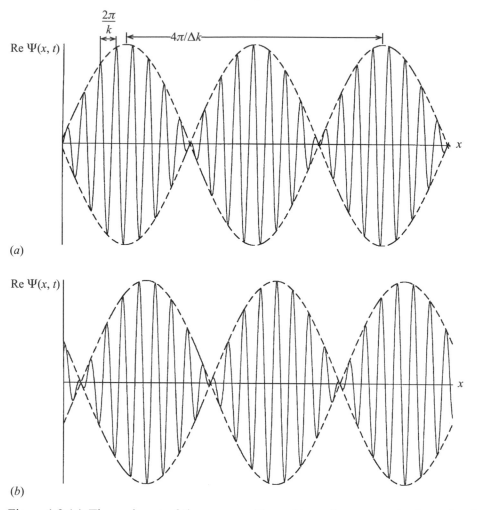

(a)

(b)

Figure 1.2 (a) The real part of the superposition of two plane waves is shown by the solid curve. The profile of the amplitude is shown by the dashed curve. (b) The positions of the curves in Figure 1.2(a) after a short time interval.

$4\pi/\Delta k$. At the points of maximum amplitude, the two original plane waves interfere constructively. At the nodes in Figure 1.2(a), the two original plane waves interfere destructively and cancel each other out.

As time increases, the plane wave $\exp[i(\bar{k}x - \bar{\omega}t)]$ moves with velocity $\bar{\omega}/\bar{k}$. If we consider a fixed point x_1 and watch the plane wave as it passes that point, we observe not only the periodic rise and fall of the amplitude of the unmodified plane wave $\exp[i(\bar{k}x - \bar{\omega}t)]$, but also the overlapping rise and fall of the amplitude due to the modulating function $2\cos[(\Delta kx - \Delta\omega t)/2]$. Without the modulating function, the plane wave would reach the same maximum and the same minimum amplitude with the passage of each cycle. The modulating function causes the maximum (or minimum) amplitude for each cycle of the plane wave to oscillate with frequency $\Delta\omega/2$.

The pattern in Figure 1.2(a) propagates along the x-axis as time progresses. After a short period of time Δt, the wave (1.8) moves to a position shown in Figure 1.2(b). Thus, the position of maximum amplitude has moved in the positive x-direction by an amount $v_g\Delta t$, where v_g is the *group velocity* of the composite wave, and is given by

$$v_g = \frac{\Delta\omega}{\Delta k} \tag{1.9}$$

The expression (1.9) for the group velocity of a composite of two plane waves is exact.

In the special case when k_2 equals $-k_1$ and ω_2 equals ω_1 in equation (1.7), the superposition of the two plane waves becomes

$$\Psi(x,\, t) = e^{i(kx-\omega t)} + e^{-i(kx+\omega t)} \tag{1.10}$$

where

$$k = k_1 = -k_2$$
$$\omega = \omega_1 = \omega_2$$

The two component plane waves in equation (1.10) travel with equal phase velocities ω/k, but in opposite directions. Using equations (A.31) and (A.32), we can express equation (1.10) in the form

$$\Psi(x,\, t) = (e^{ikx} + e^{-ikx})e^{-i\omega t}$$
$$= 2\cos kx\, e^{-i\omega t}$$
$$= 2\cos kx\,(\cos \omega t - i\sin \omega t)$$

We see that for this special case the composite wave is the product of two functions: one only of the distance x and the other only of the time t. The composite wave $\Psi(x,\, t)$ vanishes whenever $\cos kx$ is zero, i.e., when $kx = \pi/2$, $3\pi/2$, $5\pi/2$, $\dots$, regardless of the value of t. Therefore, the nodes of $\Psi(x,\, t)$ are independent of time. However, the amplitude or profile of the composite wave changes with time. The real part of $\Psi(x,\, t)$ is shown in Figure 1.3. The solid curve represents the wave when $\cos \omega t$ is a maximum, the dotted curve when $\cos \omega t$ is a minimum, and the dashed curve when $\cos \omega t$ has an intermediate value. Thus, the wave does not travel, but pulsates, increasing and decreasing in amplitude with frequency ω. The imaginary part of $\Psi(x,\, t)$ behaves in the same way. A composite wave with this behavior is known as a *standing wave*.

Re $\Psi(x, t)$

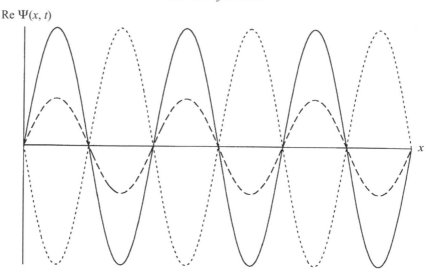

Figure 1.3 A standing harmonic wave at various times.

1.2 Wave packet

We now consider the formation of a composite wave as the superposition of a continuous spectrum of plane waves with wave numbers k confined to a narrow band of values. Such a composite wave $\Psi(x, t)$ is known as a *wave packet* and may be expressed as

$$\Psi(x, t) = \frac{1}{\sqrt{2\pi}} \int_{-\infty}^{\infty} A(k) \mathrm{e}^{\mathrm{i}(kx - \omega t)} \mathrm{d}k \qquad (1.11)$$

The weighting factor $A(k)$ for each plane wave of wave number k is negligible except when k lies within a small interval Δk. For mathematical convenience we have included a factor $(2\pi)^{-1/2}$ on the right-hand side of equation (1.11). This factor merely changes the value of $A(k)$ and has no other effect.

We note that the wave packet $\Psi(x, t)$ is the inverse Fourier transform of $A(k)$. The mathematical development and properties of Fourier transforms are presented in Appendix B. Equation (1.11) has the form of equation (B.19). According to equation (B.20), the Fourier transform $A(k)$ is related to $\Psi(x, t)$ by

$$A(k) = \frac{1}{\sqrt{2\pi}} \int_{-\infty}^{\infty} \Psi(x, t) \mathrm{e}^{-\mathrm{i}(kx - \omega t)} \mathrm{d}x \qquad (1.12)$$

It is because of the Fourier relationships between $\Psi(x, t)$ and $A(k)$ that the factor $(2\pi)^{-1/2}$ is included in equation (1.11). Although the time t appears in the integral on the right-hand side of (1.12), the function $A(k)$ does not depend on t; the time dependence of $\Psi(x, t)$ cancels the factor $\mathrm{e}^{\mathrm{i}\omega t}$. We consider below two specific examples for the functional form of $A(k)$. However, in order to evaluate the integral over k in equation (1.11), we also need to know the dependence of the angular frequency ω on the wave number k.

In general, the angular frequency $\omega(k)$ is a function of k, so that the angular frequencies in the composite wave $\Psi(x, t)$, as well as the wave numbers, vary from one plane wave to another. If $\omega(k)$ is a slowly varying function of k and the values of k are confined to a small range Δk, then $\omega(k)$ may be expanded in a Taylor series in k about some point k_0 within the interval Δk

$$\omega(k) = \omega_0 + \left(\frac{d\omega}{dk}\right)_0 (k - k_0) + \frac{1}{2}\left(\frac{d^2\omega}{dk^2}\right)_0 (k - k_0)^2 + \cdots \tag{1.13}$$

where ω_0 is the value of $\omega(k)$ at k_0 and the derivatives are also evaluated at k_0. We may neglect the quadratic and higher-order terms in the Taylor expansion (1.13) because the interval Δk and, consequently, $k - k_0$ are small. Substitution of equation (1.13) into the phase for each plane wave in (1.11) then gives

$$kx - \omega t \approx (k - k_0 + k_0)x - \omega_0 t - \left(\frac{d\omega}{dk}\right)_0 (k - k_0)t$$

$$= k_0 x - \omega_0 t + \left[x - \left(\frac{d\omega}{dk}\right)_0 t\right](k - k_0)$$

so that equation (1.11) becomes

$$\Psi(x, t) = B(x, t)e^{i(k_0 x - \omega_0 t)} \tag{1.14}$$

where

$$B(x, t) = \frac{1}{\sqrt{2\pi}}\int_{\infty}^{\infty} A(k)e^{i[x-(d\omega/dk)_0 t](k-k_0)} \, dk \tag{1.15}$$

Thus, the wave packet $\Psi(x, t)$ represents a plane wave of wave number k_0 and angular frequency ω_0 with its amplitude modulated by the factor $B(x, t)$. This modulating function $B(x, t)$ depends on x and t through the relationship $[x - (d\omega/dk)_0 t]$. This situation is analogous to the case of two plane waves as expressed in equations (1.7) and (1.8). The modulating function $B(x, t)$ moves in the positive x-direction with group velocity v_g given by

$$v_g = \left(\frac{d\omega}{dk}\right)_0 \tag{1.16}$$

In contrast to the group velocity for the two-wave case, as expressed in equation (1.9), the group velocity in (1.16) for the wave packet is not uniquely defined. The point k_0 is chosen arbitrarily and, therefore, the value at k_0 of the derivative $d\omega/dk$ varies according to that choice. However, the range of k is narrow and $\omega(k)$ changes slowly with k, so that the variation in v_g is small. Combining equations (1.15) and (1.16), we have

$$B(x, t) = \frac{1}{\sqrt{2\pi}} \int_{-\infty}^{\infty} A(k) e^{i(x - v_g t)(k - k_0)} \, dk \qquad (1.17)$$

Since the function $A(k)$ is the Fourier transform of $\Psi(x, t)$, the two functions obey Parseval's theorem as given by equation (B.28) in Appendix B

$$\int_{-\infty}^{\infty} |\Psi(x, t)|^2 dx = \int_{-\infty}^{\infty} |B(x, t)|^2 \, dx = \int_{-\infty}^{\infty} |A(k)|^2 \, dk \qquad (1.18)$$

Gaussian wave number distribution

In order to obtain a specific mathematical expression for the wave packet, we need to select some form for the function $A(k)$. In our first example we choose $A(k)$ to be the gaussian function

$$A(k) = \frac{1}{\sqrt{2\pi}\alpha} e^{-(k - k_0)^2 / 2\alpha^2} \qquad (1.19)$$

This function $A(k)$ is a maximum at wave number k_0, which is also the average value for k for this distribution of wave numbers. Substitution of equation (1.19) into (1.17) gives

$$|\Psi(x, t)| = B(x, t) = \frac{1}{\sqrt{2\pi}} e^{-\alpha^2 (x - v_g t)^2 / 2} \qquad (1.20)$$

where equation (A.8) has been used. The resulting modulating factor $B(x, t)$ is also a gaussian function–following the general result that the Fourier transform of a gaussian function is itself gaussian. We have also noted in equation (1.20) that $B(x, t)$ is always positive and is therefore equal to the absolute value $|\Psi(x, t)|$ of the wave packet. The functions $A(k)$ and $|\Psi(x, t)|$ are shown in Figure 1.4.

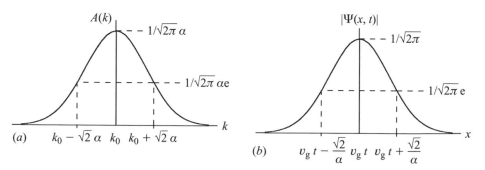

Figure 1.4 (*a*) A gaussian wave number distribution. (*b*) The modulating function corresponding to the wave number distribution in Figure 1.4(*a*).

Figure 1.5 shows the real part of the plane wave $\exp[i(k_0 x - \omega_0 t)]$ with its amplitude modulated by $B(x, t)$ of equation (1.20). The plane wave moves in the positive x-direction with phase velocity v_{ph} equal to ω_0/k_0. The maximum amplitude occurs at $x = v_g t$ and propagates in the positive x-direction with group velocity v_g equal to $(\mathrm{d}\omega/\mathrm{d}k)_0$.

The value of the function $A(k)$ falls from its maximum value of $(\sqrt{2\pi}\alpha)^{-1}$ at k_0 to $1/e$ of its maximum value when $|k - k_0|$ equals $\sqrt{2}\alpha$. Most of the area under the curve (actually 84.3%) comes from the range

$$-\sqrt{2}\alpha < (k - k_0) < \sqrt{2}\alpha$$

Thus, the distance $\sqrt{2}\alpha$ may be regarded as a measure of the width of the distribution $A(k)$ and is called the *half width*. The half width may be defined using $1/2$ or some other fraction instead of $1/e$. The reason for using $1/e$ is that the value of k at that point is easily obtained without consulting a table of numerical values. These various possible definitions give different numerical values for the half width, but all these values are of the same order of magnitude. Since the value of $|\Psi(x, t)|$ falls from its maximum value of $(2\pi)^{-1/2}$ to $1/e$ of that value when $|x - v_g t|$ equals $\sqrt{2}/\alpha$, the distance $\sqrt{2}/\alpha$ may be considered the half width of the wave packet.

When the parameter α is small, the maximum of the function $A(k)$ is high and the function drops off in value rapidly on each side of k_0, giving a small value for the half width. The half width of the wave packet, however, is large because it is proportional to $1/\alpha$. On the other hand, when the parameter α is large, the maximum of $A(k)$ is low and the function drops off slowly, giving a large half width. In this case, the half width of the wave packet becomes small.

If we regard the uncertainty Δk in the value of k as the half width of the distribution $A(k)$ and the uncertainty Δx in the position of the wave packet as its half width, then the product of these two uncertainties is

$$\Delta x \Delta k = 2$$

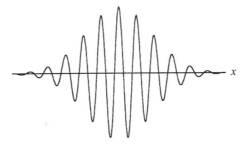

Figure 1.5 The real part of a wave packet for a gaussian wave number distribution.

Thus, the product of these two uncertainties Δx and Δk is a constant of order unity, independent of the parameter α.

Square pulse wave number distribution

As a second example, we choose $A(k)$ to have a constant value of unity for k between k_1 and k_2 and to vanish elsewhere, so that

$$A(k) = 1, \qquad k_1 \leqslant k \leqslant k_2$$
$$= 0, \qquad k < k_1, \; k > k_2 \tag{1.21}$$

as illustrated in Figure 1.6(*a*). With this choice for $A(k)$, the modulating function $B(x, t)$ in equation (1.17) becomes

$$B(x, t) = \frac{1}{\sqrt{2\pi}} \int_{k_1}^{k_2} e^{i(x - v_g t)(k - k_0)} \, dk$$

$$= \frac{1}{\sqrt{2\pi} i (x - v_g t)} [e^{i(x - v_g t)(k_2 - k_0)} - e^{i(x - v_g t)(k_1 - k_0)}]$$

$$= \frac{1}{\sqrt{2\pi} i (x - v_g t)} [e^{i(x - v_g t)\Delta k/2} - e^{-i(x - v_g t)\Delta k/2}]$$

$$= \sqrt{\frac{2}{\pi}} \frac{\sin[(x - v_g t)\Delta k/2]}{x - v_g t} \tag{1.22}$$

where k_0 is chosen to be $(k_1 + k_2)/2$, Δk is defined as $(k_2 - k_1)$, and equation (A.33) has been used. The function $B(x, t)$ is shown in Figure 1.6(*b*).

The real part of the wave packet $\Psi(x, t)$ obtained from combining equations (1.14) and (1.22) is shown in Figure 1.7. The amplitude of the plane wave $\exp[i(k_0 x - \omega_0 t)]$ is modulated by the function $B(x, t)$ of equation (1.22), which has a maximum when $(x - v_g t)$ equals zero, i.e., when $x = v_g t$. The nodes of $B(x, t)$ nearest to the maximum occur when $(x - v_g t)\Delta k/2$ equals $\pm \pi$, i.e., when x is $\pm(2\pi/\Delta k)$ from the point of maximum amplitude. If we consider the half width of the wave packet between these two nodes as a measure of the uncertainty Δx in the location of the wave packet and the width $(k_2 - k_1)$ of the square pulse $A(k)$ as a measure of the uncertainty Δk in the value of k, then the product of these two uncertainties is

$$\Delta x \Delta k = 2\pi$$

Uncertainty relation

We have shown in the two examples above that the uncertainty Δx in the position of a wave packet is inversely related to the uncertainty Δk in the wave numbers of the constituent plane waves. This relationship is generally valid and

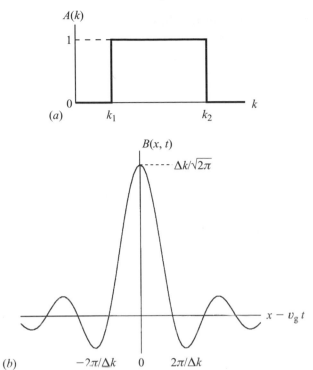

Figure 1.6 (*a*) A square pulse wave number distribution. (*b*) The modulating function corresponding to the wave number distribution in Figure 1.6(*a*).

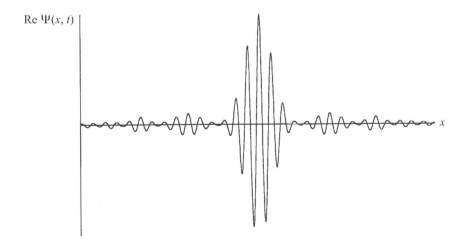

Figure 1.7 The real part of a wave packet for a square pulse wave number distribution.

is a property of Fourier transforms. In order to localize a wave packet so that the uncertainty Δx is very small, it is necessary to employ a broad spectrum of plane waves in equations (1.11) or (1.17). The function $A(k)$ must have a wide distribution of wave numbers, giving a large uncertainty Δk. If the distribution $A(k)$ is very narrow, so that the uncertainty Δk is small, then the wave packet becomes broad and the uncertainty Δx is large.

Thus, for all wave packets the product of the two uncertainties has a lower bound of order unity

$$\Delta x \Delta k \geqslant 1 \tag{1.23}$$

The lower bound applies when the narrowest possible range Δk of values for k is used in the construction of the wave packet, so that the quadratic and higher-order terms in equation (1.13) can be neglected. If a broader range of k is allowed, then the product $\Delta x \Delta k$ can be made arbitrarily large, making the right-hand side of equation (1.23) a lower bound. The actual value of the lower bound depends on how the uncertainties are defined. Equation (1.23) is known as the *uncertainty relation*.

A similar uncertainty relation applies to the variables t and ω. To show this relation, we write the wave packet (1.11) in the form of equation (B.21)

$$\Psi(x,\, t) = \frac{1}{\sqrt{2\pi}} \int_{-\infty}^{\infty} G(\omega) e^{i(kx - \omega t)} \, d\omega \tag{1.24}$$

where the weighting factor $G(\omega)$ has the form of equation (B.22)

$$G(\omega) = \frac{1}{\sqrt{2\pi}} \int_{-\infty}^{\infty} \Psi(x,\, t) e^{-i(kx - \omega t)} \, dt$$

In the evaluation of the integral in equation (1.24), the wave number k is regarded as a function of the angular frequency ω, so that in place of (1.13) we have

$$k(\omega) = k_0 + \left(\frac{dk}{d\omega}\right)_0 (\omega - \omega_0) + \cdots$$

If we neglect the quadratic and higher-order terms in this expansion, then equation (1.24) becomes

$$\Psi(x,\, t) = C(x,\, t) e^{i(k_0 x - \omega_0 t)}$$

where

$$C(x,\, t) = \frac{1}{\sqrt{2\pi}} \int_{-\infty}^{\infty} A(\omega) e^{-i[t - (dk/d\omega)_0 x](\omega - \omega_0)} \, d\omega$$

As before, the wave packet is a plane wave of wave number k_0 and angular frequency ω_0 with its amplitude modulated by a factor that moves in the positive x-direction with group velocity v_g, given by equation (1.16). Following the previous analysis, if we select a specific form for the modulating function $G(\omega)$ such as a gaussian or a square pulse distribution, we can show that the product of the uncertainty Δt in the time variable and the uncertainty $\Delta \omega$ in the angular frequency of the wave packet has a lower bound of order unity, i.e.

$$\Delta t \Delta \omega \geqslant 1 \tag{1.25}$$

This uncertainty relation is also a property of Fourier transforms and is valid for all wave packets.

1.3 Dispersion of a wave packet

In this section we investigate the change in contour of a wave packet as it propagates with time.

The general expression for a wave packet $\Psi(x, t)$ is given by equation (1.11). The weighting factor $A(k)$ in (1.11) is the inverse Fourier transform of $\Psi(x, t)$ and is given by (1.12). Since the function $A(k)$ is independent of time, we may set t equal to any arbitrary value in the integral on the right-hand side of equation (1.12). If we let t equal zero in (1.12), then that equation becomes

$$A(k) = \frac{1}{\sqrt{2\pi}} \int_{-\infty}^{\infty} \Psi(\xi, 0) e^{-ik\xi} \, d\xi \tag{1.26}$$

where we have also replaced the dummy variable of integration by ξ. Substitution of equation (1.26) into (1.11) yields

$$\Psi(x, t) = \frac{1}{2\pi} \iint_{-\infty}^{\infty} \Psi(\xi, 0) e^{i[k(x-\xi)-\omega t]} \, dk \, d\xi \tag{1.27}$$

Since the limits of integration do not depend on the variables ξ and k, the order of integration over these variables may be interchanged.

Equation (1.27) relates the wave packet $\Psi(x, t)$ at time t to the wave packet $\Psi(x, 0)$ at time $t = 0$. However, the angular frequency $\omega(k)$ is dependent on k and the functional form must be known before we can evaluate the integral over k.

If $\omega(k)$ is proportional to $|k|$ as expressed in equation (1.6), then (1.27) gives

$$\Psi(x, t) = \frac{1}{2\pi} \iint_{-\infty}^{\infty} \Psi(\xi, 0) e^{ik(x-ct-\xi)} \, dk \, d\xi$$

The integral over k may be expressed in terms of the Dirac delta function through equation (C.6) in Appendix C, so that we have

$$\Psi(x,\,t) = \int_{-\infty}^{\infty} \Psi(\xi,\,0)\delta(x - ct - \xi)\,\mathrm{d}\xi = \Psi(x - ct,\,0)$$

Thus, the wave packet $\Psi(x,\,t)$ has the same value at point x and time t that it had at point $x - ct$ at time $t = 0$. The wave packet has traveled with velocity c without a change in its contour, i.e., it has traveled without dispersion. Since the phase velocity v_{ph} is given by $\omega_0/k_0 = c$ and the group velocity v_{g} is given by $(\mathrm{d}\omega/\mathrm{d}k)_0 = c$, the two velocities are the same for an undispersed wave packet.

We next consider the more general situation where the angular frequency $\omega(k)$ is not proportional to $|k|$, but is instead expanded in the Taylor series (1.13) about $(k - k_0)$. Now, however, we retain the quadratic term, but still neglect the terms higher than quadratic, so that

$$\omega(k) \approx \omega_0 + v_{\text{g}}(k - k_0) + \gamma(k - k_0)^2$$

where equation (1.16) has been substituted for the first-order derivative and γ is an abbreviation for the second-order derivative

$$\gamma \equiv \frac{1}{2}\left(\frac{\mathrm{d}^2\omega}{\mathrm{d}k^2}\right)_0$$

The phase in equation (1.27) then becomes

$$k(x - \xi) - \omega t = (k - k_0)(x - \xi) + k_0(x - \xi) - \omega_0 t$$

$$- v_{\text{g}} t(k - k_0) - \gamma t(k - k_0)^2$$

$$= k_0 x - \omega_0 t - k_0\xi + (x - v_{\text{g}} t - \xi)(k - k_0) - \gamma t(k - k_0)^2$$

so that the wave packet (1.27) takes the form

$$\Psi_\gamma(x,\,t) = \frac{\mathrm{e}^{\mathrm{i}(k_0 x - \omega_0 t)}}{2\pi} \int\!\!\!\int_{-\infty}^{\infty} \Psi(\xi,\,0)\mathrm{e}^{-\mathrm{i}k_0\xi}\mathrm{e}^{\mathrm{i}(x - v_{\text{g}} t - \xi)(k - k_0) - \mathrm{i}\gamma t(k - k_0)^2}\,\mathrm{d}k\,\mathrm{d}\xi$$

The subscript γ has been included in the notation $\Psi_\gamma(x,\,t)$ in order to distinguish that wave packet from the one in equations (1.14) and (1.15), where the quadratic term in $\omega(k)$ is omitted. The integral over k may be evaluated using equation (A.8), giving the result

$$\Psi_\gamma(x,\,t) = \frac{\mathrm{e}^{\mathrm{i}(k_0 x - \omega_0 t)}}{2\sqrt{\mathrm{i}\pi\gamma t}} \int\!\!\!\int_{-\infty}^{\infty} \Psi(\xi,\,0)\mathrm{e}^{-\mathrm{i}k_0\xi}\mathrm{e}^{-(x - v_{\text{g}} t - \xi)^2/4\mathrm{i}\gamma t}\,\mathrm{d}\xi \qquad (1.28)$$

Equation (1.28) relates the wave packet at time t to the wave packet at time $t = 0$ if the k-dependence of the angular frequency includes terms up to k^2. The profile of the wave packet $\Psi_\gamma(x,\,t)$ changes as time progresses because of the factor $t^{-1/2}$ before the integral and the t in the exponent within the integral. If we select a specific form for the wave packet at time $t = 0$, the nature of this time dependence becomes more evident.

Gaussian wave packet

Let us suppose that $\Psi(x, 0)$ has the gaussian distribution (1.20) as its profile, so that equation (1.14) at time $t = 0$ is

$$\Psi(\xi, 0) = e^{ik_0\xi}B(\xi, 0) = \frac{1}{\sqrt{2\pi}}e^{ik_0\xi}e^{-\alpha^2\xi^2/2} \tag{1.29}$$

Substitution of equation (1.29) into (1.28) gives

$$\Psi_\gamma(x, t) = \frac{e^{i(k_0x-\omega_0t)}}{2\pi\sqrt{2i\gamma t}} \int_{-\infty}^{\infty} e^{-\alpha^2\xi^2/2}e^{-(x-v_gt-\xi)^2/4i\gamma t}\,d\xi$$

The integral may be evaluated using equation (A.8) accompanied with some tedious, but straightforward algebraic manipulations, yielding

$$\Psi_\gamma(x, t) = \frac{e^{i(k_0x-\omega_0t)}}{\sqrt{2\pi(1+2i\alpha^2\gamma t)}}e^{\alpha^2(x-v_gt)^2/2(1+2i\alpha^2\gamma t)} \tag{1.30}$$

The wave packet, then, consists of the plane wave $\exp i[k_0x - \omega_0t]$ with its amplitude modulated by

$$\frac{1}{\sqrt{2\pi(1+2i\alpha^2\gamma t)}}e^{-\alpha^2(x-v_gt)^2/2(1+2i\alpha^2\gamma t)}$$

which is a complex function that depends on the time t. When γ equals zero so that the quadratic term in $\omega(k)$ is neglected, this complex modulating function reduces to $B(x, t)$ in equation (1.20). The absolute value $|\Psi_\gamma(x, t)|$ of the wave packet (1.30) is given by

$$|\Psi_\gamma(x, t)| = \frac{1}{(2\pi)^{1/2}(1+4\alpha^4\gamma^2t^2)^{1/4}}e^{-\alpha^2(x-v_gt)^2/2(1+4\alpha^4\gamma^2t^2)} \tag{1.31}$$

We now contrast the behavior of the wave packet in equation (1.31) with that of the wave packet in (1.20). At any time t, the maximum amplitudes of both occur at $x = v_gt$ and travel in the positive x-direction with the same group velocity v_g. However, at that time t, the value of $|\Psi_\gamma(x, t)|$ is $1/e$ of its maximum value when the exponent in equation (1.31) is unity, so that the half width or uncertainty Δx for $|\Psi_\gamma(x, t)|$ is given by

$$\Delta x = |x - v_gt| = \frac{\sqrt{2}}{\alpha}\sqrt{1+4\alpha^4\gamma^2t^2}$$

Moreover, the maximum amplitude for $|\Psi_\gamma(x, t)|$ at time t is given by

$$(2\pi)^{-1/2}(1+4\alpha^4\gamma^2t^2)^{-1/4}$$

As time increases from $-\infty$ to 0, the half width of the wave packet $|\Psi_\gamma(x,\ t)|$ continuously decreases and the maximum amplitude continuously increases. At $t = 0$ the half width attains its lowest value of $\sqrt{2}/\alpha$ and the maximum amplitude attains its highest value of $1/\sqrt{2\pi}$, and both values are in agreement with the wave packet in equation (1.20). As time increases from 0 to ∞, the half width continuously increases and the maximum amplitude continuously decreases. Thus, as t^2 increases, the wave packet $|\Psi_\gamma(x,\ t)|$ remains gaussian in shape, but broadens and flattens out in such a way that the area under the square $|\Psi_\gamma(x,\ t)|^2$ of the wave packet remains constant over time at a value of $(2\sqrt{\pi}\alpha)^{-1}$, in agreement with Parseval's theorem (1.18).

The product $\Delta x \Delta k$ for this spreading wave packet $\Psi_\gamma(x,\ t)$ is

$$\Delta x \Delta k = 2\sqrt{1 + 4\alpha^4 \gamma^2 t^2}$$

and increases as $|t|$ increases. Thus, the value of the right-hand side when $t = 0$ is the lower bound for the product $\Delta x \Delta k$ and is in agreement with the uncertainty relation (1.23).

1.4 Particles and waves

To explain the photoelectric effect, Einstein (1905) postulated that light, or electromagnetic radiation, consists of a beam of particles, each of which travels at the same velocity c (the speed of light), where c has the value

$$c = 2.997\,92 \times 10^8 \text{ m s}^{-1}$$

Each particle, later named a *photon*, has a characteristic frequency ν and an energy $h\nu$, where h is Planck's constant with the value

$$h = 6.626\,08 \times 10^{-34} \text{ J s}$$

The constant h and the hypothesis that energy is quantized in integral multiples of $h\nu$ had previously been introduced by M. Planck (1900) in his study of blackbody radiation.[1] In terms of the angular frequency ω defined in equation (1.2), the energy E of a photon is

$$E = \hbar\omega \tag{1.32}$$

where $\hbar$ is defined by

$$\hbar \equiv \frac{h}{2\pi} = 1.054\,57 \times 10^{-34} \text{ J s}$$

Because the photon travels with velocity c, its motion is governed by relativity

---

[1] The history of the development of quantum concepts to explain observed physical phenomena, which occurred mainly in the first three decades of the twentieth century, is discussed in introductory texts on physical chemistry and on atomic physics. A much more detailed account is given in M. Jammer (1966) *The Conceptual Development of Quantum Mechanics* (McGraw-Hill, New York).

theory, which requires that its rest mass be zero. The magnitude of the momentum p for a particle with zero rest mass is related to the relativistic energy E by $p = E/c$, so that

$$p = \frac{E}{c} = \frac{h\nu}{c} = \frac{\hbar\omega}{c}$$

Since the velocity c equals ω/k, the momentum is related to the wave number k for a photon by

$$p = \hbar k \tag{1.33}$$

Einstein's postulate was later confirmed experimentally by A. Compton (1924).

Noting that it had been fruitful to regard light as having a corpuscular nature, L. de Broglie (1924) suggested that it might be useful to associate wave-like behavior with the motion of a particle. He postulated that a particle with linear momentum p be associated with a wave whose wavelength λ is given by

$$\lambda = \frac{2\pi}{k} = \frac{h}{p} \tag{1.34}$$

and that expressions (1.32) and (1.33) also apply to particles. The hypothesis of wave properties for particles and the de Broglie relation (equation (1.34)) have been confirmed experimentally for electrons by G. P. Thomson (1927) and by Davisson and Germer (1927), for neutrons by E. Fermi and L. Marshall (1947), and by W. H. Zinn (1947), and for helium atoms and hydrogen molecules by I. Estermann, R. Frisch, and O. Stern (1931).

The classical, non-relativistic energy E for a free particle, i.e., a particle in the absence of an external force, is expressed as the sum of the kinetic and potential energies and is given by

$$E = \frac{1}{2}mv^2 + V = \frac{p^2}{2m} + V \tag{1.35}$$

where m is the mass and v the velocity of the particle, the linear momentum p is

$$p = mv$$

and V is a constant potential energy. The force F acting on the particle is given by

$$F = -\frac{dV}{dx} = 0$$

and vanishes because V is constant. In classical mechanics the choice of the zero-level of the potential energy is arbitrary. Since the potential energy for the free particle is a constant, we may, without loss of generality, take that constant value to be zero, so that equation (1.35) becomes

$$E = \frac{p^2}{2m} \tag{1.36}$$

Following the theoretical scheme of Schrödinger, we associate a wave packet $\Psi(x, t)$ with the motion in the x-direction of this free particle. This wave packet is readily constructed from equation (1.11) by substituting (1.32) and (1.33) for ω and k, respectively

$$\Psi(x, t) = \frac{1}{\sqrt{2\pi\hbar}} \int_{-\infty}^{\infty} A(p) e^{i(px - Et)/\hbar} \, dp \tag{1.37}$$

where, for the sake of symmetry between $\Psi(x, t)$ and $A(p)$, a factor $\hbar^{-1/2}$ has been absorbed into $A(p)$. The function $A(k)$ in equation (1.12) is now $\hbar^{1/2} A(p)$, so that

$$A(p) = \frac{1}{\sqrt{2\pi\hbar}} \int_{-\infty}^{\infty} \Psi(x, t) e^{-i(px - Et)/\hbar} \, dx \tag{1.38}$$

The law of dispersion for this wave packet may be obtained by combining equations (1.32), (1.33), and (1.36) to give

$$\omega(k) = \frac{E}{\hbar} = \frac{p^2}{2m\hbar} = \frac{\hbar k^2}{2m} \tag{1.39}$$

This dispersion law with ω proportional to k^2 is different from that for undispersed light waves, where ω is proportional to k.

If $\omega(k)$ in equation (1.39) is expressed as a power series in $k - k_0$, we obtain

$$\omega(k) = \frac{\hbar k_0^2}{2m} + \frac{\hbar k_0}{m}(k - k_0) + \frac{\hbar}{2m}(k - k_0)^2 \tag{1.40}$$

This expansion is exact; there are no terms of higher order than quadratic. From equation (1.40) we see that the phase velocity v_{ph} of the wave packet is given by

$$v_{\mathrm{ph}} = \frac{\omega_0}{k_0} = \frac{\hbar k_0}{2m} \tag{1.41}$$

and the group velocity v_{g} is

$$v_{\mathrm{g}} = \left(\frac{d\omega}{dk}\right)_0 = \frac{\hbar k_0}{m} \tag{1.42}$$

while the parameter γ of equations (1.28), (1.30), and (1.31) is

$$\gamma = \frac{1}{2}\left(\frac{d^2\omega}{dk^2}\right)_0 = \frac{\hbar}{2m} \tag{1.43}$$

If we take the derivative of $\omega(k)$ in equation (1.39) with respect to k and use equation (1.33), we obtain

$$\frac{d\omega}{dk} = \frac{\hbar k}{m} = \frac{p}{m} = v$$

Thus, the velocity v of the particle is associated with the group velocity v_g of the wave packet

$$v = v_g$$

If the constant potential energy V in equation (1.35) is set at some arbitrary value other then zero, then equation (1.39) takes the form

$$\omega(k) = \frac{\hbar k^2}{2m} + \frac{V}{\hbar}$$

and the phase velocity v_{ph} becomes

$$v_{ph} = \frac{\hbar k_0}{2m} + \frac{V}{\hbar k_0}$$

Thus, both the angular frequency $\omega(k)$ and the phase velocity v_{ph} are dependent on the choice of the zero-level of the potential energy and are therefore arbitrary; neither has a physical meaning for a wave packet representing a particle.

Since the parameter γ is non-vanishing, the wave packet will disperse with time as indicated by equation (1.28). For a gaussian profile, the absolute value of the wave packet is given by equation (1.31) with γ given by (1.43). We note that γ is proportional to m^{-1}, so that as m becomes larger, γ becomes smaller. Thus, for heavy particles the wave packet spreads slowly with time.

As an example, the value of γ for an electron, which has a mass of 9.11×10^{-31} kg, is 5.78×10^{-5} m^2 s^{-1}. For a macroscopic particle whose mass is approximately a microgram, say 9.11×10^{-10} kg in order to make the calculation easier, the value of γ is 5.78×10^{-26} m^2 s^{-1}. The ratio of the macroscopic particle to the electron is 10^{21}. The time dependence in the dispersion terms in equations (1.31) occurs as the product γt. Thus, for the same extent of spreading, the macroscopic particle requires a factor of 10^{21} longer than the electron.

1.5 Heisenberg uncertainty principle

Since a free particle is represented by the wave packet $\Psi(x, t)$, we may regard the uncertainty Δx in the position of the wave packet as the uncertainty in the position of the particle. Likewise, the uncertainty Δk in the wave number is related to the uncertainty Δp in the momentum of the particle by $\Delta k = \Delta p/\hbar$. The uncertainty relation (1.23) for the particle is, then

$$\Delta x \Delta p \geqslant \hbar \qquad (1.44)$$

This relationship is known as the *Heisenberg uncertainty principle*.

The consequence of this principle is that at any instant of time the position of the particle is defined only as a range Δx and the momentum of the particle is defined only as a range Δp. The product of these two ranges or 'uncertainties' is of order $\hbar$ or larger. The exact value of the lower bound is dependent on how the uncertainties are defined. A precise definition of the uncertainties in position and momentum is given in Sections 2.3 and 3.10.

The Heisenberg uncertainty principle is a consequence of the stipulation that a quantum particle is a wave packet. The mathematical construction of a wave packet from plane waves of varying wave numbers dictates the relation (1.44). It is *not* the situation that while the position and the momentum of the particle are well-defined, they cannot be measured simultaneously to any desired degree of accuracy. The position and momentum are, in fact, not simultaneously precisely defined. The more precisely one is defined, the less precisely is the other, in accordance with equation (1.44). This situation is in contrast to classical-mechanical behavior, where both the position and the momentum can, in principle, be specified simultaneously as precisely as one wishes.

In quantum mechanics, if the momentum of a particle is precisely specified so that $p = p_0$ and $\Delta p = 0$, then the function $A(p)$ is

$$A(p) = \delta(p - p_0)$$

The wave packet (1.37) then becomes

$$\Psi(x, t) = \frac{1}{\sqrt{2\pi\hbar}} \int_{-\infty}^{\infty} \delta(p - p_0) e^{i(px - Et)/\hbar} \, dp = \frac{1}{\sqrt{2\pi\hbar}} e^{i(p_0 x - Et)/\hbar}$$

which is a plane wave with wave number $p_0/\hbar$ and angular frequency $E/\hbar$. Such a plane wave has an infinite value for the uncertainty Δx. Likewise, if the position of a particle is precisely specified, the uncertainty in its momentum is infinite.

Another Heisenberg uncertainty relation exists for the energy E of a particle and the time t at which the particle has that value for the energy. The uncertainty $\Delta\omega$ in the angular frequency of the wave packet is related to the uncertainty ΔE in the energy of the particle by $\Delta\omega = \Delta E/\hbar$, so that the relation (1.25) when applied to a free particle becomes

$$\Delta E \Delta t \geqslant \hbar \tag{1.45}$$

Again, this relation arises from the representation of a particle by a wave packet and is a property of Fourier transforms.

The relation (1.45) may also be obtained from (1.44) as follows. The uncertainty ΔE is the spread of the kinetic energies in a wave packet. If Δp is small, then ΔE is related to Δp by

$$\Delta E = \Delta\left(\frac{p^2}{2m}\right) = \frac{p}{m}\Delta p \tag{1.46}$$

The time Δt for a wave packet to pass a given point equals the uncertainty in its position x divided by the group velocity v_g

$$\Delta t = \frac{\Delta x}{v_g} = \frac{\Delta x}{v} = \frac{m}{p}\Delta x \qquad (1.47)$$

Combining equations (1.46) and (1.47), we see that $\Delta E \Delta t = \Delta x \Delta p$. Thus, the relation (1.45) follows from (1.44). The Heisenberg uncertainty relation (1.45) is treated more thoroughly in Section 3.10.

1.6 Young's double-slit experiment

The essential features of the particle–wave duality are clearly illustrated by Young's double-slit experiment. In order to explain all of the observations of this experiment, light must be regarded as having both wave-like and particle-like properties. Similar experiments on electrons indicate that they too possess both particle-like and wave-like characteristics. The consideration of the experimental results leads directly to a physical interpretation of Schrödinger's wave function, which is presented in Section 1.8.

The experimental apparatus is illustrated schematically in Figure 1.8. Monochromatic light emitted from the point source S is focused by a lens L onto a detection or observation screen D. Between L and D is an opaque screen with two closely spaced slits A and B, each of which may be independently opened or closed.

A monochromatic light beam from S passing through the opaque screen with slit A open and slit B closed gives a diffraction pattern on D with an intensity distribution I_A as shown in Figure 1.9(a). In that figure the points A and B are directly in line with slits A and B, respectively. If slit A is closed and slit B open, the intensity distribution of the diffraction pattern is given by the curve labeled I_B in Figure 1.9(a). For an experiment in which slit A is open and slit B is closed half of the time, while slit A is closed and slit B is open the other half of the time, the resulting intensity distribution is the sum of I_A and I_B, as shown in Figure 1.9(b). However, when both slits are open throughout an

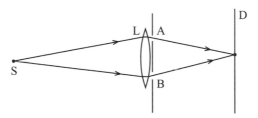

Figure 1.8 Diagram of Young's double-slit experiment.

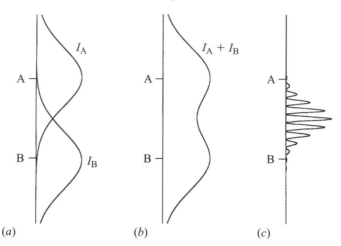

Figure 1.9 (*a*) Intensity distributions I_A from slit A alone and I_B from slit B alone. (*b*) The sum of the intensity distributions I_A and I_B. (c) The intensity interference pattern when slits A and B are open simultaneously.

experiment, an interference pattern as shown in Figure 1.9(*c*) is observed. The intensity pattern in this case is not the sum $I_A + I_B$, but rather an alternating series of bright and dark interference fringes with a bright maximum midway between points A and B. The spacing of the fringes depends on the distance between the two slits.

The wave theory for light provides a satisfactory explanation for these observations. It was, indeed, this very experiment conducted by T. Young (1802) that, in the nineteenth century, led to the replacement of Newton's particle theory of light by a wave theory.

The wave interpretation of the interference pattern observed in Young's experiment is inconsistent with the particle or photon concept of light as required by Einstein's explanation of the photoelectric effect. If the monochromatic beam of light consists of a stream of individual photons, then each photon presumably must pass through either slit A or slit B. To test this assertion, detectors are placed directly behind slits A and B and both slits are opened. The light beam used is of such low intensity that only one photon at a time is emitted by S. In this situation each photon is recorded by either one detector or the other, never by both at once. Half of the photons are observed to pass through slit A, half through slit B in random order. This result is consistent with particle behavior.

How then is a photon passing through only one slit influenced by the other slit to produce an interference pattern? A possible explanation is that somehow photons passing through slit A interact with other photons passing through slit

B and vice versa. To answer this question, Young's experiment is repeated with both slits open and with only one photon at a time emitted by S. The elapsed time between each emission is long enough to rule out any interactions among the photons. While it might be expected that, under these circumstances, the pattern in Figure 1.9(*b*) would be obtained, in fact the interference fringes of Figure 1.9(*c*) are observed. Thus, the same result is obtained regardless of the intensity of the light beam, even in the limit of diminishing intensity.

If the detection screen D is constructed so that the locations of individual photon impacts can be observed (with an array of scintillation counters, for example), then two features become apparent. The first is that only whole photons are detected; each photon strikes the screen D at only one location. The second is that the interference pattern is slowly built up as the cumulative effect of very many individual photon impacts. The behavior of any particular photon is unpredictable; it strikes the screen at a random location. The density of the impacts at each point on the screen D gives the interference fringes. Looking at it the other way around, the interference pattern is the probability distribution of the location of the photon impacts.

If only slit A is open half of the time and only slit B the other half of the time, then the interference fringes are not observed and the diffraction pattern of Figure 1.9(*b*) is obtained. The photons passing through slit A one at a time form in a statistical manner the pattern labeled I_A in Figure 1.9(*a*), while those passing through slit B yield the pattern I_B. If both slits A and B are left open, but a detector is placed at slit A so that we know for certain whether each given photon passes through slit A or through slit B, then the interference pattern is again *not* observed; only the pattern of Figure 1.9(*b*) is obtained. The act of ascertaining through which slit the photon passes has the same effect as closing the other slit.

The several variations on Young's experiment cannot be explained exclusively by a wave concept of light nor by a particle concept. Both wave and particle behavior are needed for a complete description. When the photon is allowed to pass undetected through the slits, it displays wave behavior and an interference pattern is observed. Typical of particle behavior, each photon strikes the detection screen D at a specific location. However, the location is different for each photon and the resulting pattern for many photons is in accord with a probability distribution. When the photon is observed or constrained to pass through a specific slit, whether the other slit is open or closed, the behavior is more like that of a particle and the interference fringes are not observed. It should be noted, however, that the curve I_A in Figure 1.9(*a*) is the diffraction pattern for a wave passing through a slit of width comparable to the wavelength of the wave. Thus, even with only one slit open and with the photons passing through the slit one at a time, wave behavior is observed.

Analogous experiments using electrons instead of photons have been carried out with the same results. Electrons passing through a system with double slits produce an interference pattern. If a detector determines through which slit each electron passes, then the interference pattern is not observed. As with the photon, the electron exhibits both wave-like and particle-like behavior and its location on a detection screen is randomly determined by a probability distribution.

1.7 Stern–Gerlach experiment

Another experiment that relates to the physical interpretation of the wave function was performed by O. Stern and W. Gerlach (1922). Their experiment is a dramatic illustration of a quantum-mechanical effect which is in direct conflict with the concepts of classical theory. It was the first experiment of a non-optical nature to show quantum behavior directly.

In the Stern–Gerlach experiment, a beam of silver atoms is produced by evaporating silver in a high-temperature oven and allowing the atoms to escape through a small hole. The beam is further collimated by passage through a series of slits. As shown in Figure 1.10, the beam of silver atoms then passes through a highly inhomogeneous magnetic field and condenses on a detection plate. The cross-section of the magnet is shown in Figure 1.11. One pole has a very sharp edge in order to produce a large gradient in the magnetic field. The atomic beam is directed along this edge (the z-axis) so that the silver atoms experience a gradient in magnetic field in the vertical or x-direction, but not in the horizontal or y-direction.

Silver atoms, being paramagnetic, have a magnetic moment **M**. In a magnetic field **B**, the potential energy V of each atom is

$$V = -\mathbf{M} \cdot \mathbf{B}$$

Between the poles of the magnet, the magnetic field **B** varies rapidly in the x-

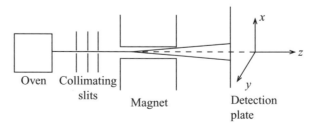

Figure 1.10 Diagram of the Stern–Gerlach experiment.

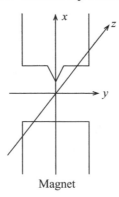

Figure 1.11 A cross-section of the magnet in Figure 1.10.

direction, resulting in a force F_x in the x-direction acting on each silver atom. This force is given by

$$F_x = -\frac{\partial V}{\partial x} = M \cos\theta \frac{\partial B}{\partial x}$$

where M and B are the magnitudes of the vectors $\mathbf{M}$ and $\mathbf{B}$ and θ is the angle between the direction of the magnetic moment and the positive x-axis. Thus, the inhomogeneous magnetic field deflects the path of a silver atom by an amount dependent on the orientation angle θ of its magnetic moment. If the angle θ is between $0°$ and $90°$, then the force is positive and the atom moves in the positive x-direction. For an angle θ between $90°$ and $180°$, the force is negative and the atom moves in the negative x-direction.

As the silver atoms escape from the oven, their magnetic moments are randomly oriented so that all possible values of the angle θ occur. According to classical mechanics, we should expect the beam of silver atoms to form, on the detection plate, a continuous vertical line, corresponding to a gaussian distribution of impacts with a maximum intensity at the center ($x = 0$). The outer limits of this line would correspond to the magnetic moment of a silver atom parallel ($\theta = 0°$) and antiparallel ($\theta = 180°$) to the magnetic field gradient ($\partial B/\partial x$). What is actually observed on the detection plate are two spots, located at each of the outer limits predicted by the classical theory. Thus, the beam of silver atoms splits into two distinct components, one corresponding to $\theta = 0°$, the other to $\theta = 180°$. There are no trajectories corresponding to intermediate values of θ. There is nothing unique or special about the vertical direction. If the magnet is rotated so that the magnetic field gradient is along the y-axis, then again only two spots are observed on the detection plate, but are now located on the horizontal axis.

The Stern–Gerlach experiment shows that the magnetic moment of each silver atom is found only in one of two orientations, either parallel or antiparallel to the magnetic field gradient, even though the magnetic moments of the atoms are randomly oriented when they emerge from the oven. Thus, the possible orientations of the atomic magnetic moment are *quantized*, i.e., only certain discrete values are observed. Since the direction of the quantization is determined by the direction of the magnetic field gradient, the experimental process itself influences the result of the measurement. This feature occurs in other experiments as well and is characteristic of quantum behavior.

If the beam of silver atoms is allowed to pass sequentially between the poles of two or three magnets, additional interesting phenomena are observed. We describe here three such related experimental arrangements. In the first arrangement the collimated beam passes through a magnetic field gradient pointing in the positive x-direction. One of the two exiting beams is blocked (say the one with antiparallel orientation), while the other (with parallel orientation) passes through a second magnetic field gradient which is parallel to the first. The atoms exiting the second magnet are deposited on a detection plate. In this case only one spot is observed, because the magnetic moments of the atoms entering the second magnetic field are all oriented parallel to the gradient and remain parallel until they strike the detection plate.

The second arrangement is the same as the first except that the gradient of the second magnetic field is along the positive y-axis, i.e., it is perpendicular to the gradient of the first magnetic field. For this arrangement, two spots of silver atoms appear on the detection plate, one to the left and one to the right of the vertical x-axis. The beam leaving the first magnet with all the atomic magnetic moments oriented in the positive x-direction is now split into two equal beams with the magnetic moments oriented parallel and antiparallel to the second magnetic field gradient.

The third arrangement adds yet another vertical inhomogeneous magnetic field to the setup of the second arrangement. In this new arrangement the collimated beam of silver atoms coming from the oven first encounters a magnetic field gradient in the positive x-direction, which splits the beam vertically into two parts. The lower beam is blocked and the upper beam passes through a magnetic field gradient in the positive y-direction. This beam is split horizontally into two parts. The left beam is blocked and the right beam is now directed through a magnetic field gradient parallel to the first one, i.e., oriented in the positive x-direction. The resulting pattern on the detection plate might be expected to be a single spot, corresponding to the magnetic moments of all atoms being aligned in the positive x-direction. What is observed in this case, however, are two spots situated on a vertical axis and corresponding to atomic magnetic moments aligned in equal numbers in both the positive and negative *x*-directions. The passage of the atoms through the second magnet apparently realigned their magnetic moments parallel and antiparallel to the positive *y*-axis and thereby destroyed the previous information regarding their alignment by the first magnet.

The original Stern–Gerlach experiment has also been carried out with the same results using sodium, potassium, copper, gold, thallium, and hydrogen atoms in place of silver atoms. Each of these atoms, including silver, has a single unpaired electron among the valence electrons surrounding its nucleus and core electrons. In hydrogen, of course, there is only one electron about the nucleus. The magnetic moment of such an atom is due to the intrinsic angular momentum, called *spin*, of this odd electron. The quantization of the magnetic moment by the inhomogeneous magnetic field is then the quantization of this electron spin angular momentum. The spin of the electron and of other particles is discussed in Chapter 7.

Since the splitting of the atomic beam in the Stern–Gerlach experiment is due to the spin of an unpaired electron, one might wonder why a beam of electrons is not used directly rather than having the electrons attached to atoms. In order for a particle to pass between the poles of a magnet and be deflected by a distance proportional to the force acting on it, the trajectory of the particle must be essentially a classical path. As discussed in Section 1.4, such a particle is described by a wave packet and wave packets disperse with time–the lighter the particle, the faster the dispersion and the greater the uncertainty in the position of the particle. The application of Heisenberg's uncertainty principle to an electron beam shows that, because of the small mass of the electron, it is meaningless to assign a magnetic moment to a free electron. As a result, the pattern on the detection plate from an electron beam would be sufficiently diffuse from interference effects that no conclusions could be drawn [2]. However, when the electron is bound unpaired in an atom, then the atom, having a sufficiently larger mass, has a magnetic moment and an essentially classical path through the Stern–Gerlach apparatus.

1.8 Physical interpretation of the wave function

Young's double-slit experiment and the Stern–Gerlach experiment, as described in the two previous sections, lead to a physical interpretation of the wave function associated with the motion of a particle. Basic to the concept of the wave function is the postulate that the wave function contains all the

[2] This point is discussed in more detail in N. F. Mott and H. S. W. Massey (1965) *The Theory of Atomic Collisions*, 3rd edition, p. 215–16, (Oxford University Press, Oxford).

information that can be known about the particle that it represents. The wave function is a complete description of the quantum behavior of the particle. For this reason, the wave function is often also called the *state* of the system.

In the double-slit experiment, the patterns observed on the detection screen are slowly built up from many individual particle impacts, whether these particles are photons or electrons. The position of the impact of any single particle cannot be predicted; only the cumulative effect of many impacts is predetermined. Accordingly, a theoretical interpretation of the experiment must involve probability distributions rather than specific particle trajectories. The probability that a particle will strike the detection screen between some point x and a neighboring point $x + dx$ is $P(x)\,dx$ and is proportional to the range dx. The larger the range dx, the greater the probability for a given particle to strike the detection screen in that range. The proportionality factor $P(x)$ is called the *probability density* and is a function of the position x. For example, the probability density $P(x)$ for the curve I_A in Figure 1.9(a) has a maximum at the point A and decreases symmetrically on each side of A.

If the motion of a particle in the double-slit experiment is to be represented by a wave function, then that wave function must determine the probability density $P(x)$. For mechanical waves in matter and for electromagnetic waves, the intensity of a wave is proportional to the square of its amplitude. By analogy, the probability density $P(x)$ is postulated to be the square of the absolute value of the wave function $\Psi(x)$

$$P(x) = |\Psi(x)|^2 = \Psi^*(x)\Psi(x)$$

On the basis of this postulate, the interference pattern observed in the double-slit experiment can be explained in terms of quantum particle behavior.

A particle, photon or electron, passing through slit A and striking the detection screen at point x has wave function $\Psi_A(x)$, while a similar particle passing through slit B has wave function $\Psi_B(x)$. Since a particle is observed to retain its identity and not divide into smaller units, its wave function $\Psi(x)$ is postulated to be the sum of the two possibilities

$$\Psi(x) = \Psi_A(x) + \Psi_B(x) \tag{1.48}$$

When only slit A is open, the particle emitted by the source S passes through slit A, thereby causing the wave function $\Psi(x)$ in equation (1.48) to change or *collapse* suddenly to $\Psi_A(x)$. The probability density $P_A(x)$ that the particle strikes point x on the detection screen is, then

$$P_A(x) = |\Psi_A(x)|^2$$

and the intensity distribution I_A in Figure 1.9(a) is obtained. When only slit B is open, the particle passes through slit B and the wave function $\Psi(x)$ collapses to $\Psi_B(x)$. The probability density $P_B(x)$ is then given by

$$P_B(x) = |\Psi_B(x)|^2$$

and curve I_B in Figure 1.9(a) is observed. If slit A is open and slit B closed half of the time, and slit A is closed and slit B open the other half of the time, then the resulting probability density on the detection screen is just

$$P_A(x) + P_B(x) = |\Psi_A(x)|^2 + |\Psi_B(x)|^2$$

giving the curve in Figure 1.9(b).

When both slits A and B are open at the same time, the interpretation changes. In this case, the probability density $P_{AB}(x)$ is

$$\begin{aligned} P_{AB}(x) &= |\Psi_A(x) + \Psi_B(x)|^2 \\ &= |\Psi_A(x)|^2 + |\Psi_B(x)|^2 + \Psi_A^*(x)\Psi_B(x) + \Psi_B^*(x)\Psi_A(x) \\ &= P_A(x) + P_B(x) + \mathscr{T}_{AB}(x) \end{aligned} \qquad (1.49)$$

where

$$\mathscr{T}_{AB}(x) = \Psi_A^*(x)\Psi_B(x) + \Psi_B^*(x)\Psi_A(x)$$

The probability density $P_{AB}(x)$ has an interference term $\mathscr{T}_{AB}(x)$ in addition to the terms $P_A(x)$ and $P_B(x)$. This interference term is real and is positive for some values of x, but negative for others. Thus, the term $\mathscr{T}_{AB}(x)$ modifies the sum $P_A(x) + P_B(x)$ to give an intensity distribution with interference fringes as shown in Figure 1.9(c).

For the experiment with both slits open and a detector placed at slit A, the interaction between the wave function and the detector must be taken into account. Any interaction between a particle and observing apparatus modifies the wave function of the particle. In this case, the wave function has the form of a wave packet which, according to equation (1.37), oscillates with time as $e^{-iEt/\hbar}$. During the time period Δt that the particle and the detector are interacting, the energy of the interacting system is uncertain by an amount ΔE, which, according to the Heisenberg energy–time uncertainty principle, equation (1.45), is related to Δt by $\Delta E \geq \hbar/\Delta t$. Thus, there is an uncertainty in the phase $Et/\hbar$ of the wave function and $\Psi_A(x)$ is replaced by $e^{i\varphi}\Psi_A(x)$, where φ is real. The value of φ varies with each particle–detector interaction and is totally unpredictable. Therefore, the wave function $\Psi(x)$ for a particle in this experiment is

$$\Psi(x) = e^{i\varphi}\Psi_A(x) + \Psi_B(x) \qquad (1.50)$$

and the resulting probability density $P_\varphi(x)$ is

$$P_\varphi(x) = |\Psi_A(x)|^2 + |\Psi_B(x)|^2 + e^{-i\varphi}\Psi_A^*(x)\Psi_B(x) + e^{i\varphi}\Psi_B^*(x)\Psi_A(x)$$

$$= P_A(x) + P_B(x) + \mathscr{T}_\varphi(x) \tag{1.51}$$

where $\mathscr{T}_\varphi(x)$ is defined by

$$\mathscr{T}_\varphi(x) = e^{-i\varphi}\Psi_A^*(x)\Psi_B(x) + e^{i\varphi}\Psi_B^*(x)\Psi_A(x)$$

The interaction with the detector at slit A has changed the interference term from $\mathscr{T}_{AB}(x)$ to $\mathscr{T}_\varphi(x)$.

For any particular particle leaving the source S and ultimately striking the detection screen D, the value of φ is determined by the interaction with the detector at slit A. However, this value is not known and cannot be controlled; for all practical purposes it is a randomly determined and unverifiable number. The value of φ does, however, influence the point x where the particle strikes the detection screen. The pattern observed on the screen is the result of a large number of impacts of particles, each with wave function $\Psi(x)$ in equation (1.50), but with random values for φ. In establishing this pattern, the term $\mathscr{T}_\varphi(x)$ in equation (1.51) averages to zero. Thus, in this experiment the probability density $P_\varphi(x)$ is just the sum of $P_A(x)$ and $P_B(x)$, giving the intensity distribution shown in Figure 1.9(b).

In comparing the two experiments with both slits open, we see that interacting with the system by placing a detector at slit A changes the wave function of the system and the experimental outcome. This feature is an essential characteristic of quantum theory. We also note that without a detector at slit A, there are two indistinguishable ways for the particle to reach the detection screen D and the two wave functions $\Psi_A(x)$ and $\Psi_B(x)$ are added together. With a detector at slit A, the two paths are distinguishable and it is the probability densities $P_A(x)$ and $P_B(x)$ that are added.

An analysis of the Stern–Gerlach experiment also contributes to the interpretation of the wave function. When an atom escapes from the high-temperature oven, its magnetic moment is randomly oriented. Before this atom interacts with the magnetic field, its wave function Ψ is the weighted sum of two possible states α and β

$$\Psi = c_\alpha \alpha + c_\beta \beta \tag{1.52}$$

where c_α and c_β are constants and are related by

$$|c_\alpha|^2 + |c_\beta|^2 = 1$$

In the presence of the inhomogeneous magnetic field, the wave function Ψ collapses to either α or β with probabilities $|c_\alpha|^2$ and $|c_\beta|^2$, respectively. The state α corresponds to the atomic magnetic moment being parallel to the magnetic field gradient, the state β being antiparallel. Regardless of the orientation of the magnetic field gradient, vertical (up or down), horizontal (left or right), or any angle in between, the wave function of the atom is always given by equation (1.52) with α parallel and β antiparallel to the magnetic field gradient. Since the atomic magnetic moments are initially randomly oriented, half of the wave functions collapse to α and half to β.

In the Stern–Gerlach experiment with two magnets having parallel magnetic field gradients–the 'first arrangement' described in Section 1.7–all the atoms entering the second magnet are in state α and therefore are all deflected in the same direction by the second magnetic field gradient. Thus, it is clear that the wave function Ψ before any interaction is permanently changed by the inter-action with the first magnet.

In the 'second arrangement' of the Stern–Gerlach experiment, the atoms emerging from the first magnet and entering the second magnet are all in the same state, say α. (Recall that the other beam of atoms in state β is blocked.) The wave function α may be regarded as the weighted sum of two states α' and β'

$$\alpha = c'_\alpha \alpha' + c'_\beta \beta'$$

where α' and β' refer to states with atomic magnetic moments parallel and antiparallel, respectively, to the second magnetic field gradient and where c'_α and c'_β are constants related by

$$|c'_\alpha|^2 + |c'_\beta|^2 = 1$$

In the 'second arrangement', the second magnetic field gradient is perpendicu-lar to the first, so that

$$|c'_\alpha|^2 - |c'_\beta|^2 = \tfrac{1}{2}$$

and

$$\alpha = \frac{1}{\sqrt{2}}(\alpha' \pm \beta')$$

The interaction of the atoms in state α with the second magnet collapses the wave function α to either α' or β' with equal probabilities.

In the 'third arrangement', the right beam of atoms emerging from the second magnet (all atoms being in state α'), passes through a third magnetic field gradient parallel to the first. In this case, the wave function α' may be expressed as the sum of states α and β

$$\alpha' = \frac{1}{\sqrt{2}}(\alpha \pm \beta)$$

The interaction between the third magnetic field gradient and each atom collapses the wave function α' to either α or β with equal probabilities.

The interpretation of the various arrangements in the Stern–Gerlach experiment reinforces the postulate that the wave function for a particle is the sum of indistinguishable paths and is modified when the paths become distinguishable by means of a measurement. The nature of the modification is the collapse of the wave function to one of its components in the sum. Moreover, this new collapsed wave function may be expressed as the sum of subsequent indistinguishable paths, but remains unchanged if no further interactions with measuring devices occur.

This statistical interpretation of the significance of the wave function was postulated by M. Born (1926), although his ideas were based on some experiments other than the double-slit and Stern–Gerlach experiments. The concepts that the wave function contains all the information known about the system it represents and that it collapses to a different state in an experimental observation were originated by W. Heisenberg (1927). These postulates regarding the meaning of the wave function are part of what has become known as the *Copenhagen interpretation* of quantum mechanics. While the Copenhagen interpretation is disputed by some scientists and philosophers, it is accepted by the majority of scientists and it provides a consistent theory which agrees with all experimental observations to date. We adopt the Copenhagen interpretation of quantum mechanics in this book.[3]

Problems

1.1 The law of dispersion for surface waves on a sheet of water of uniform depth d is[4]

$$\omega(k) = (gk \tanh dk)^{1/2}$$

where g is the acceleration due to gravity. What is the group velocity of the resultant composite wave? What is the limit for deep water ($dk \geqslant 4$)?

1.2 The phase velocity for a particular wave is $v_{\mathrm{ph}} = A/\lambda$, where A is a constant. What is the dispersion relation? What is the group velocity?

1.3 Show that

$$\int_{-\infty}^{\infty} A(k)\, \mathrm{d}k = 1$$

for the gaussian function $A(k)$ in equation (1.19).

---

[3] The historical and philosophical aspects of the Copenhagen interpretation are more extensively discussed in J. Baggott (1992) *The Meaning of Quantum Theory* (Oxford University Press, Oxford).

[4] For a derivation, see H. Lamb (1932) *Hydrodynamics*, pp. 363–81 (Cambridge University Press, Cambridge).

1.4 Show that the average value of k is k_0 for the gaussian function $A(k)$ in equation (1.19).

1.5 Show that the gaussian functions $A(k)$ and $\Psi(x, t)$ obey Parseval's theorem (1.18).

1.6 Show that the square pulse $A(k)$ in equation (1.21) and the corresponding function $\Psi(x, t)$ obey Parseval's theorem.

2

Schrödinger wave mechanics

2.1 The Schrödinger equation

In the previous chapter we introduced the wave function to represent the motion of a particle moving in the absence of an external force. In this chapter we extend the concept of a wave function to make it apply to a particle acted upon by a non-vanishing force, i.e., a particle moving under the influence of a potential which depends on position. The force F acting on the particle is related to the potential or potential energy $V(x)$ by

$$F = -\frac{dV}{dx} \tag{2.1}$$

As in Chapter 1, we initially consider only motion in the x-direction. In Section 2.7, however, we extend the formalism to include three-dimensional motion.

In Chapter 1 we associated the wave packet

$$\Psi(x,\, t) = \frac{1}{\sqrt{2\pi\hbar}} \int_{-\infty}^{\infty} A(p) e^{i(px - Et)/\hbar} \, dp \tag{2.2}$$

with the motion in the x-direction of a free particle, where the weighting factor $A(p)$ is given by

$$A(p) = \frac{1}{\sqrt{2\pi\hbar}} \int_{-\infty}^{\infty} \Psi(x,\, t) e^{-i(px - Et)/\hbar} \, dx \tag{2.3}$$

This wave packet satisfies a partial differential equation, which will be used as the basis for the further development of a quantum theory. To find this differential equation, we first differentiate equation (2.2) twice with respect to the distance variable x to obtain

$$\frac{\partial^2 \Psi}{\partial x^2} = \frac{-1}{\sqrt{2\pi\hbar^5}} \int_{-\infty}^{\infty} p^2 A(p) e^{i(px - Et)/\hbar} \, dp \tag{2.4}$$

Differentiation of (2.2) with respect to the time t gives

$$\frac{\partial \Psi}{\partial t} = \frac{-i}{\sqrt{2\pi\hbar^3}} \int_{-\infty}^{\infty} EA(p)e^{i(px-Et)/\hbar}\, dp \tag{2.5}$$

The total energy E for a free particle (i.e., for a particle moving in a region of constant potential energy V) is given by

$$E = \frac{p^2}{2m} + V$$

which may be combined with equations (2.4) and (2.5) to give

$$i\hbar \frac{\partial \Psi}{\partial t} = -\frac{\hbar^2}{2m}\frac{\partial^2 \Psi}{\partial x^2} + V\Psi$$

Schrödinger (1926) postulated that this differential equation is also valid when the potential energy is not constant, but is a function of position. In that case the partial differential equation becomes

$$i\hbar \frac{\partial \Psi(x,\,t)}{\partial t} = -\frac{\hbar^2}{2m}\frac{\partial^2 \Psi(x,\,t)}{\partial x^2} + V(x)\Psi(x,\,t) \tag{2.6}$$

which is known as the *time-dependent Schrödinger equation*. The solutions $\Psi(x,\,t)$ of equation (2.6) are the *time-dependent wave functions*. An important goal in wave mechanics is solving equation (2.6) for $\Psi(x,\,t)$ using various expressions for $V(x)$ that relate to specific physical systems.

When $V(x)$ is not constant, the solutions $\Psi(x,\,t)$ to equation (2.6) may still be expanded in the form of a wave packet,

$$\Psi(x,\,t) = \frac{1}{\sqrt{2\pi\hbar}} \int_{-\infty}^{\infty} A(p,\,t)e^{i(px-Et)/\hbar}\, dp \tag{2.7}$$

The Fourier transform $A(p,\,t)$ is then, in general, a function of both p and time t, and is given by

$$A(p,\,t) = \frac{1}{\sqrt{2\pi\hbar}} \int_{-\infty}^{\infty} \Psi(x,\,t)e^{-i(px-Et)/\hbar}\, dx \tag{2.8}$$

By way of contrast, recall that in treating the free particle as a wave packet in Chapter 1, we required that the weighting factor $A(p)$ be independent of time and we needed to specify a functional form for $A(p)$ in order to study some of the properties of the wave packet.

2.2 The wave function

Interpretation

Before discussing the methods for solving the Schrödinger equation for specific choices of $V(x)$, we consider the meaning of the wave function. Since the wave function $\Psi(x,\,t)$ is identified with a particle, we need to establish the connection between $\Psi(x,\,t)$ and the observable properties of the particle. As in the case of the free particle discussed in Chapter 1, we follow the formulation of Born (1926).

The fundamental postulate relating the wave function $\Psi(x, t)$ to the properties of the associated particle is that the quantity $|\Psi(x, t)|^2 = \Psi^*(x, t)\Psi(x, t)$ gives the *probability density* for finding the particle at point x at time t. Thus, the probability of finding the particle between x and $x + dx$ at time t is $|\Psi(x, t)|^2\, dx$. The location of a particle, at least within an arbitrarily small interval, can be determined through a physical measurement. If a series of measurements are made on a number of particles, each of which has the exact same wave function, then these particles will be found in many different locations. Thus, the wave function does not indicate the actual location at which the particle will be found, but rather provides the probability for finding the particle in any given interval. More generally, quantum theory provides the probabilities for the various possible results of an observation rather than a precise prediction of the result. This feature of quantum theory is in sharp contrast to the predictive character of classical mechanics.

According to Born's statistical interpretation, the wave function completely describes the physical system it represents. There is no information about the system that is not contained in $\Psi(x, t)$. Thus, the *state* of the system is determined by its wave function. For this reason the wave function is also called the *state function* and is sometimes referred to as the state $\Psi(x, t)$.

The product of a function and its complex conjugate is always real and is positive everywhere. Accordingly, the wave function itself may be a real or a complex function. At any point x or at any time t, the wave function may be positive or negative. In order that $|\Psi(x, t)|^2$ represents a unique probability density for every point in space and at all times, the wave function must be continuous, single-valued, and finite. Since $\Psi(x, t)$ satisfies a differential equation that is second-order in x, its first derivative is also continuous. The wave function may be multiplied by a phase factor $e^{i\alpha}$, where α is real, without changing its physical significance since

$$[e^{i\alpha}\Psi(x, t)]^*[e^{i\alpha}\Psi(x, t)] = \Psi^*(x, t)\Psi(x, t) = |\Psi(x, t)|^2$$

Normalization

The particle that is represented by the wave function must be found with probability equal to unity somewhere in the range $-\infty \leqslant x \leqslant \infty$, so that $\Psi(x, t)$ must obey the relation

$$\int_{-\infty}^{\infty} |\Psi(x, t)|^2\, dx = 1 \tag{2.9}$$

A function that obeys this equation is said to be *normalized*. If a function $\Phi(x, t)$ is not normalized, but satisfies the relation

$$\int_{-\infty}^{\infty} \Phi^*(x, t)\Phi(x, t)\,dx = N$$

then the function $\Psi(x, t)$ defined by

$$\Psi(x, t) = \frac{1}{\sqrt{N}}\Phi(x, t)$$

is normalized.

In order for $\Psi(x, t)$ to satisfy equation (2.9), the wave function must be square-integrable (also called quadratically integrable). Therefore, $\Psi(x, t)$ must go to zero faster than $1/\sqrt{|x|}$ as x approaches ($\pm$) infinity. Likewise, the derivative $\partial\Psi/\partial x$ must also go to zero as x approaches ($\pm$) infinity.

Once a wave function $\Psi(x, t)$ has been normalized, it remains normalized as time progresses. To prove this assertion, we consider the integral

$$N = \int_{-\infty}^{\infty} \Psi^*\Psi\,dx$$

and show that N is independent of time for every function Ψ that obeys the Schrödinger equation (2.6). The time derivative of N is

$$\frac{dN}{dt} = \int_{-\infty}^{\infty} \frac{\partial}{\partial t}|\Psi(x, t)|^2\,dx \quad = 0 \tag{2.10}$$

where the order of differentiation and integration has been interchanged on the right-hand side. The derivative of the probability density may be expanded as follows

$$\frac{\partial}{\partial t}|\Psi(x, t)|^2 = \frac{\partial}{\partial t}(\Psi^*\Psi) = \Psi^*\frac{\partial\Psi}{\partial t} + \Psi\frac{\partial\Psi^*}{\partial t}$$

Equation (2.6) and its complex conjugate may be written in the form

$$\frac{\partial\Psi}{\partial t} = \frac{i\hbar}{2m}\frac{\partial^2\Psi}{\partial x^2} - \frac{i}{\hbar}V\Psi$$

$$\frac{\partial\Psi^*}{\partial t} = -\frac{i\hbar}{2m}\frac{\partial^2\Psi^*}{\partial x^2} + \frac{i}{\hbar}V\Psi^* \tag{2.11}$$

so that $\partial|\Psi(x, t)|^2/\partial t$ becomes

$$\frac{\partial}{\partial t}|\Psi(x, t)|^2 = \frac{i\hbar}{2m}\left(\Psi^*\frac{\partial^2\Psi}{\partial x^2} - \Psi\frac{\partial^2\Psi^*}{\partial x^2}\right)$$

where the terms containing V cancel. We next note that

$$\frac{\partial}{\partial x}\left(\Psi^*\frac{\partial\Psi}{\partial x} - \Psi\frac{\partial\Psi^*}{\partial x}\right) = \Psi^*\frac{\partial^2\Psi}{\partial x^2} - \Psi\frac{\partial^2\Psi^*}{\partial x^2}$$

so that

$$\frac{\partial}{\partial t}|\Psi(x,\,t)|^2 = \frac{i\hbar}{2m}\frac{\partial}{\partial x}\left(\Psi^*\frac{\partial\Psi}{\partial x} - \Psi\frac{\partial\Psi^*}{\partial x}\right) \qquad (2.12)$$

Substitution of equation (2.12) into (2.10) and evaluation of the integral give

$$\frac{\mathrm{d}N}{\mathrm{d}t} = \frac{i\hbar}{2m}\int_{-\infty}^{\infty}\frac{\partial}{\partial x}\left(\Psi^*\frac{\partial\Psi}{\partial x} - \Psi\frac{\partial\Psi^*}{\partial x}\right)\mathrm{d}x = \frac{i\hbar}{2m}\left[\Psi^*\frac{\partial\Psi}{\partial x} - \Psi\frac{\partial\Psi^*}{\partial x}\right]_{-\infty}^{\infty}$$

Since $\Psi(x,\,t)$ goes to zero as x goes to $(\pm)$ infinity, the right-most term vanishes and we have

$$\frac{\mathrm{d}N}{\mathrm{d}t} = 0$$

Thus, the integral N is time-independent and the normalization of $\Psi(x,\,t)$ does not change with time.

Not all wave functions can be normalized. In such cases the quantity $|\Psi(x,\,t)|^2$ may be regarded as the *relative probability density*, so that the ratio

$$\frac{\int_{a_1}^{a_2}|\Psi(x,\,t)|^2\,\mathrm{d}x}{\int_{b_1}^{b_2}|\Psi(x,\,t)|^2\,\mathrm{d}x}$$

represents the probability that the particle will be found between a_1 and a_2 relative to the probability that it will be found between b_1 and b_2. As an example, the plane wave

$$\Psi(x,\,t) = \mathrm{e}^{\mathrm{i}(px-Et)/\hbar}$$

does not approach zero as x approaches $(\pm)$ infinity and consequently cannot be normalized. The probability density $|\Psi(x,\,t)|^2$ is unity everywhere, so that the particle is equally likely to be found in any region of a specified width.

Momentum-space wave function

The wave function $\Psi(x,\,t)$ may be represented as a Fourier integral, as shown in equation (2.7), with its Fourier transform $A(p,\,t)$ given by equation (2.8). The transform $A(p,\,t)$ is uniquely determined by $\Psi(x,\,t)$ and the wave function $\Psi(x,\,t)$ is uniquely determined by $A(p,\,t)$. Thus, knowledge of one of these functions is equivalent to knowledge of the other. Since the wave function $\Psi(x,\,t)$ completely describes the physical system that it represents, its Fourier transform $A(p,\,t)$ also possesses that property. Either function may serve as a complete description of the state of the system. As a consequence, we may interpret the quantity $|A(p,\,t)|^2$ as the probability density for the momentum at time t. By Parseval's theorem (equation (B.28)), if $\Psi(x, t)$ is normalized, then its Fourier transform $A(p, t)$ is normalized,

$$\int_{-\infty}^{\infty} |\Psi(x, t)|^2 \, \mathrm{d}x = \int_{-\infty}^{\infty} |A(p, t)|^2 \, \mathrm{d}p = 1$$

The transform $A(p, t)$ is called the *momentum-space wave function*, while $\Psi(x, t)$ is more accurately known as the *coordinate-space wave function*. When there is no confusion, however, $\Psi(x, t)$ is usually simply referred to as the *wave function*.

2.3 Expectation values of dynamical quantities

Suppose we wish to measure the position of a particle whose wave function is $\Psi(x, t)$. The Born interpretation of $|\Psi(x, t)|^2$ as the probability density for finding the associated particle at position x at time t implies that such a measurement will not yield a unique result. If we have a large number of particles, each of which is in state $\Psi(x, t)$ and we measure the position of each of these particles in separate experiments all at some time t, then we will obtain a multitude of different results. We may then calculate the average or mean value $\langle x \rangle$ of these measurements. In quantum mechanics, average values of dynamical quantities are called *expectation values*. This name is somewhat misleading, because in an experimental measurement one does not *expect* to obtain the expectation value.

By definition, the average or expectation value of x is just the sum over all possible values of x of the product of x and the probability of obtaining that value. Since x is a continuous variable, we replace the probability by the probability density and the sum by an integral to obtain

$$\langle x \rangle = \int_{-\infty}^{\infty} x |\Psi(x, t)|^2 \, \mathrm{d}x \tag{2.13}$$

More generally, the expectation value $\langle f(x) \rangle$ of any function $f(x)$ of the variable x is given by

$$\langle f(x) \rangle = \int_{-\infty}^{\infty} f(x) |\Psi(x, t)|^2 \, \mathrm{d}x \tag{2.14}$$

Since $\Psi(x, t)$ depends on the time t, the expectation values $\langle x \rangle$ and $\langle f(x) \rangle$ in equations (2.13) and (2.14) are functions of t.

The expectation value $\langle p \rangle$ of the momentum p may be obtained using the momentum-space wave function $A(p, t)$ in the same way that $\langle x \rangle$ was obtained from $\Psi(x, t)$. The appropriate expression is

$$\langle p \rangle = \int_{-\infty}^{\infty} p|A(p, t)|^2 \, dp = \int_{-\infty}^{\infty} pA^*(p, t)A(p, t) \, dp \qquad (2.15)$$

The expectation value $\langle f(p) \rangle$ of any function $f(p)$ of p is given by an expression analogous to equation (2.14)

$$\langle f(p) \rangle = \int_{-\infty}^{\infty} f(p)|A(p, t)|^2 \, dp \qquad (2.16)$$

In general, $A(p, t)$ depends on the time, so that the expectation values $\langle p \rangle$ and $\langle f(p) \rangle$ are also functions of time.

Both $\Psi(x, t)$ and $A(p, t)$ contain the same information about the system, making it possible to find $\langle p \rangle$ using the coordinate-space wave function $\Psi(x, t)$ in place of $A(p, t)$. The result of establishing such a procedure will prove useful when determining expectation values for functions of both position and momentum. We begin by taking the complex conjugate of $A(p, t)$ in equation (2.8)

$$A^*(p, t) = \frac{1}{\sqrt{2\pi\hbar}} \int_{-\infty}^{\infty} \Psi^*(x, t)e^{i(px-Et)/\hbar} \, dx$$

Substitution of $A^*(p, t)$ into the integral on the right-hand side of equation (2.15) gives

$$\langle p \rangle = \frac{1}{\sqrt{2\pi\hbar}} \iint_{-\infty}^{\infty} \Psi^*(x, t)pA(p, t)e^{i(px-Et)/\hbar} \, dx \, dp$$

$$= \int_{-\infty}^{\infty} \Psi^*(x, t)\left[\frac{1}{\sqrt{2\pi\hbar}} \int_{-\infty}^{\infty} pA(p, t)e^{i(px-Et)/\hbar} \, dp \right] dx \qquad (2.17)$$

In order to evaluate the integral over p, we observe that the derivative of $\Psi(x, t)$ in equation (2.7), with respect to the position variable x, is

$$\frac{\partial\Psi(x, t)}{\partial x} = \frac{1}{\sqrt{2\pi\hbar}} \int_{-\infty}^{\infty} \frac{i}{\hbar} pA(p, t)e^{i(px-Et)/\hbar} \, dp$$

Substitution of this observation into equation (2.21) gives the final result

$$\langle p \rangle = \int_{-\infty}^{\infty} \Psi^*(x, t)\left(\frac{\hbar}{i} \frac{\partial}{\partial x} \right)\Psi(x, t) \, dx \qquad (2.18)$$

Thus, the expectation value of the momentum can be obtained by an integration in coordinate space.

The expectation value of p^2 is given by equation (2.16) with $f(p) = p^2$. The expression analogous to (2.17) is

$$\langle p^2 \rangle = \int_{-\infty}^{\infty} \Psi^*(x, t)\left[\frac{1}{\sqrt{2\pi\hbar}} \int_{-\infty}^{\infty} p^2 A(p, t)\, e^{i(px-Et)/\hbar} \, dp \right] dx$$

From equation (2.7) it can be seen that the quantity in square brackets equals

$$\left(\frac{\hbar}{i}\right)^2 \frac{\partial^2 \Psi(x,\, t)}{\partial x^2}$$

so that

$$\langle p^2 \rangle = \int_{-\infty}^{\infty} \Psi^*(x,\, t) \left(\frac{\hbar}{i}\right)^2 \frac{\partial^2}{\partial x^2} \Psi(x,\, t)\, dx \tag{2.19}$$

Similarly, the expectation value of p^n is given by

$$\langle p^n \rangle = \int_{-\infty}^{\infty} \Psi^*(x,\, t) \left(\frac{\hbar}{i} \frac{\partial}{\partial x}\right)^n \Psi(x,\, t)\, dx \tag{2.20}$$

Each of the integrands in equations (2.18), (2.19), and (2.20) is the complex conjugate of the wave function multiplied by an operator acting on the wave function. Thus, in the coordinate-space calculation of the expectation value of the momentum p or the nth power of the momentum, we associate with p the operator $(\hbar/i)(\partial/\partial x)$. We generalize this association to apply to the expectation value of any function $f(p)$ of the momentum, so that

$$\langle f(p) \rangle = \int_{-\infty}^{\infty} \Psi^*(x,\, t) f\left(\frac{\hbar}{i} \frac{\partial}{\partial x}\right) \Psi(x,\, t)\, dx \tag{2.21}$$

Equation (2.21) is equivalent to the momentum-space equation (2.16).

We may combine equations (2.14) and (2.21) to find the expectation value of a function $f(x,\, p)$ of the position and momentum

$$\langle f(x,\, p) \rangle = \int_{-\infty}^{\infty} \Psi^*(x,\, t) f\left(x,\, \frac{\hbar}{i} \frac{\partial}{\partial x}\right) \Psi(x,\, t)\, dx \tag{2.22}$$

Ehrenfest's theorems

According to the *correspondence principle* as stated by N. Bohr (1928), the average behavior of a well-defined wave packet should agree with the classical-mechanical laws of motion for the particle that it represents. Thus, the expectation values of dynamical variables such as position, velocity, momentum, kinetic energy, potential energy, and force as calculated in quantum mechanics should obey the same relationships that the dynamical variables obey in classical theory. This feature of wave mechanics is illustrated by the derivation of two relationships known as Ehrenfest's theorems.

The first relationship is obtained by considering the time dependence of the expectation value of the position coordinate x. The time derivative of $\langle x \rangle$ in equation (2.13) is

$$\frac{d\langle x \rangle}{dt} = \frac{d}{dt} \int_{-\infty}^{\infty} x |\Psi(x, t)|^2 \, dx = \int_{-\infty}^{\infty} x \frac{\partial}{\partial t} |\Psi(x, t)|^2 \, dx$$

$$= \frac{i\hbar}{2m} \int_{-\infty}^{\infty} x \frac{\partial}{\partial x} \left(\Psi^* \frac{\partial \Psi}{\partial x} - \Psi \frac{\partial \Psi^*}{\partial x} \right) dx$$

where equation (2.12) has been used. Integration by parts of the last integral gives

$$\frac{d\langle x \rangle}{dt} = \frac{i\hbar}{2m} x \left[\Psi^* \frac{\partial \Psi}{\partial x} - \Psi \frac{\partial \Psi^*}{\partial x} \right]_{-\infty}^{\infty} - \frac{i\hbar}{2m} \int_{-\infty}^{\infty} \left(\Psi^* \frac{\partial \Psi}{\partial x} - \Psi \frac{\partial \Psi^*}{\partial x} \right) dx$$

The integrated part vanishes because $\Psi(x, t)$ goes to zero as x approaches $(\pm)$ infinity. Another integration by parts of the last term on the right-hand side yields

$$\frac{d\langle x \rangle}{dt} = \frac{1}{m} \int_{-\infty}^{\infty} \Psi^* \left(\frac{\hbar}{i} \frac{\partial}{\partial x} \right) \Psi \, dx$$

According to equation (2.18), the integral on the right-hand side of this equation is the expectation value of the momentum, so that we have

$$\langle p \rangle = m \frac{d\langle x \rangle}{dt} \tag{2.23}$$

Equation (2.23) is the quantum-mechanical analog of the classical definition of momentum, $p = mv = m(dx/dt)$. This derivation also shows that the association in quantum mechanics of the operator $(\hbar/i)(\partial/\partial x)$ with the momentum is consistent with the correspondence principle.

The second relationship is obtained from the time derivative of the expectation value of the momentum $\langle p \rangle$ in equation (2.18),

$$\frac{d\langle p \rangle}{dt} = \frac{d}{dt} \int_{-\infty}^{\infty} \Psi^* \frac{\hbar}{i} \frac{\partial \Psi}{\partial x} \, dx = \frac{\hbar}{i} \int_{-\infty}^{\infty} \left(\frac{\partial \Psi^*}{\partial t} \frac{\partial \Psi}{\partial x} + \Psi^* \frac{\partial}{\partial x} \frac{\partial \Psi}{\partial t} \right) dx$$

We next substitute equations (2.11) for the time derivatives of Ψ and Ψ^* and obtain

$$\frac{d\langle p \rangle}{dt} = \int_{-\infty}^{\infty} \left[\left(\frac{-\hbar^2}{2m} \frac{\partial^2 \Psi^*}{\partial x^2} + V\Psi^* \right) \frac{\partial \Psi}{\partial x} + \Psi^* \frac{\partial}{\partial x} \left(\frac{\hbar^2}{2m} \frac{\partial^2 \Psi}{\partial x^2} - V\Psi \right) \right] dx$$

$$= \frac{-\hbar^2}{2m} \int_{-\infty}^{\infty} \frac{\partial^2 \Psi^*}{\partial x^2} \frac{\partial \Psi}{\partial x} \, dx + \frac{\hbar^2}{2m} \int_{-\infty}^{\infty} \Psi^* \frac{\partial^3 \Psi}{\partial x^3} \, dx - \int_{-\infty}^{\infty} \Psi^* \Psi \frac{dV}{dx} \, dx$$

$$\tag{2.24}$$

where the terms in V cancel. The first integral on the right-hand side of equation (2.24) may be integrated by parts twice to give

$$\int_{-\infty}^{\infty} \frac{\partial^2 \Psi^*}{\partial x^2} \frac{\partial \Psi}{\partial x} \, dx = \left[\frac{\partial \Psi^*}{\partial x} \frac{\partial \Psi}{\partial x}\right]_{-\infty}^{\infty} - \int_{-\infty}^{\infty} \frac{\partial \Psi^*}{\partial x} \frac{\partial^2 \Psi}{\partial x^2} \, dx$$

$$= \left[\frac{\partial \Psi^*}{\partial x} \frac{\partial \Psi}{\partial x} - \Psi^* \frac{\partial^2 \Psi}{\partial x^2}\right]_{-\infty}^{\infty} + \int_{-\infty}^{\infty} \Psi^* \frac{\partial^3 \Psi}{\partial x^3} \, dx$$

The integrated part vanishes because Ψ and $\partial \Psi / \partial x$ vanish at $(\pm)$ infinity. The remaining integral cancels the second integral on the right-hand side of equation (2.24), leaving the final result

$$\frac{d\langle p \rangle}{dt} = -\left\langle \frac{dV}{dx} \right\rangle = \langle F \rangle \tag{2.25}$$

where equation (2.1) has been used. Equation (2.25) is the quantum analog of Newton's second law of motion, $F = ma$, and is in agreement with the correspondence principle.

Heisenberg uncertainty principle

Using expectation values, we can derive the Heisenberg uncertainty principle introduced in Section 1.5. If we define the uncertainties Δx and Δp as the *standard deviations* of x and p, as used in statistics, then we have

$$\Delta x = \langle (x - \langle x \rangle)^2 \rangle^{1/2}$$

$$\Delta p = \langle (p - \langle p \rangle)^2 \rangle^{1/2}$$

The expectation values of x and of p at a time t are given by equations (2.13) and (2.18), respectively. For the sake of simplicity in this derivation, we select the origins of the position and momentum coordinates at time t to be the centers of the wave packet and its Fourier transform, so that $\langle x \rangle = 0$ and $\langle p \rangle = 0$. The squares of the uncertainties Δx and Δp are then given by

$$(\Delta x)^2 = \int_{-\infty}^{\infty} x^2 \Psi^* \Psi \, dx$$

$$(\Delta p)^2 = \left(\frac{\hbar}{i}\right)^2 \int_{-\infty}^{\infty} \Psi^* \frac{\partial^2 \Psi}{\partial x^2} \, dx = \left[\left(\frac{\hbar}{i}\right)^2 \Psi^* \frac{\partial \Psi}{\partial x}\right]_{-\infty}^{\infty} - \left(\frac{\hbar}{i}\right)^2 \int_{-\infty}^{\infty} \frac{\partial \Psi^*}{\partial x} \frac{\partial \Psi}{\partial x} \, dx$$

$$= \int_{-\infty}^{\infty} \left(\frac{-\hbar}{i} \frac{\partial \Psi^*}{\partial x}\right) \left(\frac{\hbar}{i} \frac{\partial \Psi}{\partial x}\right) \, dx$$

where the integrated term for $(\Delta p)^2$ vanishes because Ψ goes to zero as x approaches $(\pm)$ infinity.

The product $(\Delta x \Delta p)^2$ is

$$(\Delta x \Delta p)^2 = \int_{-\infty}^{\infty} (x\Psi^*)(x\Psi)\,dx \int_{-\infty}^{\infty} \left(\frac{-\hbar}{i}\frac{\partial \Psi^*}{\partial x}\right)\left(\frac{\hbar}{i}\frac{\partial \Psi}{\partial x}\right)dx$$

Applying Schwarz's inequality (A.56), we obtain

$$(\Delta x \Delta p)^2 \geq \frac{1}{4}\left|\frac{\hbar}{i}\int_{-\infty}^{\infty}\left(x\Psi^*\frac{\partial \Psi}{\partial x} + x\Psi\frac{\partial \Psi^*}{\partial x}\right)dx\right|^2 = \frac{\hbar^2}{4}\left|\int_{-\infty}^{\infty} x\frac{\partial}{\partial x}(\Psi^*\Psi)\,dx\right|^2$$

$$= \frac{\hbar^2}{4}\left|\left[x\Psi^*\Psi\right]_{-\infty}^{\infty} - \int_{-\infty}^{\infty}\Psi^*\Psi\,dx\right|^2$$

The integrated part vanishes because Ψ goes to zero faster than $1/\sqrt{|x|}$, as x approaches ($\pm$) infinity and the remaining integral is unity by equation (2.9). Taking the square root, we obtain an explicit form of the Heisenberg uncertainty principle

$$\Delta x \Delta p \geq \frac{\hbar}{2} \tag{2.26}$$

This expression is consistent with the earlier form, equation (1.44), but relation (2.26) is based on a precise definition of the uncertainties, whereas relation (1.44) is not.

2.4 Time-independent Schrödinger equation

The first step in the solution of the partial differential equation (2.6) is to express the wave function $\Psi(x, t)$ as the product of two functions

$$\Psi(x, t) = \psi(x)\chi(t) \tag{2.27}$$

where $\psi(x)$ is a function of only the distance x and $\chi(t)$ is a function of only the time t. Substitution of equation (2.27) into (2.6) and division by the product $\psi(x)\chi(t)$ give

$$i\hbar\frac{1}{\chi(t)}\frac{d\chi(t)}{dt} = -\frac{\hbar^2}{2m}\frac{1}{\psi(x)}\frac{d^2\psi(x)}{dx^2} + V(x) \tag{2.28}$$

The left-hand side of equation (2.28) is a function only of t, while the right-hand side is a function only of x. Since x and t are independent variables, each side of equation (2.28) must equal a constant. If this were not true, then the left-hand side could be changed by varying t while the right-hand side remained fixed and so the equality would no longer apply. For reasons that will soon be apparent, we designate this *separation constant* by E and assume that it is a real number.

Equation (2.28) is now separable into two independent differential equations, one for each of the two independent variables x and t. The time-dependent equation is

$$i\hbar \frac{d\chi(t)}{dt} = E\chi(t)$$

which has the solution

$$\chi(t) = e^{-iEt/\hbar} \tag{2.29}$$

The integration constant in equation (2.29) has arbitrarily been set equal to unity. The spatial-dependent equation is

$$-\frac{\hbar^2}{2m}\frac{d^2\psi(x)}{dx^2} + V(x)\psi(x) = E\psi(x) \tag{2.30}$$

and is called the *time-independent Schrödinger equation*. The solution of this differential equation depends on the specification of the potential energy $V(x)$. Note that the separation of equation (2.6) into spatial and temporal parts is contingent on the potential $V(x)$ being time-independent.

The wave function $\Psi(x, t)$ is then

$$\Psi(x, t) = \psi(x)e^{-iEt/\hbar} \tag{2.31}$$

and the probability density $|\Psi(x, t)|^2$ is now given by

$$|\Psi(x, t)|^2 = \Psi^*(x, t)\Psi(x, t) = \psi^*(x)e^{iEt/\hbar}\psi(x)e^{-iEt/\hbar} = |\psi(x)|^2$$

Thus, the probability density depends only on the position variable x and does not change with time. For this reason the wave function $\Psi(x, t)$ in equation (2.31) is called a *stationary state*. If $\Psi(x, t)$ is normalized, then $\psi(x)$ is also normalized

$$\int_{-\infty}^{\infty} |\psi(x)|^2 \, dx = 1 \tag{2.32}$$

which is the reason why we set the integration constant in equation (2.29) equal to unity.

The total energy, when expressed in terms of position and momentum, is called the *Hamiltonian*, H, and is given by

$$H(x, p) = \frac{p^2}{2m} + V(x)$$

The expectation value $\langle H \rangle$ of the Hamiltonian may be obtained by applying equation (2.22)

$$\langle H \rangle = \int_{-\infty}^{\infty} \Psi^*(x, t)\left[-\frac{\hbar^2}{2m}\frac{\partial^2}{\partial x^2} + V(x)\right]\Psi(x, t)\,dx$$

For the stationary state (2.31), this expression becomes

$$\langle H \rangle = \int_{-\infty}^{\infty} \psi^*(x)\left[-\frac{\hbar^2}{2m}\frac{\partial^2}{\partial x^2} + V(x)\right]\psi(x)\,dx$$

If we substitute equation (2.30) into the integrand, we obtain

$$\langle H \rangle = E \int_{-\infty}^{\infty} \psi^*(x)\psi(x)\,\mathrm{d}x = E$$

where we have also applied equation (2.32). We have just shown that the separation constant E is the expectation value of the Hamiltonian, or the total energy for the stationary state, so that 'E' is a desirable designation. Since the energy is a real physical quantity, the assumption that E is real is justified.

In the application of Schrödinger's equation (2.30) to specific physical examples, the requirements that $\psi(x)$ be continuous, single-valued, and square-integrable restrict the acceptable solutions to an infinite set of specific functions $\psi_n(x)$, $n = 1, 2, 3, \ldots$, each with a corresponding energy value E_n. Thus, the energy is quantized, being restricted to certain values. This feature is illustrated in Section 2.5 with the example of a particle in a one-dimensional box.

Since the partial differential equation (2.6) is linear, any linear superposition of solutions is also a solution. Therefore, the most general solution of equation (2.6) for a time-independent potential energy $V(x)$ is

$$\Psi(x, t) = \sum_n c_n \psi_n(x) \mathrm{e}^{-\mathrm{i}E_n t/\hbar} \tag{2.33}$$

where the coefficients c_n are arbitrary complex constants. The wave function $\Psi(x, t)$ in equation (2.33) is not a stationary state, but rather a sum of stationary states, each with a different energy E_n.

2.5 Particle in a one-dimensional box

As an illustration of the application of the time-independent Schrödinger equation to a system with a specific form for $V(x)$, we consider a particle confined to a box with infinitely high sides. The potential energy for such a particle is given by

$$V(x) = 0, \qquad 0 \leqslant x \leqslant a$$
$$= \infty, \qquad x < 0, \quad x > a$$

and is illustrated in Figure 2.1.

Outside the potential well, the Schrödinger equation (2.30) is given by

$$-\frac{\hbar^2}{2m}\frac{\mathrm{d}^2\psi}{\mathrm{d}x^2} + \infty\psi = E\psi$$

for which the solution is simply $\psi(x) = 0$; the probability is zero for finding the particle outside the box where the potential is infinite. Inside the box, the Schrödinger equation is

$$-\frac{\hbar^2}{2m}\frac{\mathrm{d}^2\psi}{\mathrm{d}x^2} = E\psi$$

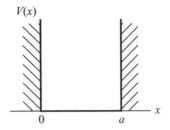

Figure 2.1 The potential energy $V(x)$ for a particle in a one-dimensional box of length a.

or

$$\frac{\mathrm{d}^2\psi}{\mathrm{d}x^2} = -\frac{4\pi^2}{\lambda^2}\psi \tag{2.34}$$

where λ is the de Broglie wavelength,

$$\lambda = \frac{2\pi\hbar}{\sqrt{2mE}} = \frac{h}{p} \tag{2.35}$$

We have implicitly assumed here that E is not negative. If E were negative, then the wave function ψ and its second derivative would have the same sign. As $|x|$ increases, the wave function $\psi(x)$ and its curvature $\mathrm{d}^2\psi/\mathrm{d}x^2$ would become larger and larger in magnitude and $\psi(x)$ would approach ($\pm$) infinity as $x \to \infty$.

The solutions to equation (2.34) are functions that are proportional to their second derivatives, namely $\sin(2\pi x/\lambda)$ and $\cos(2\pi x/\lambda)$. The functions $\exp[2\pi ix/\lambda]$ and $\exp[-2\pi ix/\lambda]$, which as equation (A.31) shows are equivalent to the trigonometric functions, are also solutions, but are more difficult to use for this system. Thus, the general solution to equation (2.34) is

$$\psi(x) = A\sin\frac{2\pi x}{\lambda} + B\cos\frac{2\pi x}{\lambda} \tag{2.36}$$

where A and B are arbitrary constants of integration.

The constants A and B are determined by the *boundary conditions* placed on the solution $\psi(x)$. Since $\psi(x)$ must be continuous, the boundary conditions require that $\psi(x)$ vanish at each end of the box so as to match the value of $\psi(x)$ outside the box, i.e., $\psi(0) = \psi(a) = 0$. At $x = 0$, the function $\psi(0)$ from (2.36) is

$$\psi(0) = A\sin 0 + B\cos 0 = B$$

so that $B = 0$ and $\psi(x)$ is now

$$\psi(x) = A\sin\frac{2\pi x}{\lambda} \tag{2.37}$$

At $x = a$, $\psi(a)$ is

$$\psi(a) = A \sin \frac{2\pi a}{\lambda} = 0$$

The constant A cannot be zero, for then $\psi(x)$ would vanish everywhere and there would be no particle. Consequently, we have $\sin(2\pi a/\lambda) = 0$ or

$$\frac{2\pi a}{\lambda} = n\pi, \qquad n = 1, 2, 3, \ldots$$

where n is any positive integer greater than zero. The solution $n = 0$ would cause $\psi(x)$ to vanish everywhere and is therefore not acceptable. Negative values of n give redundant solutions because $\sin(-\theta)$ equals $-\sin\theta$.

We have found that only distinct values for the de Broglie wavelength satisfy the requirement that the wave function represents the motion of the particle. These distinct values are denoted as λ_n and are given by

$$\lambda_n = \frac{2a}{n}, \qquad n = 1, 2, 3, \ldots \tag{2.38}$$

Consequently, from equation (2.35) only distinct values E_n of the energy are allowed

$$E_n = \frac{n^2 \pi^2 \hbar^2}{2ma^2} = \frac{n^2 h^2}{8ma^2}, \qquad n = 1, 2, 3, \ldots \tag{2.39}$$

so that the energy for a particle in a box is quantized.

The lowest allowed energy level is called the *zero-point energy* and is given by $E_1 = h^2/8ma^2$. This zero-point energy is always greater than the zero value of the constant potential energy of the system and increases as the length a of the box decreases. The non-zero value for the lowest energy level is related to the Heisenberg uncertainty principle. For the particle in a box, the uncertainty Δx in position is equal to the length a since the particle is somewhere within the box. The uncertainty Δp in momentum is equal to $2|p|$ since the momentum ranges from $-|p|$ to $|p|$. The momentum and energy are related by

$$|p| = \sqrt{2mE} = \frac{nh}{2a}$$

so that

$$\Delta x \Delta p = nh$$

is in agreement with the Heisenberg uncertainty principle (2.26). If the lowest allowed energy level were zero, then the Heisenberg uncertainty principle would be violated.

The allowed wave functions $\psi_n(x)$ for the particle in a box are obtained by substituting equation (2.38) into (2.37),

$$\psi_n(x) = A \sin \frac{n\pi x}{a}, \qquad 0 \leqslant x \leqslant a$$

The remaining constant of integration A is determined by the normalization condition (2.32),

$$\int_{-\infty}^{\infty} |\psi_n(x)|^2\, dx = |A|^2 \int_0^a \sin^2 \frac{n\pi x}{a}\, dx = |A|^2 \frac{a}{\pi} \int_0^\pi \sin^2 n\theta\, d\theta = |A|^2 \frac{a}{2} = 1$$

where equation (A.15) was used. Therefore, we have

$$|A|^2 = \frac{2}{a}$$

or

$$A = e^{i\alpha} \sqrt{\frac{2}{a}}$$

Setting the phase α equal to zero since it has no physical significance, we obtain for the normalized wave functions

$$\psi_n(x) = \sqrt{\frac{2}{a}} \sin \frac{n\pi x}{a}, \qquad 0 \le x \le a \tag{2.40}$$
$$= 0, \qquad\qquad x < 0, \quad x > a$$

The time-dependent Schrödinger equation (2.30) for the particle in a box has an infinite set of solutions $\psi_n(x)$ given by equation (2.40). The first four wave functions $\psi_n(x)$ for $n = 1$, 2, 3, and 4 and their corresponding probability densities $|\psi_n(x)|^2$ are shown in Figure 2.2. The wave function $\psi_1(x)$ corresponding to the lowest energy level E_1 is called the *ground state*. The other wave functions are called *excited states*.

If we integrate the product of two different wave functions $\psi_l(x)$ and $\psi_n(x)$, we find that

$$\int_0^a \psi_l(x)\psi_n(x)\, dx = \frac{2}{a}\int_0^a \sin\left(\frac{l\pi x}{a}\right)\sin\left(\frac{n\pi x}{a}\right) dx = \frac{2}{\pi}\int_0^\pi \sin l\theta \sin n\theta\, d\theta = 0 \tag{2.41}$$

where equation (A.15) has been introduced. This result may be combined with the normalization relation to give

$$\int_0^a \psi_l(x)\psi_n(x)\, dx = \delta_{ln} \tag{2.42}$$

where δ_{ln} is the Kronecker delta,

$$\delta_{ln} = 1, \qquad l = n$$
$$= 0, \qquad l \ne n \tag{2.43}$$

Functions that obey equation (2.41) are called *orthogonal functions*. If the orthogonal functions are also normalized, as in equation (2.42), then they are

Schrödinger wave mechanics

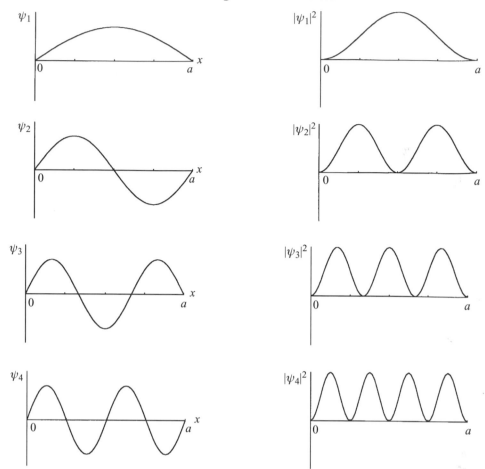

Figure 2.2 Wave functions ψ_i and probability densities $|\psi_i|^2$ for a particle in a one-dimensional box of length a.

said to be *orthonormal*. The orthogonal property of wave functions in quantum mechanics is discussed in a more general context in Section 3.3.

The stationary states $\Psi(x, t)$ for the particle in a one-dimensional box are given by substitution of equations (2.39) and (2.40) into (2.31),

$$\Psi(x, t) = \sqrt{\frac{2}{a}} \sin\left(\frac{n\pi x}{a}\right) e^{-i(n^2\pi^2\hbar/2ma^2)t} \tag{2.44}$$

The most general solution (2.33) is, then,

$$\Psi(x, t) = \sqrt{\frac{2}{a}} \sum_n c_n \sin\left(\frac{n\pi x}{a}\right) e^{-i(n^2\pi^2\hbar/2ma^2)t} \tag{2.45}$$

2.6 Tunneling

As a second example of the application of the Schrödinger equation, we consider the behavior of a particle in the presence of a potential barrier. The specific form that we choose for the potential energy $V(x)$ is given by

$$V(x) = V_0, \qquad 0 \leqslant x \leqslant a$$

$$= 0, \qquad x < 0, \quad x > a$$

and is shown in Figure 2.3. The region where $x < 0$ is labeled I, where $0 \leqslant x \leqslant a$ is labeled II, and where $x > a$ is labeled III.

Suppose a particle of mass m and energy E coming from the left approaches the potential barrier. According to classical mechanics, if E is less than the barrier height V_0, the particle will be reflected by the barrier; it cannot pass through the barrier and appear in region III. In quantum theory, as we shall see, the particle can penetrate the barrier and appear on the other side. This effect is called *tunneling*.

In regions I and III, where $V(x)$ is zero, the Schrödinger equation (2.30) is

$$\frac{d^2\psi(x)}{dx^2} = -\frac{2mE}{\hbar^2}\psi(x) \tag{2.46}$$

The general solutions to equation (2.46) for these regions are

$$\psi_{\mathrm{I}} = A e^{i\alpha x} + B e^{-i\alpha x} \tag{2.47 a}$$

$$\psi_{\mathrm{III}} = F e^{i\alpha x} + G e^{-i\alpha x} \tag{2.47 b}$$

where A, B, F, and G are arbitrary constants of integration and α is the abbreviation

$$\alpha = \frac{\sqrt{2mE}}{\hbar} \tag{2.48}$$

In region II, where $V(x) = V_0 > E$, the Schrödinger equation (2.30) becomes

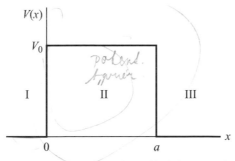

Figure 2.3 Potential energy barrier of height V_0 and width a.

$$\frac{\mathrm{d}^2\psi(x)}{\mathrm{d}x^2} = \frac{2m}{\hbar^2}(V_0 - E)\psi(x) \tag{2.49}$$

for which the general solution is

$$\psi_{\mathrm{II}} = Ce^{\beta x} + De^{-\beta x} \tag{2.50}$$

where C and D are integration constants and β is the abbreviation

$$\beta = \frac{\sqrt{2m(V_0 - E)}}{\hbar} \tag{2.51}$$

The term $\exp[i\alpha x]$ in equations (2.47) indicates travel in the positive x-direction, while $\exp[-i\alpha x]$ refers to travel in the opposite direction. The coefficient A is, then, the amplitude of the incident wave, B is the amplitude of the reflected wave, and F is the amplitude of the transmitted wave. In region III, the particle moves in the positive x-direction, so that G is zero. The relative probability of tunneling is given by the *transmission coefficient T*

$$T = \frac{|F|^2}{|A|^2} \tag{2.52}$$

and the relative probability of reflection is given by the *reflection coefficient R*

$$R = \frac{|B|^2}{|A|^2} \tag{2.53}$$

The wave function for the particle is obtained by joining the three parts ψ_{I}, ψ_{II}, and ψ_{III} such that the resulting wave function $\psi(x)$ and its first derivative $\psi'(x)$ are continuous. Thus, the following boundary conditions apply

$$\psi_{\mathrm{I}}(0) = \psi_{\mathrm{II}}(0), \qquad \psi_{\mathrm{I}}'(0) = \psi_{\mathrm{II}}'(0) \tag{2.54}$$

$$\psi_{\mathrm{II}}(a) = \psi_{\mathrm{III}}(a), \qquad \psi_{\mathrm{II}}'(a) = \psi_{\mathrm{III}}'(a) \tag{2.55}$$

These four relations are sufficient to determine any four of the constants A, B, C, D, F in terms of the fifth. If the particle were confined to a finite region of space, then its wave function could be normalized, thereby determining the fifth and final constant. However, in this example, the position of the particle may range from $-\infty$ to ∞. Accordingly, the wave function cannot be normalized, the remaining constant cannot be evaluated, and only relative probabilities such as the transmission and reflection coefficients can be determined.

We first evaluate the transmission coefficient T in equation (2.52). Applying equations (2.55) to (2.47 b) and (2.50), we obtain

$$Ce^{\beta a} + De^{-\beta a} = Fe^{i\alpha a}$$

$$\beta(Ce^{\beta a} - De^{-\beta a}) = i\alpha Fe^{i\alpha a}$$

from which it follows that

$$C = F\left(\frac{\beta + i\alpha}{2\beta}\right)e^{(i\alpha - \beta)a}$$

$$D = F\left(\frac{\beta - i\alpha}{2\beta}\right)e^{(i\alpha + \beta)a}$$

(2.56)

Application of equation (2.54) to (2.47 a) and (2.50) gives

$$A + B = C + D$$

$$i\alpha(A - B) = \beta(C - D)$$

(2.57)

Elimination of B from the pair of equations (2.57) and substitution of equations (2.56) for C and for D yield

$$A = \frac{1}{2i\alpha}[(\beta + i\alpha)C - (\beta - i\alpha)D]$$

$$= F\frac{e^{i\alpha a}}{4i\alpha\beta}[(\beta + i\alpha)^2 e^{-\beta a} - (\beta - i\alpha)^2 e^{\beta a}]$$

At this point it is easier to form $|A|^2$ before any further algebraic simplifications

$$|A|^2 = A^* A$$

$$- |F|^2 \frac{1}{16\alpha^2\beta^2}[(\beta^2 + \alpha^2)^2 e^{-2\beta a} + (\beta^2 + \alpha^2)^2 e^{2\beta a} - (\beta - i\alpha)^4 - (\beta + i\alpha)^4]$$

$$= |F|^2 \frac{1}{16\alpha^2\beta^2}[(\alpha^2 + \beta^2)^2 (e^{\beta a} - e^{-\beta a})^2 + 16\alpha^2\beta^2]$$

$$= |F|^2 \left[1 + \frac{(\alpha^2 + \beta^2)^2}{4\alpha^2\beta^2} \sinh^2 \beta a\right]$$

where equation (A.46) has been used. Combining this result with equations (2.48), (2.51), and (2.52), we obtain

$$T = \left[1 + \frac{V_0^2}{4E(V_0 - E)} \sinh^2(\sqrt{2m(V_0 - E)}\, a/\hbar)\right]^{-1}$$

(2.58)

To find the reflection coefficient R, we eliminate A from the pair of equations (2.57) and substitute equations (2.56) for C and for D to obtain

$$B = \frac{1}{2i\alpha}[-(\beta - i\alpha)C + (\beta + i\alpha)D] = F\frac{e^{i\alpha a}}{4i\alpha\beta}[(\alpha^2 + \beta^2)e^{\beta a} - (\alpha^2 + \beta^2)e^{-\beta a}]$$

$$= F\frac{e^{i\alpha a}}{2i\alpha\beta}(\alpha^2 + \beta^2)\sinh \beta a$$

where again equation (A.46) has been used. Combining this result with equations (2.52) and (2.53), we find that

$$R = T\frac{|B|^2}{|F|^2} = T\frac{(\alpha^2 + \beta^2)^2}{4\alpha^2\beta^2}\sinh^2\beta a$$

Substitution of equations (2.48), (2.51), and (2.58) yields

$$R = \frac{\dfrac{V_0^2}{4E(V_0 - E)}\sinh^2(\sqrt{2m(V_0 - E)}\,a/\hbar)}{1 + \dfrac{V_0^2}{4E(V_0 - E)}\sinh^2(\sqrt{2m(V_0 - E)}\,a/\hbar)} \qquad (2.59)$$

The transmission coefficient T in equation (2.58) is the relative probability that a particle impinging on the potential barrier tunnels through the barrier. The reflection coefficient R in equation (2.59) is the relative probability that the particle bounces off the barrier and moves in the negative x-direction. Since the particle must do one or the other of these two possibilities, the sum of T and R should equal unity

$$T + R = 1$$

which we observe from equations (2.58) and (2.59) to be the case.

We also note that the (relative) probability for the particle being in the region $0 \leqslant x \leqslant a$ is not zero. In this region, the potential energy is greater than the total particle energy, making the kinetic energy of the particle negative. This concept is contrary to classical theory and does not have a physical significance. For this reason we cannot observe the particle experimentally within the potential barrier. Further, we note that because the particle is not confined to a finite region, the boundary conditions on the wave function have not imposed any restrictions on the energy E. Thus, the energy in this example is not quantized.

In this analysis we considered the relative probabilities for tunneling and reflection for a single particle. The conclusions apply equally well to a beam of particles, each of mass m and total energy E, traveling initially in the positive x-direction. In that case, the transmission coefficient T in equation (2.58) gives the fraction of incoming particles that tunnel through the barrier, and the reflection coefficient R in equation (2.59) gives the fraction that are reflected by the barrier.

If the potential barrier is thick (a is large), the potential barrier is high compared with the particle energy E ($V_0 \gg E$), the mass m of the particle is large, or any combination of these characteristics, then we have

$$\sinh\beta a = \frac{1}{2}(e^{\beta a} - e^{-\beta a}) \approx \frac{e^{\beta a}}{2}$$

so that T and R become

$$T \approx \frac{16E(V_0 - E)}{V_0^2} e^{-2a\sqrt{2m(V_0 - E)}/\hbar}$$

$$R \approx 1 - \frac{16E(V_0 - E)}{V_0^2} e^{-2a\sqrt{2m(V_0 - E)}/\hbar}$$

In the limit as $a \to \infty$, as $V_0 \to \infty$, as $m \to \infty$, or any combination, the transmission coefficient T approaches zero and the reflection coefficient R approaches unity, which are the classical-mechanical values. We also note that in the limit $\hbar \to 0$, the classical values for T and R are obtained.

Examples of tunneling in physical phenomena occur in the spontaneous emission of an alpha particle by a nucleus, oxidation–reduction reactions, electrode reactions, and the umbrella inversion of the ammonia molecule. For these cases, the potential is not as simple as the one used here, but must be selected to approximate as closely as possible the actual potential. However, the basic qualitative results of the treatment here serve to explain the general concept of tunneling.

2.7 Particles in three dimensions

Up to this point we have considered particle motion only in the x-direction. The generalization of Schrödinger wave mechanics to three dimensions is straightforward. In this section we summarize the basic ideas and equations of wave mechanics as expressed in their three-dimensional form.

The position of any point in three-dimensional cartesian space is denoted by the vector $\mathbf{r}$ with components x, y, z, so that

$$\mathbf{r} = \mathbf{i}x + \mathbf{j}y + \mathbf{k}z \tag{2.60}$$

where $\mathbf{i}$, $\mathbf{j}$, $\mathbf{k}$ are, respectively, the unit vectors along the x, y, z cartesian coordinate axes. The linear momentum $\mathbf{p}$ of a particle of mass m is given by

$$\mathbf{p} = m\frac{d\mathbf{r}}{dt} = m\left(\mathbf{i}\frac{dx}{dt} + \mathbf{j}\frac{dy}{dt} + \mathbf{k}\frac{dz}{dt}\right) = \mathbf{i}p_x + \mathbf{j}p_y + \mathbf{k}p_z \tag{2.61}$$

The x-component, p_x, of the momentum now needs to carry a subscript, whereas before it was denoted simply as p. The *scalar* or *dot product* of $\mathbf{r}$ and $\mathbf{p}$ is

$$\mathbf{r} \cdot \mathbf{p} = \mathbf{p} \cdot \mathbf{r} = xp_x + yp_y + zp_z$$

and the *magnitude* p of the vector $\mathbf{p}$ is

$$p = (\mathbf{p} \cdot \mathbf{p})^{1/2} = (p_x^2 + p_y^2 + p_z^2)^{1/2}$$

The classical Hamiltonian $H(\mathbf{p}, \mathbf{r})$ takes the form

$$H(\mathbf{p},\,\mathbf{r}) = \frac{p^2}{2m} + V(r) = \frac{1}{2m}(p_x^2 + p_y^2 + p_z^2) + V(\mathbf{r}) \qquad (2.62)$$

When expressed in three dimensions, the de Broglie relation is

$$\mathbf{p} = \hbar\mathbf{k} \qquad (2.63)$$

where $\mathbf{k}$ is the vector wave number with components k_x, k_y, k_z. The de Broglie wavelength λ is still given by

$$\lambda = \frac{2\pi}{k} = \frac{h}{p}$$

where now k and p are the magnitudes of the corresponding vectors. The wave packet representing a particle in three dimensions is

$$\Psi(\mathbf{r},\,t) = \frac{1}{(2\pi\hbar)^{3/2}} \int A(\mathbf{p},\,t)e^{\mathrm{i}(\mathbf{p}\cdot\mathbf{r}-Et)/\hbar}\,\mathrm{d}\mathbf{p} \qquad (2.64)$$

As shown by equations (B.19), (B.20), and (B.27), the momentum-space wave function $A(\mathbf{p},\,t)$ is a generalized Fourier transform of $\Psi(\mathbf{r},\,t)$,

$$A(\mathbf{p},\,t) = \frac{1}{(2\pi\hbar)^{3/2}} \int \Psi(\mathbf{r},\,t)e^{-\mathrm{i}(\mathbf{p}\cdot\mathbf{r}-Et)/\hbar}\,\mathrm{d}\mathbf{r} \qquad (2.65)$$

The volume elements $\mathrm{d}\mathbf{r}$ and $\mathrm{d}\mathbf{p}$ are defined as

$$\mathrm{d}\mathbf{r} = \mathrm{d}x\,\mathrm{d}y\,\mathrm{d}z$$
$$\mathrm{d}\mathbf{p} = \mathrm{d}p_x\,\mathrm{d}p_y\,\mathrm{d}p_z$$

and the integrations extend over the complete range of each variable.

For a particle moving in three-dimensional space, the quantity

$$\Psi^*(\mathbf{r},\,t)\Psi(\mathbf{r},\,t)\,\mathrm{d}\mathbf{r} = \Psi^*(x,\,y,\,z,\,t)\Psi(x,\,y,\,z,\,t)\,\mathrm{d}x\,\mathrm{d}y\,\mathrm{d}z$$

is the probability at time t of finding the particle with its x-coordinate between x and $x + \mathrm{d}x$, its y-coordinate between y and $y + \mathrm{d}y$, and its z-coordinate between z and $z + \mathrm{d}z$. The product $\Psi^*(\mathbf{r},\,t)\Psi(\mathbf{r},\,t)$ is, then, the probability density at the point $\mathbf{r}$ at time t. If the particle is under the influence of an external potential field $V(\mathbf{r})$, the wave function $\Psi(\mathbf{r},\,t)$ may be normalized

$$\int \Psi^*(\mathbf{r},\,t)\Psi(\mathbf{r},\,t)\,\mathrm{d}\mathbf{r} = 1 \qquad (2.66)$$

The quantum-mechanical operators corresponding to the components of $\mathbf{p}$ are

$$\hat{p}_x = \frac{\hbar}{\mathrm{i}}\frac{\partial}{\partial x}, \qquad \hat{p}_y = \frac{\hbar}{\mathrm{i}}\frac{\partial}{\partial y}, \qquad \hat{p}_z = \frac{\hbar}{\mathrm{i}}\frac{\partial}{\partial z}$$

or, in vector notation

$$\hat{\mathbf{p}} = \frac{\hbar}{\mathrm{i}}\boldsymbol{\nabla} \qquad (2.67)$$

where the *gradient* operator ∇ is defined as

$$\nabla \equiv \mathbf{i}\frac{\partial}{\partial x} + \mathbf{j}\frac{\partial}{\partial y} + \mathbf{k}\frac{\partial}{\partial z}$$

Using these relations, we may express the Hamiltonian operator in three dimensions as

$$\hat{H} = \frac{-\hbar^2}{2m}\nabla^2 + V(\mathbf{r})$$

where the *laplacian* operator ∇^2 is defined by

$$\nabla^2 \equiv \nabla \cdot \nabla = \frac{\partial^2}{\partial x^2} + \frac{\partial^2}{\partial y^2} + \frac{\partial^2}{\partial z^2}$$

The time-dependent Schrödinger equation is

$$i\hbar\frac{\partial \Psi(\mathbf{r},\,t)}{\partial t} = \hat{H}\Psi(\mathbf{r},\,t)$$

$$= \frac{-\hbar^2}{2m}\nabla^2\Psi(\mathbf{r},\,t) + V(\mathbf{r})\Psi(\mathbf{r},\,t) \tag{2.68}$$

The stationary-state solutions to this differential equation are

$$\Psi_n(\mathbf{r},\,t) = \psi_n(\mathbf{r})e^{-iE_n t/\hbar} \tag{2.69}$$

where the spatial functions $\psi_n(\mathbf{r})$ are solutions of the time-independent Schrödinger equation

$$\frac{-\hbar^2}{2m}\nabla^2\psi_n(\mathbf{r}) + V(\mathbf{r})\psi_n(\mathbf{r}) = E_n\psi_n(\mathbf{r}) \tag{2.70}$$

The most general solution to equation (2.68) is

$$\Psi(\mathbf{r},\,t) = \sum_n c_n\psi_n(\mathbf{r})e^{-iE_n t/\hbar} \tag{2.71}$$

where c_n are arbitrary complex constants.

The expectation value of a function $f(\mathbf{r},\,\mathbf{p})$ of position and momentum is given by

$$\langle f(\mathbf{r},\,\mathbf{p})\rangle = \int \Psi^*(\mathbf{r},\,t) f\left(\mathbf{r},\,\frac{\hbar}{i}\nabla\right)\Psi(\mathbf{r},\,t)\,d\mathbf{r} \tag{2.72}$$

Equivalently, expectation values of three-dimensional dynamical quantities may be evaluated for each dimension and then combined, if appropriate, into vector notation. For example, the two Ehrenfest theorems in three dimensions are

$$\langle \mathbf{p} \rangle = m \frac{d\langle \mathbf{r} \rangle}{dt}$$

$$\frac{d\langle \mathbf{p} \rangle}{dt} = -\langle \boldsymbol{\nabla} V \rangle = \langle \mathbf{F} \rangle$$

where $\mathbf{F}$ is the vector force acting on the particle. The Heisenberg uncertainty principle becomes

$$\Delta x \Delta p_x \geqslant \frac{\hbar}{2}, \quad \Delta y \Delta p_y \geqslant \frac{\hbar}{2}, \quad \Delta z \Delta p_z \geqslant \frac{\hbar}{2}$$

Multi-particle system

For a system of N *distinguishable* particles in three-dimensional space, the classical Hamiltonian is

$$H(\mathbf{p}_1, \mathbf{p}_2, \ldots, \mathbf{p}_N, \mathbf{r}_1, \mathbf{r}_2, \ldots, \mathbf{r}_N) = \frac{p_1^2}{2m_1} + \frac{p_2^2}{2m_2} + \cdots + \frac{p_N^2}{2m_N}$$

$$+ V(\mathbf{r}_1, \mathbf{r}_2, \ldots, \mathbf{r}_N)$$

where $\mathbf{r}_k$ and $\mathbf{p}_k$ are the position and momentum vectors of particle k. Thus, the quantum-mechanical Hamiltonian operator is

$$\hat{H} = \frac{-\hbar^2}{2} \left(\frac{1}{m_1} \nabla_1^2 + \frac{1}{m_2} \nabla_2^2 + \cdots + \frac{1}{m_N} \nabla_N^2 \right) + V(\mathbf{r}_1, \mathbf{r}_2, \ldots, \mathbf{r}_N) \quad (2.73)$$

where ∇_k^2 is the laplacian with respect to the position of particle k.

The wave function for this system is a function of the N position vectors: $\Psi(\mathbf{r}_1, \mathbf{r}_2, \ldots, \mathbf{r}_N, t)$. Thus, although the N particles are moving in three-dimensional space, the wave function is $3N$-dimensional. The physical interpretation of the wave function is analogous to that for the three-dimensional case. The quantity

$$\Psi^*(\mathbf{r}_1, \mathbf{r}_2, \ldots, \mathbf{r}_N, t)\Psi(\mathbf{r}_1, \mathbf{r}_2, \ldots, \mathbf{r}_N, t)\, d\mathbf{r}_1\, d\mathbf{r}_2 \ldots d\mathbf{r}_N$$

$$= \Psi^*(x_1, y_1, z_1, x_2, \ldots, z_N)\Psi(x_1, y_1, z_1, x_2, \ldots, z_N)\, dx_1\, dy_1\, dz_1\, dx_2 \ldots dz_N$$

is the probability at time t that, simultaneously, particle 1 is between x_1, y_1, z_1 and $x_1 + dx_1$, $y_1 + dy_1$, $z_1 + dz_1$, particle 2 is between x_2, y_2, z_2 and $x_2 + dx_2$, $y_2 + dy_2$, $z_2 + dz_2$, $\ldots$, and particle N is between x_N, y_N, z_N and $x_N + dx_N$, $y_N + dy_N$, $z_N + dz_N$. The normalization condition is

$$\int \Psi^*(\mathbf{r}_1, \mathbf{r}_2, \ldots, \mathbf{r}_N, t)\Psi(\mathbf{r}_1, \mathbf{r}_2, \ldots, \mathbf{r}_N, t)\, d\mathbf{r}_1\, d\mathbf{r}_2 \ldots d\mathbf{r}_N = 1 \quad (2.74)$$

This discussion applies only to systems with *distinguishable* particles; for example, systems where each particle has a different mass. The treatment of wave functions for systems with *indistinguishable* particles is more complicated and is discussed in Chapter 8. Such systems include atoms or molecules with more than one electron, and molecules with two or more identical nuclei.

2.8 Particle in a three-dimensional box

A simple example of a three-dimensional system is a particle confined to a rectangular container with sides of lengths a, b, and c. Within the box there is no force acting on the particle, so that the potential $V(\mathbf{r})$ is given by

$$V(\mathbf{r}) = 0, \qquad 0 \leqslant x \leqslant a, \ 0 \leqslant y \leqslant b, \ 0 \leqslant z \leqslant c$$

$$= \infty, \qquad x < 0, \ x > a; \ y < 0, \ y > b; \ z < 0, \ z > c$$

The wave function $\psi(\mathbf{r})$ outside the box vanishes because the potential is infinite there. Inside the box, the wave function obeys the Schrödinger equation (2.70) with the potential energy set equal to zero

$$\frac{-\hbar^2}{2m} \left(\frac{\partial^2 \psi(\mathbf{r})}{\partial x^2} + \frac{\partial^2 \psi(\mathbf{r})}{\partial y^2} + \frac{\partial^2 \psi(\mathbf{r})}{\partial z^2} \right) = E\psi(\mathbf{r}) \tag{2.75}$$

The standard procedure for solving a partial differential equation of this type is to assume that the function $\psi(\mathbf{r})$ may be written as the product of three functions, one for each of the three variables

$$\psi(\mathbf{r}) = \psi(x, y, z) = X(x)Y(y)Z(z) \tag{2.76}$$

Thus, $X(x)$ is a function only of the variable x, $Y(y)$ only of y, and $Z(z)$ only of z. Substitution of equation (2.76) into (2.75) and division by the product XYZ give

$$\frac{-\hbar^2}{2mX} \frac{d^2 X}{dx^2} + \frac{-\hbar^2}{2mY} \frac{d^2 Y}{dy^2} + \frac{-\hbar^2}{2mZ} \frac{d^2 Z}{dz^2} = E \tag{2.77}$$

The first term on the left-hand side of equation (2.77) depends only on the variable x, the second only on y, and the third only on z. No matter what the values of x, or y, or z, the sum of these three terms is always equal to the same constant E. The only way that this condition can be met is for each of the three terms to equal some constant, say E_x, E_y, and E_z, respectively. The partial differential equation (2.77) can then be separated into three equations, one for each variable

$$\frac{d^2 X}{dx^2} + \frac{2m}{\hbar^2} E_x X = 0, \qquad \frac{d^2 Y}{dy^2} + \frac{2m}{\hbar^2} E_y Y = 0, \qquad \frac{d^2 Z}{dz^2} + \frac{2m}{\hbar^2} E_z Z = 0$$

$$\tag{2.78}$$

where

$$E_x + E_y + E_z = E \tag{2.79}$$

Thus, the three-dimensional problem has been reduced to three one-dimensional problems.

The differential equations (2.78) are identical in form to equation (2.34) and the boundary conditions are the same as before. Consequently, the solutions inside the box are given by equation (2.40) as

$$X(x) = \sqrt{\frac{2}{a}} \sin \frac{n_x \pi x}{a}, \qquad n_x = 1, 2, 3, \ldots$$

$$Y(y) = \sqrt{\frac{2}{b}} \sin \frac{n_y \pi y}{b}, \qquad n_y = 1, 2, 3, \ldots \qquad (2.80)$$

$$Z(z) = \sqrt{\frac{2}{c}} \sin \frac{n_z \pi z}{c}, \qquad n_z = 1, 2, 3, \ldots$$

and the constants E_x, E_y, E_z are given by equation (2.39)

$$E_x = \frac{n_x^2 h^2}{8ma^2}, \qquad n_x = 1, 2, 3, \ldots$$

$$E_y = \frac{n_y^2 h^2}{8mb^2}, \qquad n_y = 1, 2, 3, \ldots \qquad (2.81)$$

$$E_z = \frac{n_z^2 h^2}{8mc^2}, \qquad n_z = 1, 2, 3, \ldots$$

The quantum numbers n_x, n_y, n_z take on positive integer values independently of each other. Combining equations (2.76) and (2.80) gives the wave functions inside the three-dimensional box

$$\psi_{n_x, n_y, n_z}(\mathbf{r}) = \sqrt{\frac{8}{v}} \sin \frac{n_x \pi x}{a} \sin \frac{n_y \pi y}{b} \sin \frac{n_z \pi z}{c} \qquad (2.82)$$

where $v = abc$ is the volume of the box. The energy levels for the particle are obtained by substitution of equations (2.81) into (2.79)

$$E_{n_x, n_y, n_z} = \frac{h^2}{8m} \left(\frac{n_x^2}{a^2} + \frac{n_y^2}{b^2} + \frac{n_z^2}{c^2} \right) \qquad (2.83)$$

Degeneracy of energy levels

If the box is cubic, we have $a = b = c$ and the energy levels become

$$E_{n_x, n_y, n_z} = \frac{h^2}{8ma^2} (n_x^2 + n_y^2 + n_z^2) \qquad (2.84)$$

The lowest or zero-point energy is $E_{1,1,1} = 3h^2/8ma^2$, which is three times the zero-point energy for a particle in a one-dimensional box of the same length. The second or next-highest value for the energy is obtained by setting one of

Table 2.1. *Energy levels for a particle in a three-dimensional box with $a = b = c$*

Energy	Degeneracy	Values of n_x, n_y, n_z
$3(h^2/8ma^2)$	1	1,1,1
$6(h^2/8ma^2)$	3	2,1,1 1,2,1 1,1,2
$9(h^2/8ma^2)$	3	2,2,1 2,1,2 1,2,2
$11(h^2/8ma^2)$	3	3,1,1 1,3,1 1,1,3
$12(h^2/8ma^2)$	1	2,2,2
$14(h^2/8ma^2)$	6	3,2,1 3,1,2 2,3,1 2,1,3 1,3,2 1,2,3

Table 2.2. *Energy levels for a particle in a three-dimensional box with $b = a/2$, $c = a/3$*

Energy	Degeneracy	Values of n_x, n_y, n_z
$14(h^2/8ma^2)$	1	1,1,1
$17(h^2/8ma^2)$	1	2,1,1
$22(h^2/8ma^2)$	1	3,1,1
$26(h^2/8ma^2)$	1	1,2,1
$29(h^2/8ma^2)$	2	2,2,1 4,1,1
$34(h^2/8ma^2)$	1	3,2,1
$38(h^2/8ma^2)$	1	5,1,1
$41(h^2/8ma^2)$	2	1,1,2 4,2,1

the integers n_x, n_y, n_z equal to 2 and the remaining ones equal to unity. Thus, there are three ways of obtaining the value $6h^2/8ma^2$, namely, $E_{2,1,1}$, $E_{1,2,1}$, and $E_{1,1,2}$. Each of these three possibilities corresponds to a different wave function, respectively, $\psi_{2,1,1}(\mathbf{r})$, $\psi_{1,2,1}(\mathbf{r})$, and $\psi_{1,1,2}(\mathbf{r})$. An energy level that corresponds to more than one wave function is said to be *degenerate*. The second energy level in this case is threefold or triply degenerate. The zero-point energy level is *non-degenerate*. The energies and degeneracies for the first six energy levels are listed in Table 2.1.

The degeneracies of the energy levels in this example are the result of symmetry in the lengths of the sides of the box. If, instead of the box being cubic, the lengths of b and c in terms of a were $b = a/2$, $c = a/3$, then the values of the energy levels and their degeneracies are different, as shown in Table 2.2 for the lowest eight levels.

Degeneracy is discussed in more detail in Chapter 3.

Problems

2.1 Consider a particle in a one-dimensional box of length a and in quantum state n. What is the probability that the particle is in the left quarter of the box $(0 \leqslant x \leqslant a/4)$? For which state n is the probability a maximum? What is the probability that the particle is in the left half of the box $(0 \leqslant x \leqslant a/2)$?

2.2 Consider a particle of mass m in a one-dimensional potential such that

$$V(x) = 0, \qquad -a/2 \leqslant x \leqslant a/2$$
$$= \infty, \qquad x < -a/2, \quad x > a/2$$

Solve the time-independent Schrödinger equation for this particle to obtain the energy levels and the normalized wave functions. (Note that the boundary conditions are different from those in Section 2.5.)

2.3 Consider a particle of mass m confined to move on a circle of radius a. Express the Hamiltonian operator in plane polar coordinates and then determine the energy levels and wave functions.

2.4 Consider a particle of mass m and energy E approaching from the left a potential barrier of height V_0, as shown in Figure 2.3 and discussed in Section 2.6. However, suppose now that E is greater than V_0 $(E > V_0)$. Obtain expressions for the reflection and transmission coefficients for this case. Show that T equals unity when $E - V_0 = n^2\pi^2\hbar^2/2ma^2$ for $n = 1, 2, \ldots$ Show that between these periodic maxima T has minima which lie progressively closer to unity as E increases.

2.5 Find the expression for the transmission coefficient T for Problem 2.4 when the energy E of the particle is equal to the potential barrier height V_0.

3

General principles of quantum theory

3.1 Linear operators

The wave mechanics discussed in Chapter 2 is a linear theory. In order to develop the theory in a more formal manner, we need to discuss the properties of linear operators. An operator $\hat{A}$ is a mathematical entity that transforms a function ψ into another function ϕ

$$\phi = \hat{A}\psi \tag{3.1}$$

Throughout this book a circumflex is used to denote operators. For example, multiplying the function $\psi(x)$ by the variable x to give a new function $\phi(x)$ may be regarded as operating on the function $\psi(x)$ with the operator $\hat{x}$, where $\hat{x}$ means multiply by x: $\phi(x) = \hat{x}\psi(x) = x\psi(x)$. Generally, when the operation is simple multiplication, the circumflex on the operator is omitted. The operator $\hat{D}_x$, defined as d/dx, acting on $\psi(x)$ gives the first derivative of $\psi(x)$ with respect to x, so that in this case

$$\phi = \hat{D}_x\psi = \frac{d\psi}{dx}$$

The operator $\hat{A}$ may involve a more complex procedure, such as taking the integral of ψ with respect to x either implicitly or between a pair of limits.

The operator $\hat{A}$ is *linear* if it satisfies two criteria

$$\hat{A}(\psi_1 + \psi_2) = \hat{A}\psi_1 + \hat{A}\psi_2 \tag{3.2a}$$

$$\hat{A}(c\psi) = c\hat{A}\psi \tag{3.2b}$$

where c is any complex constant. In the three examples given above, the operators are linear. Some nonlinear operators are 'exp' (take the exponential of) and $[\,]^2$ (take the square of), since

65

$$e^{x+y} = e^x e^y \neq e^x + e^y$$

$$e^{cx} \neq ce^x$$

$$[x + y]^2 = x^2 + 2xy + y^2 \neq x^2 + y^2$$

$$[c(x + y)]^2 \neq c[x + y]^2$$

The operator $\hat{C}$ is the *sum* of the operators $\hat{A}$ and $\hat{B}$ if

$$\hat{C}\psi = (\hat{A} + \hat{B})\psi = \hat{A}\psi + \hat{B}\psi$$

The operator $\hat{C}$ is the *product* of the operators $\hat{A}$ and $\hat{B}$ if

$$\hat{C}\psi = \hat{A}\hat{B}\psi = \hat{A}(\hat{B}\psi)$$

where first $\hat{B}$ operates on ψ and then $\hat{A}$ operates on the resulting function. Operators obey the *associative law* of multiplication, namely

$$\hat{A}(\hat{B}\hat{C}) = (\hat{A}\hat{B})\hat{C}$$

Operators may be combined. Thus, the *square* $\hat{A}^2$ of an operator $\hat{A}$ is just the product $\hat{A}\hat{A}$

$$\hat{A}^2\psi = \hat{A}\hat{A}\psi = \hat{A}(\hat{A}\psi)$$

Similar definitions apply to higher powers of $\hat{A}$. As another example, the differential equation

$$\frac{d^2 y}{dx^2} + k^2 y = 0$$

may be written as $(\hat{D}_x^2 + k^2)y = 0$, where the operator $(\hat{D}_x^2 + k^2)$ is the sum of the two product operators $\hat{D}_x^2$ and k^2.

In multiplication, the order of $\hat{A}$ and $\hat{B}$ is important because $\hat{A}\hat{B}\psi$ is not necessarily equal to $\hat{B}\hat{A}\psi$. For example, if $\hat{A} = x$ and $\hat{B} = \hat{D}_x$, then we have $\hat{A}\hat{B}\psi = x\hat{D}_x\psi = x(d\psi/dx)$ while, on the other hand, $\hat{B}\hat{A}\psi = \hat{D}_x(x\psi) = \psi + x(d\psi/dx)$. The *commutator* of $\hat{A}$ and $\hat{B}$, written as $[\hat{A}, \hat{B}]$, is an operator defined as

$$[\hat{A}, \hat{B}] = \hat{A}\hat{B} - \hat{B}\hat{A} \tag{3.3}$$

from which it follows that $[\hat{A}, \hat{B}] = -[\hat{B}, \hat{A}]$. If $\hat{A}\hat{B}\psi = \hat{B}\hat{A}\psi$, then we have $\hat{A}\hat{B} = \hat{B}\hat{A}$ and $[\hat{A}, \hat{B}] = 0$; in this case we say that $\hat{A}$ and $\hat{B}$ *commute*. By expansion of each side of the following expressions, we can readily prove the relationships

$$[\hat{A}, \hat{B}\hat{C}] = [\hat{A}, \hat{B}]\hat{C} + \hat{B}[\hat{A}, \hat{C}] \tag{3.4a}$$

$$[\hat{A}\hat{B}, \hat{C}] = [\hat{A}, \hat{C}]\hat{B} + \hat{A}[\hat{B}, \hat{C}] \tag{3.4b}$$

The operator $\hat{A}$ is the *reciprocal* of $\hat{B}$ if $\hat{A}\hat{B} = \hat{B}\hat{A} = 1$, where 1 may be regarded as the unit operator, i.e., 'multiply by unity'. We may write $\hat{A} = \hat{B}^{-1}$

and $\hat{B} = \hat{A}^{-1}$. If the operator $\hat{A}$ possesses a reciprocal, it is *non-singular*, in which case the expression $\phi = \hat{A}\psi$ may be solved for ψ, giving $\psi = \hat{A}^{-1}\phi$. If $\hat{A}$ possesses no reciprocal, it is *singular* and the expression $\phi = \hat{A}\psi$ may not be inverted.

3.2 Eigenfunctions and eigenvalues

Consider a finite set of functions f_i and the relationship

$$c_1 f_1 + c_2 f_2 + \cdots + c_n f_n = 0$$

where c_1, c_2, ... are complex constants. If an equation of this form exists, then the functions are linearly dependent. However, if no such relationship exists, except for the trivial one with $c_1 = c_2 = \cdots = c_n = 0$, then the functions are *linearly independent*. This definition can be extended to include an infinite set of functions.

In general, the function ϕ obtained by the application of the operator $\hat{A}$ on an arbitrary function ψ, as expressed in equation (3.1), is linearly independent of ψ. However, for some particular function ψ_1, it is possible that

$$\hat{A}\psi_1 = \alpha_1\psi_1$$

where α_1 is a complex number. In such a case ψ_1 is said to be an *eigenfunction* of $\hat{A}$ and α_1 is the corresponding *eigenvalue*. For a given operator $\hat{A}$, many eigenfunctions may exist, so that

$$\hat{A}\psi_i = \alpha_i\psi_i \tag{3.5}$$

where ψ_i are the eigenfunctions, which may even be infinite in number, and α_i are the corresponding eigenvalues. Each eigenfunction of $\hat{A}$ is unique, that is to say, is linearly independent of the other eigenfunctions.

Sometimes two or more eigenfunctions have the same eigenvalue. In that situation the eigenvalue is said to be *degenerate*. When two, three, ..., n eigenfunctions have the same eigenvalue, the eigenvalue is *doubly, triply, ...*, *n-fold degenerate*. When an eigenvalue corresponds only to a single eigenfunction, the eigenvalue is *non-degenerate*.

A simple example of an eigenvalue equation involves the operator $\hat{D}_x$ mentioned in Section 3.1. When $\hat{D}_x$ operates on e^{kx}, the result is

$$\hat{D}_x e^{kx} = \frac{d}{dx} e^{kx} = k e^{kx}$$

Thus, the exponentials e^{kx} are eigenfunctions of $\hat{D}_x$ with corresponding eigenvalues k. Since both the real part and the imaginary part of k can have any values from $-\infty$ to $+\infty$, there are an infinite number of eigenfunctions and these eigenfunctions form a continuum of functions.

Another example is the operator $\hat{D}_x^2$ acting on either $\sin nx$ or $\cos nx$, where n is a positive integer ($n \geqslant 1$), for which we obtain

$$\hat{D}_x^2 \sin nx = -n^2 \sin nx$$

$$\hat{D}_x^2 \cos nx = -n^2 \cos nx$$

The functions $\sin nx$ and $\cos nx$ are eigenfunctions of $\hat{D}_x^2$ with eigenvalues $-n^2$. Although there are an infinite number of eigenfunctions in this example, these eigenfunctions form a discrete, rather than a continuous, set.

In order that the eigenfunctions ψ_i have physical significance in their application to quantum theory, they are chosen from a special class of functions, namely, those which are continuous, have continuous derivatives, are single-valued, and are square integrable. We refer to functions with these properties as *well-behaved functions*. Throughout this book we implicitly assume that all functions are well-behaved.

Scalar product and orthogonality

The *scalar product* of two functions $\psi(x)$ and $\phi(x)$ is defined as

$$\int_{-\infty}^{\infty} \phi^*(x)\psi(x)\,dx$$

For functions of the three cartesian coordinates x, y, z, the scalar product of $\psi(x, y, z)$ and $\phi(x, y, z)$ is

$$\int_{-\infty}^{\infty} \phi^*(x, y, z)\psi(x, y, z)\,dx\,dy\,dz$$

For the functions $\psi(r, \theta, \varphi)$ and $\phi(r, \theta, \varphi)$ of the spherical coordinates r, θ, φ, the scalar product is

$$\int_0^{2\pi}\int_0^{\pi}\int_0^{\infty} \phi^*(r, \theta, \varphi)\psi(r, \theta, \varphi)r^2 \sin\theta\,dr\,d\theta\,d\varphi$$

In order to express equations in general terms, we adopt the notation $\int d\tau$ to indicate integration over the full range of all the coordinates of the system being considered and write the scalar product in the form

$$\int \phi^*\psi\,d\tau$$

For further convenience we also introduce a notation devised by Dirac and write the scalar product of ψ and ϕ as $\langle \phi \,|\, \psi \rangle$, so that

$$\langle \phi \,|\, \psi \rangle \equiv \int \phi^*\psi\,d\tau$$

The significance of this notation is discussed in Section 3.6. From the definition of the scalar product and of the notation $\langle \phi \,|\, \psi \rangle$, we note that

$$\langle \phi \mid \psi \rangle^* = \langle \psi \mid \phi \rangle$$

$$\langle \phi \mid c\psi \rangle = c\langle \phi \mid \psi \rangle$$

$$\langle c\phi \mid \psi \rangle = c^*\langle \phi \mid \psi \rangle$$

where c is an arbitrary complex constant. Since the integral $\langle \psi \mid \psi \rangle^*$ equals $\langle \psi \mid \psi \rangle$, the scalar product $\langle \psi \mid \psi \rangle$ is real.

If the scalar product of ψ and ϕ vanishes, i.e., if $\langle \phi \mid \psi \rangle = 0$, then ψ and ϕ are said to be *orthogonal*. If the eigenfunctions ψ_i of an operator $\hat{A}$ obey the expressions

$$\langle \psi_j \mid \psi_i \rangle = 0 \quad \text{all } i, j \text{ with } i \neq j$$

the functions ψ_i form an orthogonal set. Furthermore, if the scalar product of ψ_i with itself is unity, the function ψ_i is said to be *normalized*. A set of functions which are both orthogonal to one another and normalized are said to be *orthonormal*

$$\langle \psi_j \mid \psi_i \rangle = \delta_{ij} \tag{3.6}$$

where δ_{ij} is the *Kronecker delta function*,

$$\begin{aligned} \delta_{ij} &= 1, \quad i = j \\ &= 0, \quad i \neq j \end{aligned} \tag{3.7}$$

3.3 Hermitian operators

The linear operator $\hat{A}$ is *hermitian* with respect to the set of functions ψ_i of the variables $q_1, q_2, \ldots$ if it possesses the property that

$$\int \psi_j^* \hat{A} \psi_i \, d\tau = \int \psi_i (\hat{A} \psi_j)^* \, d\tau \tag{3.8}$$

The integration is over the entire range of all the variables. The differential $d\tau$ has the form

$$d\tau = w(q_1, q_2, \ldots) \, dq_1 \, dq_2 \ldots$$

where $w(q_1, q_2, \ldots)$ is a *weighting function* that depends on the choice of the coordinates $q_1, q_2, \ldots$ For cartesian coordinates the weighting function $w(x, y, z)$ equals unity; for spherical coordinates, $w(r, \theta, \varphi)$ equals $r^2 \sin \theta$. Special variables introduced to simplify specific problems have their own weighting functions, which may differ from unity (see for example Section 6.3). Equation (3.8) may also be expressed in Dirac notation

$$\langle \psi_j \mid \hat{A} \psi_i \rangle = \langle \hat{A} \psi_j \mid \psi_i \rangle \tag{3.9}$$

in which the brackets indicate integration over all the variables using their weighting function.

For illustration, we consider some examples involving only one variable, namely, the cartesian coordinate x, for which $w(x) = 1$. An operator that results in multiplying by a real function $f(x)$ is hermitian, since in this case $f(x)^* = f(x)$ and equation (3.8) is an identity. Likewise, the momentum operator $\hat{p} = (\hbar/\mathrm{i})(\mathrm{d}/\mathrm{d}x)$, which was introduced in Section 2.3, is hermitian since

$$\int_{-\infty}^{\infty} \psi_j^* \hat{p}\psi_i \, \mathrm{d}x = \int_{-\infty}^{\infty} \psi_j^* \frac{\hbar}{\mathrm{i}} \frac{\mathrm{d}\psi_i}{\mathrm{d}x} \, \mathrm{d}x = \frac{\hbar}{\mathrm{i}} \psi_j^* \psi_i \Big|_{-\infty}^{\infty} - \frac{\hbar}{\mathrm{i}} \int_{-\infty}^{\infty} \psi_i \frac{\mathrm{d}\psi_j^*}{\mathrm{d}x} \, \mathrm{d}x$$

The integrated part is zero if the functions ψ_i vanish at infinity, which they must in order to be well-behaved. The remaining integral is $\int \psi_i \hat{p}^* \psi_j^* \, \mathrm{d}x$, so that we have

$$\int_{-\infty}^{\infty} \psi_j^* \hat{p}\psi_i \, \mathrm{d}x = \int_{-\infty}^{\infty} \psi_i (\hat{p}\psi_j)^* \, \mathrm{d}x$$

The imaginary unit i contained in the operator $\hat{p}$ is essential for the hermitian character of that operator. The operator $\hat{D}_x = \mathrm{d}/\mathrm{d}x$ is not hermitian because

$$\int_{-\infty}^{\infty} \psi_j^* \frac{\mathrm{d}\psi_i}{\mathrm{d}x} \, \mathrm{d}x = -\int_{-\infty}^{\infty} \psi_i \frac{\mathrm{d}\psi_j^*}{\mathrm{d}x} \, \mathrm{d}x \tag{3.10}$$

where again the integrated part vanishes. The negative sign on the right-hand side of equation (3.10) indicates that the operator is not hermitian. The operator $\hat{D}_x^2$, however, is hermitian.

The hermitian character of an operator depends not only on the operator itself, but also on the functions on which it acts and on the range of integration. An operator may be hermitian with respect to one set of functions, but not with respect to another set. It may be hermitian with respect to a set of functions defined over one range of variables, but not with respect to the same set over a different range. For example, the hermiticity of the momentum operator $\hat{p}$ is dependent on the vanishing of the functions ψ_i at infinity.

The product of two hermitian operators may or may not be hermitian. Consider the product $\hat{A}\hat{B}$ where $\hat{A}$ and $\hat{B}$ are separately hermitian with respect to a set of functions ψ_i, so that

$$\langle \psi_j | \hat{A}\hat{B}\psi_i \rangle = \langle \hat{A}\psi_j | \hat{B}\psi_i \rangle = \langle \hat{B}\hat{A}\psi_j | \psi_i \rangle \tag{3.11}$$

where we have assumed that the functions $\hat{A}\psi_i$ and $\hat{B}\psi_i$ also lie in the hermitian domain of $\hat{A}$ and $\hat{B}$. The product $\hat{A}\hat{B}$ is hermitian if, and only if, $\hat{A}$ and $\hat{B}$ commute. Using the same procedure, one can easily demonstrate that if $\hat{A}$ and $\hat{B}$ do not commute, then the operators $(\hat{A}\hat{B} + \hat{B}\hat{A})$ and $\mathrm{i}[\hat{A}, \hat{B}]$ are hermitian.

By setting $\hat{B}$ equal to $\hat{A}$ in the product $\hat{A}\hat{B}$ in equation (3.11), we see that the square of a hermitian operator is hermitian. This result can be generalized to any integral power of $\hat{A}$. Since $|\hat{A}\psi|^2$ is always positive, the integral $\langle \hat{A}\psi \,|\, \hat{A}\psi \rangle$ is positive and consequently

$$\langle \psi \,|\, \hat{A}^2\psi \rangle \geqslant 0 \tag{3.12}$$

Eigenvalues

The eigenvalues of a hermitian operator are real. To prove this statement, we consider the eigenvalue equation

$$\hat{A}\psi = \alpha\psi \tag{3.13}$$

where $\hat{A}$ is hermitian, ψ is an eigenfunction of $\hat{A}$, and α is the corresponding eigenvalue. Multiplying by ψ^* and integrating give

$$\langle \psi \,|\, \hat{A}\psi \rangle = \alpha \langle \psi \,|\, \psi \rangle \tag{3.14}$$

Multiplication of the complex conjugate of equation (3.13) by ψ and integrating give

$$\langle \hat{A}\psi \,|\, \psi \rangle = \alpha^* \langle \psi \,|\, \psi \rangle \tag{3.15}$$

Because $\hat{A}$ is hermitian, the left-hand sides of equations (3.14) and (3.15) are equal, so that

$$(\alpha - \alpha^*)\langle \psi \,|\, \psi \rangle = 0 \tag{3.16}$$

Since the integral in equation (3.16) is not equal to zero, we conclude that $\alpha = \alpha^*$ and thus α is real.

Orthogonality theorem

If ψ_1 and ψ_2 are eigenfunctions of a hermitian operator $\hat{A}$ with different eigenvalues α_1 and α_2, then ψ_1 and ψ_2 are orthogonal. To prove this theorem, we begin with the integral

$$\langle \psi_2 \,|\, \hat{A}\psi_1 \rangle = \alpha_1 \langle \psi_2 \,|\, \psi_1 \rangle \tag{3.17}$$

Since $\hat{A}$ is hermitian and α_2 is real, the left-hand side may be written as

$$\langle \psi_2 \,|\, \hat{A}\psi_1 \rangle = \langle \hat{A}\psi_2 \,|\, \psi_1 \rangle = \alpha_2 \langle \psi_2 \,|\, \psi_1 \rangle$$

Thus, equation (3.17) becomes

$$(\alpha_2 - \alpha_1)\langle \psi_2 \,|\, \psi_1 \rangle = 0$$

Since $\alpha_1 \neq \alpha_2$, the functions ψ_1 and ψ_2 are orthogonal.

Since the Dirac notation suppresses the variables involved in the integration, we re-express the orthogonality relation in integral notation

$$\int \psi_2^*(q_1, q_2, \ldots)\psi_1(q_1, q_2, \ldots)w(q_1, q_2, \ldots)\,dq_1\,dq_2\ldots = 0$$

This expression serves as a reminder that, in general, the eigenfunctions of a hermitian operator involve several variables and that the weighting function must be used. The functions are, therefore, orthogonal *with respect to the weighting function* $w(q_1, q_2, \ldots)$.

If the weighting function is real and positive, then we can define ϕ_1 and ϕ_2 as

$$\phi_1 = \sqrt{w}\psi_1, \quad \phi_2 = \sqrt{w}\psi_2$$

The functions ϕ_1 and ϕ_2 are then mutually orthogonal with respect to a weighting function of unity. Moreover, if the operator $\hat{A}$ is hermitian with respect to ψ_1 and ψ_2 with a weighting function w, then $\hat{A}$ is hermitian with respect to ϕ_1 and ϕ_2 with a weighting function equal to unity.

If two or more linearly independent eigenfunctions have the same eigen-value, so that the eigenvalue is degenerate, the orthogonality theorem does not apply. However, it is possible to construct eigenfunctions that are mutually orthogonal. Suppose there are two independent eigenfunctions ψ_1 and ψ_2 of the operator $\hat{A}$ with the same eigenvalue α. Any linear combination $c_1\psi_1 + c_2\psi_2$, where c_1 and c_2 are any pair of complex numbers, is also an eigenfunction of $\hat{A}$ with the same eigenvalue, so that

$$\hat{A}(c_1\psi_1 + c_2\psi_2) = c_1\hat{A}\psi_1 + c_2\hat{A}\psi_2 = \alpha(c_1\psi_1 + c_2\psi_2)$$

From any pair ψ_1, ψ_2 which initially are not orthogonal, we can construct by selecting appropriate values for c_1 and c_2 a new pair which are orthogonal. By selecting different sets of values for c_1, c_2, we may obtain infinitely many new pairs of eigenfunctions which are mutually orthogonal.

As an illustration, suppose the members of a set of functions $\psi_1, \psi_2, \ldots, \psi_n$ are not orthogonal. We define a new set of functions $\phi_1, \phi_2, \ldots, \phi_n$ by the relations

$$\phi_1 = \psi_1$$

$$\phi_2 = a\phi_1 + \psi_2$$

$$\phi_3 = b_1\phi_1 + b_2\phi_2 + \psi_3$$

$$\vdots$$

If we require that ϕ_2 be orthogonal to ϕ_1 by setting $\langle \phi_1 | \phi_2 \rangle = 0$, then the constant a is given by

$$a = -\langle \psi_1 | \psi_2 \rangle / \langle \psi_1 | \psi_1 \rangle = -\langle \phi_1 | \psi_2 \rangle / \langle \phi_1 | \phi_1 \rangle$$

and ϕ_2 is determined. We next require ϕ_3 to be orthogonal to ϕ_1 and to ϕ_2, which gives

$$b_1 = -\langle \phi_1 \mid \psi_3 \rangle / \langle \phi_1 \mid \phi_1 \rangle$$

$$b_2 = -\langle \phi_2 \mid \psi_3 \rangle / \langle \phi_2 \mid \phi_2 \rangle$$

In general, we have

$$\phi_s = \psi_s + \sum_{i=1}^{s-1} k_{si} \phi_i$$

$$k_{si} = -\langle \phi_i \mid \psi_s \rangle / \langle \phi_i \mid \phi_i \rangle$$

This construction is known as the *Schmidt orthogonalization procedure.* Since the initial selection for ϕ_1 can be any of the original functions ψ_i or any linear combination of them, an infinite number of orthogonal sets ϕ_i can be obtained by the Schmidt procedure.

We conclude that all eigenfunctions of a hermitian operator are either mutually orthogonal or, if belonging to a degenerate eigenvalue, can be chosen to be mutually orthogonal. Throughout the remainder of this book, we treat all the eigenfunctions of a hermitian operator as an orthogonal set.

Extended orthogonality theorem

The orthogonality theorem can also be extended to cover a somewhat more general form of the eigenvalue equation. For the sake of convenience, we present in detail the case of a single variable, although the treatment can be generalized to any number of variables. Suppose that instead of the eigenvalue equation (3.5), we have for a hermitian operator $\hat{A}$ of one variable

$$\hat{A}\psi_i(x) = \alpha_i w(x)\psi_i(x) \tag{3.18}$$

where the function $w(x)$ is real, positive, and the same for all values of i. Therefore, equation (3.18) can also be written as

$$\hat{A}^* \psi_j^*(x) = \alpha_j^* w(x)\psi_j^*(x) \tag{3.19}$$

Multiplication of equation (3.18) by $\psi_j^*(x)$ and integration over x give

$$\int \psi_j^*(x)\hat{A}\psi_i(x)\,dx = \alpha_i \int \psi_j^*(x)\psi_i(x)w(x)\,dx \tag{3.20}$$

Now, the operator $\hat{A}$ is hermitian with respect to the functions ψ_i with a weighting function equaling unity, so that the integral on the left-hand side of equation (3.20) becomes

$$\int \psi_j^*(x)\hat{A}\psi_i(x)\,dx = \int \psi_i(x)\hat{A}^* \psi_j^*(x)\,dx = \alpha_j^* \int \psi_j^*(x)\psi_i(x)w(x)\,dx$$

where equation (3.19) has been used as well. Accordingly, equation (3.20) becomes

$$(\alpha_i - \alpha_j^*) \int \psi_j^*(x)\psi_i(x)w(x)\,dx = 0 \tag{3.21}$$

When $j = i$, the integral in equation (3.21) cannot vanish because the product $\psi_i^*\psi_i$ and the function $w(x)$ are always positive. Therefore, we have $\alpha_i = \alpha_i^*$ and the eigenvalues α_i are real. For the situation where $i \neq j$ and $\alpha_i \neq \alpha_j^*$, the integral in equation (3.21) must vanish,

$$\int \psi_j^*(x)\psi_i(x)w(x)\,dx = 0 \tag{3.22}$$

Thus, the set of functions $\psi_i(x)$ for non-degenerate eigenvalues are *mutually orthogonal when integrated with a weighting function* $w(x)$. Eigenfunctions corresponding to degenerate eigenvalues can be made orthogonal as discussed earlier.

The discussion above may be generalized to more than one variable. In the general case, equation (3.18) is replaced by

$$\hat{A}\psi_i(q_1, q_2, \ldots) = \alpha_i w(q_1, q_2, \ldots)\psi_i(q_1, q_2, \ldots) \tag{3.23}$$

and equation (3.22) by

$$\int \psi_j^*(q_1, q_2, \ldots)\psi_i(q_1, q_2, \ldots)w(q_1, q_2, \ldots)\,dq_1\,dq_2\ldots = 0 \tag{3.24}$$

Equation (3.18) can also be transformed into the more usual form, equation (3.5). We first define a set of functions $\phi_i(x)$ as

$$\phi_i(x) = [w(x)]^{1/2}\psi_i(x) = \psi_i(x)/u(x) \tag{3.25}$$

where

$$u(x) = [w(x)]^{-1/2} \tag{3.26}$$

The function $u(x)$ is real because $w(x)$ is always positive and $u(x)$ is positive because we take the positive square root. If $w(x)$ approaches infinity at any point within the range of hermiticity of $\hat{A}$ (as x approaches infinity, for example), then $\psi_i(x)$ must approach zero such that the ratio $\phi_i(x)$ approaches zero. Equation (3.18) is now multiplied by $u(x)$ and $\psi_i(x)$ is replaced by $u(x)\phi_i(x)$

$$u(x)\hat{A}u(x)\phi_i(x) = \alpha_i w(x)[u(x)]^2\phi_i(x)$$

If we define an operator $\hat{B}$ by the relation $\hat{B} = u(x)\hat{A}u(x)$ and apply equation (3.26), we obtain

$$\hat{B}\phi_i(x) = \alpha_i\phi_i(x)$$

which has the form of equation (3.5). We observe that

$$\int \psi_j^* \hat{A} \psi_i \, dx = \int \phi_j^* u \hat{A} u \phi_i \, dx = \int \phi_j^* \hat{B} \phi_i \, dx$$

$$\int (\hat{A} \psi_j)^* \psi_i \, dx = \int (\hat{A} u \phi_j)^* u \phi_i \, dx = \int (\hat{B} \phi_j)^* \phi_i \, dx$$

Since $\hat{A}$ is hermitian with respect to the ψ_is, the two integrals on the left of each equation equal each other, from which it follows that

$$\int \phi_j^* \hat{B} \phi_i \, dx = \int (\hat{B} \phi_j)^* \phi_i \, dx$$

and $\hat{B}$ is therefore hermitian with respect to the ϕ_is.

3.4 Eigenfunction expansions

Consider a set of orthonormal eigenfunctions ψ_i of a hermitian operator. Any arbitrary function f of the same variables as ψ_i defined over the same range of these variables may be expanded in terms of the members of set ψ_i

$$f = \sum_i a_i \psi_i \tag{3.27}$$

where the a_is are constants. The summation in equation (3.27) converges to the function f if the set of eigenfunctions is *complete*. By *complete* we mean that no other function g exists with the property that $\langle g \,|\, \psi_i \rangle = 0$ for any value of i, where g and ψ_i are functions of the same variables and are defined over the same variable range. As a general rule, the eigenfunctions of a hermitian operator are not only orthogonal, but are also complete. A mathematical criterion for completeness is presented at the end of this section.

The coefficients a_i are evaluated by multiplying (3.27) by the complex conjugate ψ_j^* of one of the eigenfunctions, integrating over the range of the variables, and noting that the ψ_is are orthonormal

$$\langle \psi_j \,|\, f \rangle = \left\langle \psi_j \,\middle|\, \sum_i a_i \psi_i \right\rangle = \sum_i a_i \langle \psi_j \,|\, \psi_i \rangle = a_j$$

Replacing the dummy index j by i, we have

$$a_i = \langle \psi_i \,|\, f \rangle \tag{3.28}$$

Substitution of equation (3.28) back into (3.27) gives

$$f = \sum_i \langle \psi_i \,|\, f \rangle \psi_i \tag{3.29}$$

Completeness

We now evaluate $\langle f \,|\, f \rangle$ in which f and f^* are expanded as in equation (3.27), with the two independent summations given different dummy indices

$$\langle f \,|\, f \rangle = \left\langle \sum_j a_j \psi_j \,\bigg|\, \sum_i a_i \psi_i \right\rangle = \sum_j \sum_i a_j^* a_j \langle \psi_j \,|\, \psi_i \rangle = \sum_i |a_i|^2$$

Without loss of generality we may assume that the function f is normalized, so that $\langle f \,|\, f \rangle = 1$ and

$$\sum_i |a_i|^2 = 1 \qquad\qquad (3.30)$$

Equation (3.30) may be used as a criterion for completeness. If an eigenfunction ψ_n with a non-vanishing coefficient a_n were missing from the summation in equation (3.27), then the series would still converge, but it would be incomplete and would therefore not converge to f. The corresponding coefficient a_n would be missing from the left-hand side of equation (3.30). Since each term in the summation in equation (3.30) is positive, the sum without a_n would be less than unity. Only if the expansion set ψ_i in equation (3.27) is complete will (3.30) be satisfied.

The completeness criterion can also be expressed in another form. For this purpose we need to introduce the variables explicitly. For simplicity we assume first that f is a function of only one variable x. In this case, equation (3.29) is

$$f(x) = \sum_i \left[\int \psi_i^*(x') f(x') \, dx' \right] \psi_i(x)$$

where x' is the dummy variable of integration. Interchanging the order of summation and integration gives

$$f(x) = \int \left[\sum_i \psi_i^*(x') \psi_i(x) \right] f(x') \, dx'$$

Thus, the summation is equal to the Dirac delta function (see Appendix C)

$$\sum_i \psi_i^*(x') \psi_i(x) = \delta(x - x') \qquad\qquad (3.31)$$

This expression, known as the *completeness relation* and sometimes as the *closure relation*, is valid only if the set of eigenfunctions is complete, and may be used as a mathematical test for completeness. Notice that the completeness relation (3.31) is not related to the choice of the arbitrary function f, whereas the criterion (3.30) is related.

The completeness relation for the multi-variable case is slightly more complex. When expressed explicitly in terms of its variables, equation (3.29) is

$$f(q_1, q_2, \ldots) = \sum_i \left[\int \psi_i^*(q_1', q_2', \ldots) f(q_1', q_2', \ldots) w(q_1', q_2', \ldots) \, dq_1', dq_2', \ldots \right]$$

$$\times \, \psi_i(q_1, q_2, \ldots)$$

Interchanging the order of summation and integration gives

$$f(q_1, q_2, \ldots) = \int \left[\sum_i \psi_i^*(q_1', q_2', \ldots) \psi_i(q_1, q_2, \ldots) \right]$$

$$\times \, f(q_1', q_2', \ldots) w(q_1', q_2', \ldots) \, dq_1 \, dq_2 \ldots$$

so that the completeness relation takes the form

$$w(q_1', q_2', \ldots) \sum_i \psi_i^*(q_1', q_2', \ldots) \psi_i(q_1, q_2, \ldots) = \delta(q_1 - q_1') \delta(q_2 - q_2') \cdots$$

$$(3.32)$$

3.5 Simultaneous eigenfunctions

Suppose the members of a complete set of functions ψ_i are simultaneously eigenfunctions of two hermitian operators $\hat{A}$ and $\hat{B}$ with eigenvalues α_i and β_i, respectively

$$\hat{A}\psi_i = \alpha_i \psi_i$$

$$\hat{B}\psi_i = \beta_i \psi_i$$

If we operate on the first eigenvalue equation with $\hat{B}$ and on the second with $\hat{A}$, we obtain

$$\hat{B}\hat{A}\psi_i = \alpha_i \hat{B}\psi_i = \alpha_i \beta_i \psi_i$$

$$\hat{A}\hat{B}\psi_i = \beta_i \hat{A}\psi_i = \alpha_i \beta_i \psi_i$$

from which it follows that

$$(\hat{A}\hat{B} - \hat{B}\hat{A})\psi_i = [A, B]\psi_i = 0$$

Thus, the functions ψ_i are eigenfunctions of the commutator $[\hat{A}, \hat{B}]$ with eigenvalues equal to zero. An operator that gives zero when applied to any member of a complete set of functions is itself zero, so that $\hat{A}$ and $\hat{B}$ commute. We have just shown that *if the operators $\hat{A}$ and $\hat{B}$ have a complete set of simultaneous eigenfunctions, then $\hat{A}$ and $\hat{B}$ commute.*

We now prove the converse, namely, that *eigenfunctions of commuting operators can always be constructed to be simultaneous eigenfunctions.* Suppose that $\hat{A}\psi_i = \alpha_i \psi_i$ and that $[\hat{A}, \hat{B}] = 0$. Since $\hat{A}$ and $\hat{B}$ commute, we have

$$\hat{A}\hat{B}\psi_i = \hat{B}\hat{A}\psi_i = \hat{B}(\alpha_i\psi_i) = \alpha_i\hat{B}\psi_i$$

Therefore, the function $\hat{B}\psi_i$ is an eigenfunction of $\hat{A}$ with eigenvalue α_i.

There are now two possibilities; the eigenvalue α_i of $\hat{A}$ is either non-degenerate or degenerate. If α_i is non-degenerate, then it corresponds to only one independent eigenfunction ψ_i, so that the function $\hat{B}\psi_i$ is proportional to ψ_i

$$\hat{B}\psi_i = \beta_i\psi_i$$

where β_i is the proportionality constant and therefore the eigenvalue of $\hat{B}$ corresponding to ψ_i. Thus, the function ψ_i is a simultaneous eigenfunction of both $\hat{A}$ and $\hat{B}$.

On the other hand, suppose the eigenvalue α_i is degenerate. For simplicity, we consider the case of a doubly degenerate eigenvalue α_i; the extension to n-fold degeneracy is straightforward. The function ψ_i is then any linear combination of two linearly independent, orthonormal eigenfunctions ψ_{i1} and ψ_{i2} of $\hat{A}$ corresponding to the eigenvalue α_i

$$\psi_i = c_1\psi_{i1} + c_2\psi_{i2}$$

We need to determine the coefficients c_1, c_2 such that $\hat{B}\psi_i = \beta_i\psi_i$, that is

$$c_1\hat{B}\psi_{i1} + c_2\hat{B}\psi_{i2} = \beta_i(c_1\psi_{i1} + c_2\psi_{i2})$$

If we take the scalar product of this equation first with ψ_{i1} and then with ψ_{i2}, we obtain

$$c_1(B_{11} - \beta_i) + c_2B_{12} = 0$$
$$c_1B_{21} + c_2(B_{22} - \beta_i) = 0$$

where we have introduced the simplified notation

$$B_{jk} \equiv \langle\psi_{ij}\,|\,\hat{B}\psi_{ik}\rangle$$

These simultaneous linear homogeneous equations determine c_1 and c_2 and have a non-trivial solution if the determinant of the coefficients of c_1, c_2 vanishes

$$\begin{vmatrix} B_{11} - \beta_i & B_{12} \\ B_{21} & B_{22} - \beta_i \end{vmatrix} = 0$$

or

$$\beta_i^2 - (B_{11} + B_{22})\beta_i + B_{11}B_{22} - B_{12}B_{21} = 0$$

This quadratic equation has two roots $\beta_i^{(1)}$ and $\beta_i^{(2)}$, which lead to two corresponding sets of constants $c_1^{(1)}$, $c_2^{(1)}$ and $c_1^{(2)}$, $c_2^{(2)}$. Thus, there are two distinct functions $\psi_i^{(1)}$ and $\psi_i^{(2)}$

$$\psi_i^{(1)} = c_1^{(1)}\psi_{i1} + c_2^{(1)}\psi_{i2}$$

$$\psi_i^{(2)} = c_1^{(2)}\psi_{i1} + c_2^{(2)}\psi_{i2}$$

which satisfy the relations

$$\hat{B}\psi_i^{(1)} = \beta_i^{(1)}\psi_i^{(1)}$$

$$\hat{B}\psi_i^{(2)} = \beta_i^{(2)}\psi_i^{(2)}$$

and are, therefore, simultaneous eigenfunctions of the commuting operators $\hat{A}$ and $\hat{B}$.

This analysis can be extended to three or more operators. If three operators $\hat{A}$, $\hat{B}$, and $\hat{C}$ have a complete set of simultaneous eigenfunctions, then the argument above shows that $\hat{A}$ and $\hat{B}$ commute, $\hat{B}$ and $\hat{C}$ commute, and $\hat{A}$ and $\hat{C}$ commute. Furthermore, the converse is also true. If $\hat{A}$ commutes with both $\hat{B}$ and $\hat{C}$, and $\hat{B}$ commutes with $\hat{C}$, then the three operators possess simultaneous eigenfunctions. To show this, suppose that the three operators commute with one another. We know that since $\hat{A}$ and $\hat{B}$ commute, they possess simultaneous eigenfunctions ψ_i such that

$$\hat{A}\psi_i = \alpha_i\psi_i$$

$$\hat{B}\psi_i = \beta_i\psi_i$$

We next operate on each of these expressions with $\hat{C}$, giving

$$\hat{C}\hat{A}\psi_i = \hat{A}(\hat{C}\psi_i) = \hat{C}(\alpha_i\psi_i) = \alpha_i(\hat{C}\psi_i)$$

$$\hat{C}\hat{B}\psi_i = \hat{B}(\hat{C}\psi_i) = \hat{C}(\beta_i\psi_i) = \beta_i(\hat{C}\psi_i)$$

Thus, the function $\hat{C}\psi_i$ is an eigenfunction of both $\hat{A}$ and $\hat{B}$ with eigenvalues α_i and β_i, respectively. If α_i and β_i are non-degenerate, then there is only one eigenfunction ψ_i corresponding to them and the function $\hat{C}\psi_i$ is proportional to ψ_i

$$\hat{C}\psi_i = \gamma_i\psi_i$$

and, consequently, $\hat{A}$, $\hat{B}$, and $\hat{C}$ possess simultaneous eigenfunctions. For degenerate eigenvalues α_i and/or β_i, simultaneous eigenfunctions may be constructed using a procedure parallel to the one described above for the doubly degenerate two-operator case.

We note here that if $\hat{A}$ commutes with $\hat{B}$ and $\hat{B}$ commutes with $\hat{C}$, but $\hat{A}$ does *not* commute with $\hat{C}$, then $\hat{A}$ and $\hat{B}$ possess simultaneous eigenfunctions, $\hat{B}$ and $\hat{C}$ possess simultaneous eigenfunctions, but $\hat{A}$ and $\hat{C}$ do not. The set of simultaneous eigenfunctions of $\hat{A}$ and $\hat{B}$ will differ from the set for $\hat{B}$ and $\hat{C}$. An example of this situation is discussed in Chapter 5.

In some of the derivations presented in this section, operators need not be hermitian. However, we are only interested in the properties of hermitian operators because quantum mechanics requires them. Therefore, we have implicitly assumed that all the operators are hermitian and we have not bothered to comment on the parts where hermiticity is not required.

3.6 Hilbert space and Dirac notation

This section introduces the basic mathematics of linear vector spaces as an alternative conceptual scheme for quantum-mechanical wave functions. The concept of vector spaces was developed before quantum mechanics, but Dirac applied it to wave functions and introduced a particularly useful and widely accepted notation. Much of the literature on quantum mechanics uses Dirac's ideas and notation.

A set of complete orthonormal functions $\psi_i(x)$ of a single variable x may be regarded as the basis vectors of a linear vector space of either finite or infinite dimensions, depending on whether the complete set contains a finite or infinite number of members. The situation is analogous to three-dimensional cartesian space formed by three orthogonal unit vectors. In quantum mechanics we usually (see Section 7.2 for an exception) encounter complete sets with an infinite number of members and, therefore, are usually concerned with linear vector spaces of infinite dimensionality. Such a linear vector space is called a *Hilbert space*. The functions $\psi_i(x)$ used as the basis vectors may constitute a discrete set or a continuous set. While a vector space composed of a discrete set of basis vectors is easier to visualize (even if the space is of infinite dimensionality) than one composed of a continuous set, there is no mathematical reason to exclude continuous basis vectors from the concept of Hilbert space. In Dirac notation, the basis vectors in Hilbert space are called *ket vectors* or just *kets* and are represented by the symbol $|\psi_i\rangle$ or sometimes simply by $|i\rangle$. These ket vectors determine a *ket space*.

When a ket $|\psi_i\rangle$ is multiplied by a constant c, the result $c\,|\psi_i\rangle = |c\psi_i\rangle$ is a ket in the same direction as $|\psi_i\rangle$; only the magnitude of the ket vector is changed. However, when an operator $\hat{A}$ acts on a ket $|\psi_i\rangle$, the result is another ket $|\phi_i\rangle$

$$|\phi_i\rangle = \hat{A}|\psi_i\rangle = |\hat{A}\psi_i\rangle$$

In general, the ket $|\phi_i\rangle$ is not in the same direction as $|\psi_i\rangle$ nor in the same direction as any other ket $|\psi_j\rangle$, but rather has projections along several or all basis kets. If an operator $\hat{A}$ acts on all kets $|\psi_i\rangle$ of the basis set, and the resulting set of kets $|\phi_i\rangle = |\hat{A}\psi_i\rangle$ are orthonormal, then the net result of the operation is a rotation of the basis set $|\psi_i\rangle$ about the origin to a new basis set $|\phi_i\rangle$. In the situation where $\hat{A}$ acting on $|\psi_i\rangle$ gives a constant times $|\psi_i\rangle$ (cf. equation (3.5))

$$\hat{A}|\psi_i\rangle = |\hat{A}\psi_i\rangle = \alpha_i|\psi_i\rangle$$

the ket $|\hat{A}\psi_i\rangle$ is along the direction of $|\psi_i\rangle$ and the kets $|\psi_i\rangle$ are said to be *eigenkets* of the operator $\hat{A}$.

Although the expressions $\hat{A}|\psi_i\rangle$ and $|\hat{A}\psi_i\rangle$ are completely equivalent, there is a subtle distinction between them. The first, $\hat{A}|\psi_i\rangle$, indicates the operator $\hat{A}$ being applied to the ket $|\psi_i\rangle$. The quantity $|\hat{A}\psi_i\rangle$ is the ket which results from that application.

Bra vectors

The functions $\psi_i(x)$ are, in general, complex functions. As a consequence, ket space is a complex vector space, making it mathematically necessary to introduce a corresponding set of vectors which are the *adjoints* of the ket vectors. The adjoint (sometimes also called the *complex conjugate transpose*) of a complex vector is the generalization of the complex conjugate of a complex number. In Dirac notation these adjoint vectors are called *bra vectors* or *bras* and are denoted by $\langle\psi_i|$ or $\langle i|$. Thus, the bra $\langle\psi_i|$ is the adjoint $|\psi_i\rangle^\dagger$ of the ket $|\psi_i\rangle$ and, conversely, the ket $|\psi_i\rangle$ is the adjoint $\langle\psi_i|^\dagger$ of the bra $\langle\psi_i|$

$$|\psi_i\rangle^\dagger = \langle\psi_i|$$

$$\langle\psi_i|^\dagger = |\psi_i\rangle$$

These bra vectors determine a *bra space*, just as the kets determine ket space.

The *scalar product* or *inner product* of a bra $\langle\phi|$ and a ket $|\psi\rangle$ is written in Dirac notation as $\langle\phi|\psi\rangle$ and is defined as

$$\langle\phi|\psi\rangle = \int \phi^*(x)\psi(x)\,\mathrm{d}x$$

The bracket (*bra-c-ket*) in $\langle\phi|\psi\rangle$ provides the names for the component vectors. This notation was introduced in Section 3.2 as a shorthand for the scalar product integral. The scalar product of a ket $|\psi\rangle$ with its corresponding bra $\langle\psi|$ gives a real, positive number and is the analog of multiplying a complex number by its complex conjugate. The scalar product of a bra $\langle\psi_j|$ and the ket $|\hat{A}\psi_i\rangle$ is expressed in Dirac notation as $\langle\psi_j|\hat{A}|\psi_i\rangle$ or as $\langle j|\hat{A}|i\rangle$. These scalar products are also known as the *matrix elements* of $\hat{A}$ and are sometimes denoted by A_{ij}.

To every ket in ket space, there corresponds a bra in bra space. For the ket

$c|\psi_i\rangle$, the corresponding bra is $c^*\langle\psi_i|$. We can also write $c|\psi_i\rangle$ as $|c\psi_i\rangle$, in which case the corresponding bra is $\langle c\psi_i|$, so that

$$\langle c\psi_i| = c^*\langle\psi_i|$$

For every linear operator $\hat{A}$ that transforms $|\psi_i\rangle$ in ket space into $|\phi_i\rangle = |\hat{A}\psi_i\rangle$, there is a corresponding linear operator $\hat{A}^\dagger$ in bra space which transforms $\langle\psi_i|$ into $\langle\phi_i| = \langle\hat{A}\psi_i|$. This operator $\hat{A}^\dagger$ is called the *adjoint* of $\hat{A}$. In bra space the transformation is expressed as

$$\langle\hat{A}\psi_i| = \langle\psi_i|\hat{A}^\dagger$$

Thus, for bras the operator acts on the vector to its left, whereas for kets the operator acts on the vector to its right.

To find the relationship between $\hat{A}$ and its adjoint $\hat{A}^\dagger$, we take the scalar product of $\langle\hat{A}\psi_j|$ and $|\psi_i\rangle$

$$\langle\hat{A}\psi_j|\psi_i\rangle = \langle\psi_j|\hat{A}^\dagger|\psi_i\rangle \tag{3.33a}$$

or in integral notation

$$\int(\hat{A}\psi_j)^*\psi_i\,\mathrm{d}x = \int\psi_j^*\hat{A}^\dagger\psi_i\,\mathrm{d}x \tag{3.33b}$$

A comparison with equation (3.8) shows that if $\hat{A}$ is hermitian, then we have $\hat{A}^\dagger = \hat{A}$ and $\hat{A}$ is said to be *self-adjoint*. The two terms, *hermitian* and *self-adjoint*, are synonymous. To find the adjoint of a non-hermitian operator, we apply equations (3.33). For example, we see from equation (3.10) that the adjoint of the operator $\mathrm{d}/\mathrm{d}x$ is $-\mathrm{d}/\mathrm{d}x$.

Since the scalar product $\langle\psi|\phi\rangle$ is equal to $\langle\phi|\psi\rangle^*$, we see that

$$\langle\hat{A}\psi_j|\psi_i\rangle = \langle\psi_i|\hat{A}|\psi_j\rangle^* \tag{3.34}$$

Combining equations (3.33a) and (3.34) gives

$$\langle\psi_j|\hat{A}^\dagger|\psi_i\rangle = \langle\psi_i|\hat{A}|\psi_j\rangle^* \tag{3.35}$$

If we replace $\hat{A}$ in equation (3.35) by the operator $\hat{A}^\dagger$, we obtain

$$\langle\psi_j|(\hat{A}^\dagger)^\dagger|\psi_i\rangle = \langle\psi_i|\hat{A}^\dagger|\psi_j\rangle^* \tag{3.36}$$

where $(\hat{A}^\dagger)^\dagger$ is the adjoint of the operator $\hat{A}^\dagger$. Equation (3.35) may be rewritten as

$$\langle\psi_i|\hat{A}^\dagger|\psi_j\rangle^* = \langle\psi_j|\hat{A}|\psi_i\rangle$$

and when compared with (3.36), we see that

$$\langle\psi_j|(\hat{A}^\dagger)^\dagger|\psi_i\rangle = \langle\psi_j|\hat{A}|\psi_i\rangle$$

We conclude that

$$(\hat{A}^\dagger)^\dagger = \hat{A} \tag{3.37}$$

From equation (3.35) we can also show that

$$(c\hat{A})^\dagger = c^* \hat{A}^\dagger \tag{3.38}$$

where c is any complex constant, and that

$$(\hat{A} + \hat{B})^\dagger = \hat{A}^\dagger + \hat{B}^\dagger \tag{3.39}$$

To obtain the adjoint of the product $\hat{A}\hat{B}$ of two operators, we apply equation (3.33a), first to $\hat{A}\hat{B}$, then to $\hat{A}$, and finally to $\hat{B}$

$$\langle \psi_j | (\hat{A}\hat{B})^\dagger | \psi_i \rangle = \langle \hat{A}\hat{B}\psi_j | \psi_i \rangle = \langle \hat{B}\psi_j | \hat{A}^\dagger | \psi_i \rangle = \langle \psi_j | \hat{B}^\dagger \hat{A}^\dagger | \psi_i \rangle$$

Thus, we have the relation

$$(\hat{A}\hat{B})^\dagger = \hat{B}^\dagger \hat{A}^\dagger \tag{3.40}$$

If $\hat{A}$ and $\hat{B}$ are hermitian (self-adjoint), then we have $(\hat{A}\hat{B})^\dagger = \hat{B}\hat{A}$ and further, if $\hat{A}$ and $\hat{B}$ commute, then the product $\hat{A}\hat{B}$ is hermitian or self-adjoint.

The *outer product* of a bra $\langle \phi |$ and a ket $| \psi \rangle$ is $| \psi \rangle \langle \phi |$ and behaves as an operator. If we let this outer product operate on another ket $| \chi \rangle$, we obtain the expression $| \psi \rangle \langle \phi | \chi \rangle$, which can be regarded in two ways. The scalar product $\langle \phi | \chi \rangle$ is a complex number multiplying the ket $| \psi \rangle$, so that the complete expression is a ket parallel to $| \psi \rangle$. Alternatively, the operator $| \psi \rangle \langle \phi |$ acts on the ket $| \chi \rangle$ and transfroms $| \chi \rangle$ into a ket proportional to $| \psi \rangle$.

To find the adjoint of the outer product $| \chi \rangle \langle \phi |$ of the ket $| \chi \rangle$ and the bra $\langle \phi |$, we let $\hat{A}$ in equation (3.35) be equal to $| \chi \rangle \langle \phi |$ and obtain

$$\langle \psi_j | (| \chi \rangle \langle \phi |)^\dagger | \psi_i \rangle = \langle \psi_i | (| \chi \rangle \langle \phi |) | \psi_j \rangle^* = \langle \psi_i | \chi \rangle^* \langle \phi | \psi_j \rangle^*$$

$$= \langle \chi | \psi_i \rangle \langle \psi_j | \phi \rangle = \langle \psi_j | \phi \rangle \langle \chi | \psi_i \rangle = \langle \psi_j | (| \phi \rangle \langle \chi |) | \psi_i \rangle$$

Setting equal the operators in the left-most and right-most integrals, we find that

$$(| \chi \rangle \langle \phi |)^\dagger = | \phi \rangle \langle \chi | \tag{3.41}$$

Projection operator

We define the operator $\hat{P}_i$ as the outer product of $| \psi_i \rangle$ and its corresponding bra

$$\hat{P}_i = | \psi_i \rangle \langle \psi_i | \equiv | i \rangle \langle i | \tag{3.42}$$

and apply $\hat{P}_i$ to an arbitrary ket $| \psi \rangle$

$$\hat{P}_i | \phi \rangle = | i \rangle \langle i | \phi \rangle$$

Thus, the result of $\hat{P}_i$ acting on $| \phi \rangle$ is a ket proportional to $| i \rangle$, the proportionality constant being the scalar product $\langle \psi_i | \phi \rangle$. The operator $\hat{P}_i$, then, projects $| \phi \rangle$ onto $| \psi_i \rangle$ and for that reason is known as the *projection operator*. The operator $\hat{P}_i^2$ is given by

$$\hat{P}_i^2 = \hat{P}_i \hat{P}_i = | i \rangle \langle i | i \rangle \langle i | = | i \rangle \langle i | = \hat{P}_i$$

where we have noted that the kets $| i \rangle$ are normalized. Likewise, the operator $\hat{P}_i^n$

for $n > 2$ also equals $\hat{P}_i$. This property is consistent with the interpretation of $\hat{P}_i$ as a projection operator since the result of projecting $|\phi\rangle$ onto $|i\rangle$ should be the same whether the projection is carried out once, twice, or multiple times. The operator $\hat{P}_i$ is hermitian, so that the projection of $|\phi\rangle$ on $|\psi_i\rangle$ is equal to the projection of $|\psi_i\rangle$ on $|\phi\rangle$. To show that $\hat{P}_i$ is hermitian, we let $|\chi\rangle = |\phi\rangle = |i\rangle$ in equation (3.41) and obtain $\hat{P}_i^\dagger = \hat{P}_i$.

The expansion of a function $f(x)$ in terms of the orthonormal set $\psi_i(x)$, as shown in equation (3.27), may be expressed in terms of kets as

$$|f\rangle = \sum_i a_i |\psi_i\rangle = \sum_i a_i |i\rangle$$

where $|f\rangle$ is regarded as a vector in ket space. The constants a_i are the projections of $|f\rangle$ on the 'unit ket vectors' $|i\rangle$ and are given by equation (3.28)

$$a_i = \langle i|f\rangle$$

Combining these two equations gives equation (3.29), which when expressed in Dirac notation is

$$|f\rangle = \sum_i |i\rangle\langle i|f\rangle$$

Since $f(x)$ is an arbitrary function of x, the operator $\sum_i |i\rangle\langle i|$ must equal the identity operator, so that

$$\sum_i |i\rangle\langle i| = 1 \tag{3.43}$$

From the definition of $\hat{P}_i$ in equation (3.42), we see that

$$\sum_i \hat{P}_i = 1$$

Since the operator $\sum_i |i\rangle\langle i|$ equals unity, it may be inserted at any point in an equation. Accordingly, we insert it between the bra and the ket in the scalar product of $|f\rangle$ with itself

$$\langle f|f\rangle = \left\langle f \left| \left(\sum_i |i\rangle\langle i| \right) \right| f \right\rangle = 1$$

where we have assumed $|f\rangle$ is normalized. This expression may be written as

$$\langle f|f\rangle = \sum_i \langle f|i\rangle\langle i|f\rangle = \sum_i |\langle i|f\rangle|^2 = \sum_i |a_i|^2 = 1$$

Thus, the expression (3.43) is related to the completeness criterion (3.30) and is called, therefore, the completeness relation.

3.7 Postulates of quantum mechanics

In this section we state the postulates of quantum mechanics in terms of the properties of linear operators. By way of an introduction to quantum theory, the basic principles have already been presented in Chapters 1 and 2. The purpose of that introduction is to provide a rationale for the quantum concepts by showing how the particle–wave duality leads to the postulate of a wave function based on the properties of a wave packet. Although this approach, based in part on historical development, helps to explain why certain quantum concepts were proposed, the basic principles of quantum mechanics cannot be obtained by any process of deduction. They must be stated as postulates to be accepted because the conclusions drawn from them agree with experiment without exception.

We first state the postulates succinctly and then elaborate on each of them with particular regard to the mathematical properties of linear operators. The postulates are as follows.

1. The state of a physical system is defined by a normalized function Ψ of the spatial coordinates and the time. This function contains all the information that exists on the state of the system.
2. Every physical observable A is represented by a linear hermitian operator $\hat{A}$.
3. Every individual measurement of a physical observable A yields an eigenvalue of the corresponding operator $\hat{A}$. The average value or expectation value $\langle A \rangle$ from a series of measurements of A for systems, each of which is in the exact same state Ψ, is given by $\langle A \rangle = \langle \Psi | A | \Psi \rangle$.
4. If a measurement of a physical observable A for a system in state Ψ gives the eigenvalue λ_n of $\hat{A}$, then the state of the system immediately after the measurement is the eigenfunction (if λ_n is non-degenerate) or a linear combination of eigenfunctions (if λ_n is degenerate) corresponding to λ_n.
5. The time dependence of the state function Ψ is determined by the time-dependent Schrödinger differential equation

$$i\hbar \frac{\partial \Psi}{\partial t} = \hat{H}\Psi$$

where $\hat{H}$ is the Hamiltonian operator for the system.

This list of postulates is not complete in that two quantum concepts are not covered, spin and identical particles. In Section 1.7 we mentioned in passing that an electron has an intrinsic angular momentum called spin. Other particles also possess spin. The quantum-mechanical treatment of spin is postponed until Chapter 7. Moreover, the state function for a system of two or more identical and therefore indistinguishable particles requires special consideration and is discussed in Chapter 8.

State function

According to the first postulate, the state of a physical system is completely described by a state function $\Psi(\mathbf{q}, t)$ or ket $|\Psi\rangle$, which depends on spatial coordinates $\mathbf{q}$ and the time t. This function is sometimes also called a state vector or a wave function. The coordinate vector $\mathbf{q}$ has components $q_1, q_2, \ldots$, so that the state function may also be written as $\Psi(q_1, q_2, \ldots, t)$. For a particle or system that moves in only one dimension (say along the x-axis), the vector $\mathbf{q}$ has only one component and the state vector Ψ is a function of x and t: $\Psi(x, t)$. For a particle or system in three dimensions, the components of $\mathbf{q}$ are x, y, z and Ψ is a function of the position vector $\mathbf{r}$ and t: $\Psi(\mathbf{r}, t)$. The state function is single-valued, a continuous function of each of its variables, and square or quadratically integrable.

For a one-dimensional system, the quantity $\Psi^*(x, t)\Psi(x, t)$ is the probability density for finding the system at position x at time t. In three dimensions, the quantity $\Psi^*(\mathbf{r}, t)\Psi(\mathbf{r}, t)$ is the probability density for finding the system at point $\mathbf{r}$ at time t. For a multi-variable system, the product $\Psi^*(q_1, q_2, \ldots, t)\Psi(q_1, q_2, \ldots, t)$ is the probability density that the system has coordinates $q_1, q_2, \ldots$ at time t. We show below that this interpretation of $\Psi^*\Psi$ follows from postulate 3. We usually assume that the state function is normalized

$$\int \Psi^*(q_1, q_2, \ldots, t)\Psi(q_1, q_2, \ldots, t)w(q_1, q_2, \ldots)\,\mathrm{d}q_1\,\mathrm{d}q_2 \ldots = 1$$

or in Dirac notation

$$\langle \Psi | \Psi \rangle = 1$$

where the limits of integration are over all allowed values of $q_1, q_2, \ldots$

Physical quantities or observables

The second postulate states that a physical quantity or observable is represented in quantum mechanics by a hermitian operator. To every classically defined function $A(\mathbf{r}, \mathbf{p})$ of position and momentum there corresponds a quantum-mechanical linear hermitian operator $\hat{A}(\mathbf{r}, (\hbar/\mathrm{i})\nabla)$. Thus, to obtain the quantum-mechanical operator, the momentum $\mathbf{p}$ in the classical function is replaced by the operator $\hat{\mathbf{p}}$

$$\hat{\mathbf{p}} = \frac{\hbar}{\mathrm{i}} \nabla \tag{3.44}$$

or, in terms of components

$$\hat{p}_x = \frac{\hbar}{\mathrm{i}} \frac{\partial}{\partial x}, \quad \hat{p}_y = \frac{\hbar}{\mathrm{i}} \frac{\partial}{\partial y}, \quad \hat{p}_z = \frac{\hbar}{\mathrm{i}} \frac{\partial}{\partial z}$$

For multi-particle systems with cartesian coordinates $\mathbf{r}_1, \mathbf{r}_2, \ldots$, the classical function $A(\mathbf{r}_1, \mathbf{r}_2, \ldots, \mathbf{p}_1, \mathbf{p}_2, \ldots)$ possesses the corresponding operator $\hat{A}(\mathbf{r}_1, \mathbf{r}_2, \ldots, (\hbar/\mathrm{i})\mathbf{\nabla}_1, (\hbar/\mathrm{i})\mathbf{\nabla}_2, \ldots)$ where $\mathbf{\nabla}_k$ is the gradient with respect to $\mathbf{r}_k$. For non-cartesian coordinates, the construction of the quantum-mechanical operator $\hat{A}$ is more complex and is not presented here.

The classical function A is an *observable*, meaning that it is a physically measurable property of the system. For example, for a one-particle system the Hamiltonian operator $\hat{H}$ corresponding to the classical Hamiltonian function

$$H(\mathbf{r}, \mathbf{p}) = \frac{p^2}{2m} + V(\mathbf{r})$$

where $p^2 = \mathbf{p} \cdot \mathbf{p} = p_x^2 + p_y^2 + p_z^2$, is

$$\hat{H} = -\frac{\hbar^2}{2m}\nabla^2 + V(\mathbf{r})$$

The linear operator $\hat{H}$ is easily shown to be hermitian.

Measurement of observable properties

The third postulate relates to the measurement of observable properties. Every individual measurement of a physical observable A yields an eigenvalue λ_i of the operator $\hat{A}$. The eigenvalues are given by

$$\hat{A}|i\rangle = \lambda_i|i\rangle \tag{3.45}$$

where $|i\rangle$ are the orthonormal eigenkets of $\hat{A}$. Since $\hat{A}$ is hermitian, the eigenvalues are all real. It is essential for the theory that $\hat{A}$ is hermitian because any measured quantity must, of course, be a real number. If the spectrum of $\hat{A}$ is discrete, then the eigenvalues λ_i are discrete and the measurements of A are quantized. If, on the other hand, the eigenfunctions $|i\rangle$ form a continuous, infinite set, then the eigenvalues λ_i are continuous and the measured values of A are not quantized. The set of eigenkets $|i\rangle$ of the dynamical operator $\hat{A}$ are assumed to be complete. In some cases it is possible to show explicitly that $|i\rangle$ forms a complete set, but in other cases we must assume that property.

The *expectation value* or *mean value* $\langle A \rangle$ of the physical observable A at time t for a system in a normalized state Ψ is given by

$$\langle A \rangle = \langle \Psi|\hat{A}|\Psi \rangle \tag{3.46}$$

If Ψ is not normalized, then the appropriate expression is

$$\langle A \rangle = \frac{\langle \Psi|\hat{A}|\Psi \rangle}{\langle \Psi|\Psi \rangle}$$

Some examples of expectation values are as follows

$$\langle x \rangle = \langle \Psi | x | \Psi \rangle$$

$$\langle p_x \rangle = \left\langle \Psi \left| \frac{\hbar}{i} \frac{\partial}{\partial x} \right| \Psi \right\rangle$$

$$\langle \mathbf{r} \rangle = \langle \Psi | \mathbf{r} | \Psi \rangle$$

$$\langle \mathbf{p} \rangle = \left\langle \Psi \left| \frac{\hbar}{i} \nabla \right| \Psi \right\rangle$$

$$E = \langle H \rangle = \left\langle \Psi \left| -\frac{\hbar^2}{2m} \nabla^2 + V(\mathbf{r}) \right| \Psi \right\rangle$$

The expectation value $\langle A \rangle$ is not the result of a single measurement of the property A, but rather the average of a large number (in the limit, an infinite number) of measurements of A on systems, each of which is in the same state Ψ. Each individual measurement yields one of the eigenvalues λ_i, and $\langle A \rangle$ is then the average of the observed array of eigenvalues. For example, if the eigenvalue λ_1 is observed four times, the eigenvalue λ_2 three times, the eigenvalue λ_3 once, and no other eigenvalues are observed, then the expectation value $\langle A \rangle$ is given by

$$\langle A \rangle = \frac{4\lambda_1 + 3\lambda_2 + \lambda_3}{8}$$

In practice, many more than eight observations would be required to obtain a reliable value for $\langle A \rangle$.

In general, the expectation value $\langle A \rangle$ of the observable A may be written for a discrete set of eigenfunctions as

$$\langle A \rangle = \sum_i P_i \lambda_i \tag{3.47}$$

where P_i is the probability of obtaining the value λ_i. If the state function Ψ for a system happens to coincide with one of the eigenstates $|i\rangle$, then only the eigenvalue λ_i would be observed each time a measurement of A is made and therefore the expectation value $\langle A \rangle$ would equal λ_i

$$\langle A \rangle = \langle i | \hat{A} | i \rangle = \langle i | \lambda_i | i \rangle = \lambda_i$$

It is important not to confuse the expectation value $\langle A \rangle$ with the time average of A for a single system.

For an arbitrary state Ψ at a fixed time t, the ket $|\Psi\rangle$ may be expanded in terms of the complete set of eigenkets of $\hat{A}$. In order to make the following discussion clearer, we now introduce a slightly more complicated notation. Each eigenvalue λ_i will now be distinct, so that $\lambda_i \neq \lambda_j$ for $i \neq j$. We let g_i be the degeneracy of the eigenvalue λ_i and let $|i\alpha\rangle$, $\alpha = 1, 2, \ldots, g_i$, be the orthonormal eigenkets of $\hat{A}$. We assume that the subset of kets corresponding to each eigenvalue λ_i has been made orthogonal by the Schmidt procedure outlined in Section 3.3.

If the eigenkets $|i\alpha\rangle$ constitute a discrete set, we may expand the state vector $|\Psi\rangle$ as

$$|\Psi\rangle = \sum_i \sum_{\alpha=1}^{g_i} c_{i\alpha} |i\alpha\rangle \tag{3.48}$$

where the expansion coefficients $c_{i\alpha}$ are

$$c_{i\alpha} = \langle i\alpha | \Psi \rangle \tag{3.49}$$

The expansion of the bra vector $\langle \Psi |$ is, therefore, given by

$$\langle \Psi | = \sum_j \sum_{\beta=1}^{g_j} c_{j\beta}^* \langle j\beta | \tag{3.50}$$

where the dummy indices i and α have been replaced by j and β.

The expectation value of $\hat{A}$ is obtained by substituting equations (3.48) and (3.50) into (3.46)

$$\langle A \rangle = \sum_j \sum_{\beta-1}^{g_j} \sum_i \sum_{\alpha=1}^{g_i} c_{j\beta}^\dagger c_{i\alpha} \langle j\beta | \hat{A} | i\alpha \rangle - \sum_j \sum_{\beta=1}^{g_j} \sum_i \sum_{\alpha=1}^{g_i} c_{j\beta}^* c_{i\alpha} \lambda_i \langle j\beta | i\alpha \rangle$$

$$= \sum_i \sum_{\alpha=1}^{g_i} |c_{i\alpha}|^2 \lambda_i \tag{3.51}$$

where we have noted that the kets $|i\alpha\rangle$ are orthonormal, so that

$$\langle j\beta | i\alpha \rangle = \delta_{ij} \delta_{\alpha\beta}$$

A comparison of equations (3.47) and (3.51) relates the probability P_i to the expansion coefficients $c_{i\alpha}$

$$P_i = \sum_{\alpha=1}^{g_i} |c_{i\alpha}|^2 = \sum_{\alpha=1}^{g_i} |\langle i\alpha | \Psi \rangle|^2 \tag{3.52}$$

where equation (3.49) has also been introduced. For the case where λ_i is non-degenerate, the index α is not needed and equation (3.52) reduces to

$$P_i = |c_i|^2 = |\langle i | \Psi \rangle|^2$$

For a continuous spectrum of eigenkets with non-degenerate eigenvalues, it is more convenient to write the eigenvalue equation (3.45) in the form

$$\hat{A}|\lambda\rangle = \lambda|\lambda\rangle$$

where λ is now a continuous variable and $|\lambda\rangle$ is the eigenfunction whose eigenvalue is λ. The expansion of the state vector Ψ becomes

$$|\Psi\rangle = \int c(\lambda)|\lambda\rangle \, d\lambda$$

where

$$c(\lambda) = \langle\lambda|\Psi\rangle$$

and the expectation value of A takes the form

$$\langle A \rangle = \int |c(\lambda)|^2 \, \lambda \, d\lambda \tag{3.53}$$

If dP_λ is the probability of obtaining a value of A between λ and $\lambda + d\lambda$, then equation (3.47) is replaced by

$$\langle A \rangle = \int \lambda \, dP_\lambda$$

and we see that

$$dP_\lambda = |c(\lambda)|^2 \, d\lambda = |\langle\lambda|\Psi\rangle|^2 \, d\lambda$$

The probability dP_λ is often written in the form

$$dP_\lambda = \rho(\lambda) \, d\lambda$$

where $\rho(\lambda)$ is the *probability density* of obtaining the result λ and is given by

$$\rho(\lambda) = |c(\lambda)|^2 = |\langle\lambda|\Psi\rangle|^2$$

In terms of the probability density, equation (3.53) becomes

$$\langle A \rangle = \int \lambda \rho(\lambda) \, d\lambda \tag{3.54}$$

In some applications to physical systems, the eigenkets of $\hat{A}$ possess a partially discrete and a partially continuous spectrum, in which case equations (3.51) and (3.53) must be combined.

The scalar product $\langle\Psi|\Psi\rangle$ may be evaluated from equations (3.48) and (3.50) as

$$\langle\Psi|\Psi\rangle = \sum_j \sum_{\beta=1}^{g_j} \sum_i \sum_{\alpha=1}^{g_i} c_{j\beta}^* c_{i\alpha} \langle j\beta|i\alpha\rangle = \sum_i \sum_{\alpha=1}^{g_i} |c_{i\alpha}|^2$$

$$= \sum_i \sum_{\alpha=1}^{g_i} |\langle i\alpha|\Psi\rangle|^2 = \sum_i P_i$$

Since the state vector Ψ is normalized, this expression gives

$$\sum_i P_i = 1$$

Thus, the sum of the probabilities P_i equals unity as it must from the definition of probability. For a continuous set of eigenkets, this relationship is replaced by

$$\int dP_\lambda = \int \rho(\lambda)\, d\lambda = 1$$

As an example, we consider a particle in a one-dimensional box as discussed in Section 2.5. Suppose that the state function $\Psi(x)$ for this particle is time-independent and is given by

$$\Psi(x) = C \sin^5\left(\frac{\pi x}{a}\right), \qquad 0 \leqslant x \leqslant a$$

where C is a constant which normalizes $\Psi(x)$. The eigenfunctions $|n\rangle$ and eigenvalues E_n of the Hamiltonian operator $\hat{H}$ are

$$|n\rangle = \sqrt{\frac{2}{a}}\sin\left(\frac{n\pi x}{a}\right), \qquad E_n = \frac{n^2 h^2}{8ma^2}, \qquad n = 1, 2, \ldots$$

Obviously, the state function $\Psi(x)$ is not an eigenfunction of $\hat{H}$. Following the general procedure described above, we expand $\Psi(x)$ in terms of the eigenfunctions $|n\rangle$. This expansion is the same as an expansion in a Fourier series, as described in Appendix B. As a shortcut we may use equations (A.39) and (A.40) to obtain the identity

$$\sin^5\theta = \frac{1}{16}(10\sin\theta - 5\sin 3\theta + \sin 5\theta)$$

so that the expansion of $\Psi(x)$ is

$$\Psi(x) = \frac{C}{16}\left[10\sin\left(\frac{\pi x}{a}\right) - 5\sin\left(\frac{3\pi x}{a}\right) + \sin\left(\frac{5\pi x}{a}\right)\right]$$

$$- \frac{C}{16}\sqrt{\frac{a}{2}}(10|1\rangle - 5|3\rangle + |5\rangle)$$

A measurement of the energy of a particle in state $\Psi(x)$ yields one of three values and no other value. The values and their probabilities are

$$E_1 = \frac{h^2}{8ma^2}, \qquad P_1 = \frac{10^2}{10^2 + 5^2 + 1^2} = \frac{100}{126} = 0.794$$

$$E_3 = \frac{9h^2}{8ma^2}, \qquad P_3 = \frac{5^2}{126} = 0.198$$

$$E_5 = \frac{25h^2}{8ma^2}, \qquad P_5 = \frac{1^2}{126} = 0.008$$

The sum of the probabilities is unity,

$$P_1 + P_3 + P_5 = 0.794 + 0.198 + 0.008 = 1$$

The interpretation that the quantity $\Psi^*(q_1, q_2, \ldots, t)\Psi(q_1, q_2, \ldots, t)$ is the probability density that the coordinates of the system at time t are

q_1, q_2, ... may be shown by comparing equations (3.46) and (3.54) for A equal to the coordinate vector $\mathbf{q}$

$$\langle \mathbf{q} \rangle = \langle \Psi | \mathbf{q} | \Psi \rangle = \int \Psi^*(\mathbf{q}) \Psi(\mathbf{q}) \mathbf{q} \, d\tau$$

$$\langle \mathbf{q} \rangle = \int \rho(\mathbf{q}) \mathbf{q} \, d\tau$$

For these two expressions to be mutually consistent, we must have

$$\rho(\mathbf{q}) = \Psi^*(\mathbf{q}) \Psi(\mathbf{q})$$

Thus, this interpretation of $\Psi^* \Psi$ follows from postulate 3 and for this reason is not included in the statement of postulate 1.

Collapse of the state function

The measurement of a physical observable A gives one of the eigenvalues λ_n of the operator $\hat{A}$. As stated by the fourth postulate, a consequence of this measurement is the sudden change in the state function of the system from its original form Ψ to an eigenfunction or linear combination of eigenfunctions of $\hat{A}$ corresponding to λ_n.

At a fixed time t just before the measurement takes place, the ket $|\Psi\rangle$ may be expanded in terms of the eigenkets $|i\alpha\rangle$ of $\hat{A}$, as shown in equation (3.48). If the measurement gives a non-degenerate eigenvalue λ_n, then immediately after the measurement the system is in state $|n\rangle$. The state function Ψ is said to *collapse* to the function $|n\rangle$. A second measurement of A on this same system, if taken immediately after the first, always yields the same result λ_n. If the eigenvalue λ_n is degenerate, then right after the measurement the state function is some linear combination of the eigenkets $|n\alpha\rangle$, $\alpha = 1, 2, \ldots, g_n$. A second, immediate measurement of A still yields λ_n as the result.

From postulates 4 and 5, we see that the state function Ψ can change with time for two different reasons. A discontinuous change in Ψ occurs when some property of the system is measured. The state of the system changes suddenly from Ψ to an eigenfunction or linear combination of eigenfunctions associated with the observed eigenvalue. An isolated system, on the other hand, undergoes a continuous change with time in accordance with the time-dependent Schrödinger equation.

Time evolution of the state function

The fifth postulate stipulates that the time evolution of the state function Ψ is determined by the time-dependent Schrödinger equation

$$i\hbar \frac{\partial \Psi}{\partial t} = \hat{H}\Psi \tag{3.55}$$

where $\hat{H}$ is the Hamiltonian operator of the system and, in general, changes with time. However, in this book we only consider systems for which the Hamiltonian operator is time-independent. To solve the time-dependent Schrödinger equation, we express the state function $\Psi(\mathbf{q}, t)$ as the product of two functions

$$\Psi(\mathbf{q}, t) = \psi(\mathbf{q})\chi(t) \tag{3.56}$$

where $\psi(\mathbf{q})$ depends only on the spatial variables and $\chi(t)$ depends only on the time. In Section 2.4 we discuss the procedure for separating the partial differential equation (3.55) into two differential equations, one involving only the spatial variables and the other only the time. The state function $\Psi(\mathbf{q}, t)$ is then shown to be

$$\Psi(\mathbf{q}, t) = \psi(\mathbf{q})e^{-iEt/\hbar} \tag{3.57}$$

where E is the separation constant. Since it follows from equation (3.57) that

$$|\Psi(\mathbf{q}, t)|^2 = |\psi(\mathbf{q})|^2$$

the probability density is independent of the time t and $\Psi(\mathbf{q}, t)$ is a *stationary state*.

The spatial differential equation, known as the time-independent Schrödinger equation, is

$$\hat{H}\psi(\mathbf{q}) = E\psi(\mathbf{q})$$

Thus, the spatial function $\psi(\mathbf{q})$ is actually a set of eigenfunctions $\psi_n(\mathbf{q})$ of the Hamiltonian operator $\hat{H}$ with eigenvalues E_n. The time-independent Schrödinger equation takes the form

$$\hat{H}\psi_n(\mathbf{q}) = E_n\psi_n(\mathbf{q}) \tag{3.58}$$

and the general solution of the time-dependent Schrödinger equation is

$$\Psi(\mathbf{q}, t) = \sum_n c_n\psi_n(\mathbf{q})e^{-iE_nt/\hbar} \tag{3.59}$$

where c_n are arbitrary complex constants.

The appearance of the Hamiltonian operator in equation (3.55) as stipulated by postulate 5 gives that operator a special status in quantum mechanics. Knowledge of the eigenfunctions and eigenvalues of the Hamiltonian operator for a given system is sufficient to determine the stationary states of the system and the expectation values of any other dynamical variables.

We next address the question as to whether equation (3.59) is actually the most general solution of the time-dependent Schrödinger equation. Are there other solutions that are not expressible in the form of equation (3.59)? To answer that question, we assume that $\Psi(\mathbf{q}, t)$ is any arbitrary solution of the parital differential equation (3.55). We suppose further that the set of functions $\psi_n(\mathbf{q})$ which satisfy the eigenvalue equation (3.58) is complete. Then we can, in general, expand $\Psi(\mathbf{q}, t)$ in terms of the complete set $\psi_n(\mathbf{q})$ and obtain

$$\Psi(\mathbf{q}, t) = \sum_n a_n(t)\psi_n(\mathbf{q}) \tag{3.60}$$

The coefficients $a_n(t)$ in the expansion are given by

$$a_n(t) = \langle \psi_n(\mathbf{q})|\Psi(\mathbf{q}, t)\rangle \tag{3.61}$$

and are functions of the time t, but not of the coordinates $\mathbf{q}$. We substitute the expansion (3.60) into the differential equation (3.55) to obtain

$$\sum_n \left[\frac{\hbar}{i}\frac{\partial a_n(t)}{\partial t} + a_n(t)\hat{H}\right]\psi_n(\mathbf{q}) = \sum_n \left[\frac{\hbar}{i}\frac{\partial a_n(t)}{\partial t} + E_n a_n(t)\right]\psi_n(\mathbf{q}) = 0 \tag{3.62}$$

where we have also noted that the functions $\psi_n(\mathbf{q})$ are eigenfunctions of $\hat{H}$ in accordance with equation (3.58). We next multiply equation (3.62) by $\psi_k^*(\mathbf{q})$, the complex conjugate of one of the eigenfunctions of the orthogonal set, and integrate over the spatial variables

$$\sum_n \left[\frac{\hbar}{i}\frac{\partial a_n(t)}{\partial t} + E_n a_n(t)\right]\langle\psi_k(\mathbf{q})|\psi_n(\mathbf{q})\rangle$$

$$= \sum_n \left[\frac{\hbar}{i}\frac{\partial a_n(t)}{\partial t} + E_n a_n(t)\right]\delta_{kn} = \frac{\hbar}{i}\frac{\partial a_k(t)}{\partial t} + E_k a_k(t) = 0$$

Replacing the dummy index k by n, we obtain the result

$$a_n(t) = c_n e^{-iE_n t/\hbar} \tag{3.63}$$

where c_n is a constant independent of both $\mathbf{q}$ and t. Substitution of equation (3.63) into (3.60) gives equation (3.59), showing that equation (3.59) is indeed the most general form for a solution of the time-dependent Schrödinger equation. All solutions may be expressed as the sum over stationary states.

3.8 Parity operator

The parity operator $\hat{\Pi}$ is defined by the relation

$$\hat{\Pi}\psi(\mathbf{q}) = \psi(-\mathbf{q}) \tag{3.64}$$

Thus, the parity operator reverses the sign of each cartesian coordinate. This operator is equivalent to an inversion of the coordinate system through the origin. In one and three dimensions, equation (3.64) takes the form

$$\hat{\Pi}\psi(x) = \psi(-x), \quad \hat{\Pi}\psi(\mathbf{r}) = \hat{\Pi}\psi(x, y, z) = \hat{\Pi}\psi(-x, -y, -z) = \psi(-\mathbf{r})$$

The operator $\hat{\Pi}^2$ is equal to unity since

$$\hat{\Pi}^2\psi(\mathbf{q}) = \hat{\Pi}(\hat{\Pi}\psi(\mathbf{q})) = \hat{\Pi}\psi(-\mathbf{q}) = \psi(\mathbf{q})$$

Further, we see that

$$\hat{\Pi}^n\psi(\mathbf{q}) = \psi(\mathbf{q}), \quad n \text{ even}$$
$$= \psi(-\mathbf{q}), \quad n \text{ odd}$$

or

$$\hat{\Pi}^n = 1, \quad n \text{ even}$$
$$= \hat{\Pi}, \quad n \text{ odd}$$

The operator $\hat{\Pi}$ is linear and hermitian. In the one-dimensional case, the hermiticity of $\hat{\Pi}$ is demonstrated as follows

$$\langle\phi|\hat{\Pi}|\psi\rangle = \int_{-\infty}^{\infty} \phi^*(x)\psi(-x)\,dx = -\int_{\infty}^{-\infty} \phi^*(-x')\psi(x')\,dx'$$
$$= \int_{-\infty}^{\infty} \psi(x')\hat{\Pi}\phi^*(x')\,dx' = \langle\hat{\Pi}\phi|\psi\rangle$$

where x in the second integral is replaced by $x' = -x$ to obtain the third integral. By applying the same procedure to each coordinate, we can show that $\hat{\Pi}$ is hermitian with respect to multi-dimensional functions.

The eigenvalues λ of the parity operator $\hat{\Pi}$ are given by

$$\hat{\Pi}\psi_\lambda(\mathbf{q}) = \lambda\psi_\lambda(\mathbf{q}) \tag{3.65}$$

where $\psi_\lambda(\mathbf{q})$ are the corresponding eigenfunctions. If we apply $\hat{\Pi}$ to both sides of equation (3.65), we obtain

$$\hat{\Pi}^2\psi_\lambda(\mathbf{q}) = \lambda\hat{\Pi}\psi_\lambda(\mathbf{q}) = \lambda^2\psi_\lambda(\mathbf{q})$$

Since $\hat{\Pi}^2 = 1$, we see that $\lambda^2 = 1$ and that the eigenvalues λ, which must be real because $\hat{\Pi}$ is hermitian, are equal to either $+1$ or -1. To find the eigenfunctions $\psi_\lambda(\mathbf{q})$, we note that equation (3.65) now becomes

$$\psi_\lambda(-\mathbf{q}) = \pm\psi_\lambda(\mathbf{q})$$

For $\lambda = 1$, the eigenfunctions of $\hat{\Pi}$ are even functions of $\mathbf{q}$, while for $\lambda = -1$, they are odd functions of $\mathbf{q}$. An even function of $\mathbf{q}$ is said to be of *even parity*, while *odd parity* refers to an odd function of $\mathbf{q}$. Thus, the eigenfunctions of $\hat{\Pi}$ are any well-behaved functions that are either of even or odd parity in their cartesian variables.

We show next that the parity operator $\hat{\Pi}$ commutes with the Hamiltonian operator $\hat{H}$ if the potential energy $V(\mathbf{q})$ is an even function of $\mathbf{q}$. The kinetic energy term in the Hamiltonian operator is given by

$$-\frac{\hbar^2}{2m}\nabla^2 = -\frac{\hbar^2}{2m}\left(\frac{\partial^2}{\partial q_1^2} + \frac{\partial^2}{\partial q_2^2} + \cdots\right)$$

and is an even function of each q_k. If the potential energy $V(\mathbf{q})$ is also an even function of each q_k, then we have $\hat{H}(\mathbf{q}) = \hat{H}(-\mathbf{q})$ and

$$[\hat{H}, \hat{\Pi}]f(\mathbf{q}) = \hat{H}(\mathbf{q})\hat{\Pi}f(\mathbf{q}) - \hat{\Pi}\hat{H}(\mathbf{q})f(\mathbf{q}) = \hat{H}(\mathbf{q})f(-\mathbf{q}) - \hat{H}(-\mathbf{q})f(-\mathbf{q}) = 0$$

Since the function $f(\mathbf{q})$ is arbitrary, the commutator of $\hat{H}$ and $\hat{\Pi}$ vanishes. Thus, these operators have simultaneous eigenfunctions for systems with $V(\mathbf{q}) = V(-\mathbf{q})$.

If the potential energy of a system is an even function of the coordinates and if $\psi(\mathbf{q})$ is a solution of the time-independent Schrödinger equation, then the function $\psi(-\mathbf{q})$ is also a solution. When the eigenvalues of the Hamiltonian operator are non-degenerate, these two solutions are not independent of each other, but are proportional

$$\psi(-\mathbf{q}) = c\psi(\mathbf{q})$$

These eigenfunctions are also eigenfunctions of the parity operator, leading to the conclusion that $c = \pm 1$. Consequently, some eigenfunctions will be of even parity while all the others will be of odd parity.

For a degenerate energy eigenvalue, the several corresponding eigenfunctions of $\hat{H}$ may not initially have a definite parity. However, each eigenfunction may be written as the sum of an even part $\psi_e(\mathbf{q})$ and an odd part $\psi_o(\mathbf{q})$

$$\psi(\mathbf{q}) = \psi_e(\mathbf{q}) + \psi_o(\mathbf{q})$$

where

$$\psi_e(\mathbf{q}) = \tfrac{1}{2}[\psi(\mathbf{q}) + \psi(-\mathbf{q})] = \psi_e(-\mathbf{q})$$

$$\psi_o(\mathbf{q}) = \tfrac{1}{2}[\psi(\mathbf{q}) - \psi(-\mathbf{q})] = -\psi_o(-\mathbf{q})$$

Since any linear combination of $\psi(\mathbf{q})$ and $\psi(-\mathbf{q})$ satisfies Schrödinger's equation, the functions $\psi_e(\mathbf{q})$ and $\psi_o(\mathbf{q})$ are eigenfunctions of $\hat{H}$. Furthermore, the functions $\psi_e(\mathbf{q})$ and $\psi_o(\mathbf{q})$ are also eigenfunctions of the parity operator $\hat{\Pi}$, the first with eigenvalue $+1$ and the second with eigenvalue -1.

3.9 Hellmann–Feynman theorem

A useful expression for evaluating expectation values is known as the *Hellmann–Feynman theorem*. This theorem is based on the observation that the Hamiltonian operator for a system depends on at least one parameter λ, which can be considered for mathematical purposes to be a continuous variable. For example, depending on the particular system, this parameter λ may be the mass of an electron or a nucleus, the electronic charge, the nuclear charge parameter Z, a constant in the potential energy, a quantum number, or even Planck's constant. The eigenfunctions and eigenvalues of $\hat{H}(\lambda)$ also depend on this parameter, so that the time-independent Schrödinger equation (3.58) may be written as

$$\hat{H}(\lambda)\psi_n(\lambda) = E_n(\lambda)\psi_n(\lambda) \tag{3.66}$$

The expectation value of $\hat{H}(\lambda)$ is, then

$$E_n(\lambda) = \langle \psi_n(\lambda)|\hat{H}(\lambda)|\psi_n(\lambda)\rangle \tag{3.67}$$

where we assume that $\psi_n(\lambda)$ is normalized

$$\langle \psi_n(\lambda)|\psi_n(\lambda)\rangle = 1 \tag{3.68}$$

To obtain the Hellmann–Feynman theorem, we differentiate equation (3.67) with respect to λ

$$\frac{\mathrm{d}}{\mathrm{d}\lambda} E_n(\lambda) = \left\langle \psi_n(\lambda)\left|\frac{\mathrm{d}}{\mathrm{d}\lambda}\hat{H}(\lambda)\right|\psi_n(\lambda)\right\rangle$$

$$+ \left\langle \frac{\mathrm{d}}{\mathrm{d}\lambda}\psi_n(\lambda)\left|\hat{H}(\lambda)\right|\psi_n(\lambda)\right\rangle + \left\langle \psi_n(\lambda)\left|\hat{H}(\lambda)\right|\frac{\mathrm{d}}{\mathrm{d}\lambda}\psi_n(\lambda)\right\rangle \tag{3.69}$$

Applying the hermitian property of $\hat{H}(\lambda)$ to the third integral on the right-hand side of equation (3.69) and then applying (3.66) to the second and third terms, we obtain

$$\frac{\mathrm{d}}{\mathrm{d}\lambda} E_n(\lambda) = \left\langle \psi_n(\lambda)\left|\frac{\mathrm{d}}{\mathrm{d}\lambda}\hat{H}(\lambda)\right|\psi_n(\lambda)\right\rangle$$

$$+ E_n(\lambda)\left[\left\langle \frac{\mathrm{d}}{\mathrm{d}\lambda}\psi_n(\lambda)\Big|\psi_n(\lambda)\right\rangle + \left\langle \psi_n(\lambda)\Big|\frac{\mathrm{d}}{\mathrm{d}\lambda}\psi_n(\lambda)\right\rangle\right] \tag{3.70}$$

The derivative of equation (3.68) with respect to λ is

$$\left\langle \frac{\mathrm{d}}{\mathrm{d}\lambda}\psi_n(\lambda)\Big|\psi_n(\lambda)\right\rangle + \left\langle \psi_n(\lambda)\Big|\frac{\mathrm{d}}{\mathrm{d}\lambda}\psi_n(\lambda)\right\rangle = 0$$

showing that the last term on the right-hand side of (3.70) vanishes. We thereby obtain the Hellmann–Feynman theorem

$$\frac{\mathrm{d}}{\mathrm{d}\lambda} E_n(\lambda) = \left\langle \psi_n(\lambda)\left|\frac{\mathrm{d}}{\mathrm{d}\lambda}\hat{H}(\lambda)\right|\psi_n(\lambda)\right\rangle \tag{3.71}$$

3.10 Time dependence of the expectation value

The expectation value $\langle A \rangle$ of the dynamical quantity or observable A is, in general, a function of the time t. To determine how $\langle A \rangle$ changes with time, we take the time derivative of equation (3.46)

$$\frac{d\langle A \rangle}{dt} = \frac{d}{dt}\langle \Psi | \hat{A} | \Psi \rangle = \left\langle \frac{\partial \Psi}{\partial t} \middle| \hat{A} \middle| \Psi \right\rangle + \left\langle \Psi \middle| \hat{A} \middle| \frac{\partial \Psi}{\partial t} \right\rangle + \left\langle \Psi \middle| \frac{\partial \hat{A}}{\partial t} \middle| \Psi \right\rangle$$

Equation (3.55) may be substituted for the time derivatives of the wave function to give

$$\frac{d\langle A \rangle}{dt} = \frac{i}{\hbar}\langle \hat{H}\Psi | \hat{A} | \Psi \rangle - \frac{i}{\hbar}\langle \Psi | \hat{A}\hat{H} | \Psi \rangle + \left\langle \Psi \middle| \frac{\partial \hat{A}}{\partial t} \middle| \Psi \right\rangle$$

$$= \frac{i}{\hbar}\langle \Psi | \hat{H}\hat{A} | \Psi \rangle - \frac{i}{\hbar}\langle \Psi | \hat{A}\hat{H} | \Psi \rangle + \left\langle \Psi \middle| \frac{\partial \hat{A}}{\partial t} \middle| \Psi \right\rangle$$

$$= \frac{i}{\hbar}\langle \Psi | [\hat{H}, \hat{A}] | \Psi \rangle + \left\langle \Psi \middle| \frac{\partial \hat{A}}{\partial t} \middle| \Psi \right\rangle$$

$$= \frac{i}{\hbar}\langle [\hat{H}, \hat{A}] \rangle + \left\langle \frac{\partial \hat{A}}{\partial t} \right\rangle$$

where the hermiticity of $\hat{H}$ and the definition (equation (3.3)) of the commutator have been used. If the operator $\hat{A}$ is not an explicit function of time, then the last term on the right-hand side vanishes and we have

$$\frac{d\langle A \rangle}{dt} = \frac{i}{\hbar}\langle [\hat{H}, \hat{A}] \rangle \tag{3.72}$$

If we set $\hat{A}$ equal to unity, then the commutator $[\hat{H}, \hat{A}]$ vanishes and equation (3.72) becomes

$$\frac{d\langle A \rangle}{dt} = 0$$

or

$$\frac{d}{dt}\langle \Psi | \hat{A} | \Psi \rangle = \frac{d}{dt}\langle \Psi | \Psi \rangle = 0$$

We thereby obtain the result in Section 2.2 that if Ψ is normalized, it remains normalized as time progresses.

If the operator $\hat{A}$ in equation (3.72) is set equal to $\hat{H}$, then again the commutator vanishes and we have

$$\frac{d\langle A \rangle}{dt} = \frac{d\langle H \rangle}{dt} = \frac{dE}{dt} = 0$$

Thus, the energy E of the system, which is equal to the expectation value of the Hamiltonian, is conserved if the Hamiltonian does not depend explicitly on time.

By setting the operator $\hat{A}$ in equation (3.72) equal first to the position variable x, then the variable y, and finally the variable z, we can show that

$$m\frac{d\langle x\rangle}{dt} = \langle p_x\rangle, \quad m\frac{d\langle y\rangle}{dt} = \langle p_y\rangle, \quad m\frac{d\langle z\rangle}{dt} = \langle p_z\rangle$$

or, in vector notation

$$m\frac{d\langle \mathbf{r}\rangle}{dt} = \langle \mathbf{p}\rangle$$

which is one of the Ehrenfest theorems discussed in Section 2.3. The other Ehrenfest theorem,

$$\frac{d\langle \mathbf{p}\rangle}{dt} = -\langle \mathbf{\nabla} V(\mathbf{r})\rangle$$

may be obtained from equation (3.72) by setting $\hat{A}$ successively equal to $\hat{p}_x$, $\hat{p}_y$, and $\hat{p}_z$.

3.11 Heisenberg uncertainty principle

We have shown in Section 3.5 that commuting hermitian operators have simultaneous eigenfunctions and, therefore, that the physical quantities associated with those operators can be observed simultaneously. On the other hand, if the hermitian operators $\hat{A}$ and $\hat{B}$ do not commute, then the physical observables A and B cannot both be precisely determined at the same time. We begin by demonstrating this conclusion.

Suppose that $\hat{A}$ and $\hat{B}$ do not commute. Let α_i and β_i be the eigenvalues of $\hat{A}$ and $\hat{B}$, respectively, with corresponding eigenstates $|\alpha_i\rangle$ and $|\beta_i\rangle$

$$\hat{A}|\alpha_i\rangle = \alpha_i|\alpha_i\rangle \tag{3.73a}$$

$$\hat{B}|\beta_i\rangle = \beta_i|\beta_i\rangle \tag{3.73b}$$

Some or all of the eigenvalues may be degenerate, but each eigenfunction has a unique index i. Suppose further that the system is in state $|\alpha_j\rangle$, one of the eigenstates of $\hat{A}$. If we measure the physical observable A, we obtain the result α_j. What happens if we simultaneously measure the physical observable B? To answer this question we need to calculate the expectation value $\langle B\rangle$ for this system

$$\langle B\rangle = \langle \alpha_j|\hat{B}|\alpha_j\rangle \tag{3.74}$$

If we expand the state function $|\alpha_j\rangle$ in terms of the complete, orthonormal set $|\beta_i\rangle$

$$|\alpha_j\rangle = \sum_i c_i|\beta_i\rangle$$

where c_i are the expansion coefficients, and substitute the expansion into equation (3.74), we obtain

$$\langle B \rangle = \sum_i \sum_k c_k^* c_i \langle \beta_k | \hat{B} | \beta_i \rangle = \sum_i \sum_k c_k^* c_i \beta_i \delta_{ki} = \sum_i |c_i|^2 \beta_i$$

where (3.73b) has been used. Thus, a measurement of B yields one of the many values β_i with a probability $|c_i|^2$. There is no way to predict which of the values β_i will be obtained and, therefore, the observables A and B cannot both be determined concurrently.

For a system in an arbitrary state Ψ, neither of the physical observables A and B can be precisely determined simultaneously if $\hat{A}$ and $\hat{B}$ do not commute. Let ΔA and ΔB represent the width of the spread of values for A and B, respectively. We define the *variance* $(\Delta A)^2$ by the relation

$$(\Delta A)^2 = \langle (\hat{A} - \langle A \rangle)^2 \rangle \tag{3.75}$$

that is, as the expectation value of the square of the deviation of A from its mean value. The positive square root ΔA is the *standard deviation* and is called the *uncertainty* in A. Noting that $\langle A \rangle$ is a real number, we can obtain an alternative expression for $(\Delta A)^2$ as follows:

$$(\Delta A)^2 = \langle (\hat{A} - \langle A \rangle)^2 \rangle = \langle \hat{A}^2 - 2\langle A \rangle \hat{A} + \langle A \rangle^2 \rangle$$

$$= \langle \hat{A}^2 \rangle - 2\langle A \rangle \langle A \rangle + \langle A \rangle^2 = \langle \hat{A}^2 \rangle - \langle A \rangle^2 \tag{3.76}$$

Expressions analogous to equations (3.75) and (3.76) apply for $(\Delta B)^2$.

Since $\hat{A}$ and $\hat{B}$ do not commute, we define the operator $\hat{C}$ by the relation

$$[\hat{A}, \hat{B}] = \hat{A}\hat{B} - \hat{B}\hat{A} = i\hat{C} \tag{3.77}$$

The operator $\hat{C}$ is hermitian as discussed in Section 3.3, so that its expectation value $\langle C \rangle$ is real. The commutator of $\hat{A} - \langle A \rangle$ and $\hat{B} - \langle B \rangle$ may be expanded as follows

$$[\hat{A} - \langle A \rangle, \hat{B} - \langle B \rangle] = (\hat{A} - \langle A \rangle)(\hat{B} - \langle B \rangle) - (\hat{B} - \langle B \rangle)(\hat{A} - \langle A \rangle)$$

$$= \hat{A}\hat{B} - \hat{B}\hat{A} = i\hat{C} \tag{3.78}$$

where the cross terms cancel since $\langle A \rangle$ and $\langle B \rangle$ are numbers and commute with the operators $\hat{A}$ and $\hat{B}$. We use equation (3.78) later in this section.

We now introduce the operator

$$\hat{A} - \langle A \rangle + i\lambda(\hat{B} - \langle B \rangle)$$

where λ is a real constant, and let this operator act on the state function Ψ

$$[\hat{A} - \langle A \rangle + i\lambda(\hat{B} - \langle B \rangle)]\Psi$$

The scalar product of the resulting function with itself is, of course, always positive, so that

$$\langle [\hat{A} - \langle A \rangle + i\lambda(\hat{B} - \langle B \rangle)]\Psi | [\hat{A} - \langle A \rangle + i\lambda(\hat{B} - \langle B \rangle)]\Psi \rangle \geq 0 \tag{3.79}$$

Expansion of this expression gives

$$\langle(\hat{A} - \langle A\rangle)\Psi|(\hat{A} - \langle A\rangle)\Psi\rangle + \lambda^2\langle(\hat{B} - \langle B\rangle)\Psi|(\hat{B} - \langle B\rangle)\Psi\rangle$$

$$+ i\lambda\langle(\hat{A} - \langle A\rangle)\Psi|(\hat{B} - \langle B\rangle)\Psi\rangle - i\lambda\langle(\hat{B} - \langle B\rangle)\Psi|(\hat{A} - \langle A\rangle)\Psi\rangle \geqslant 0$$

or, since $\hat{A}$ and $\hat{B}$ are hermitian

$$\langle\Psi|(\hat{A} - \langle A\rangle)^2|\Psi\rangle + \lambda^2\langle\Psi|(\hat{B} - \langle B\rangle)^2|\Psi\rangle$$

$$+ i\lambda\langle\Psi|[\hat{A} - \langle A\rangle, \hat{B} - \langle B\rangle]|\Psi\rangle \geqslant 0$$

Applying equations (3.75) and (3.78), we have

$$(\Delta A)^2 + \lambda^2(\Delta B)^2 - \lambda\langle C\rangle \geqslant 0$$

If we complete the square of the terms involving λ, we obtain

$$(\Delta A)^2 + (\Delta B)^2\left(\lambda - \frac{\langle C\rangle}{2(\Delta B)^2}\right)^2 - \frac{\langle C\rangle^2}{4(\Delta B)^2} \geqslant 0$$

Since λ is arbitrary, we select its value so as to eliminate the second term

$$\lambda = \frac{\langle C\rangle}{2(\Delta B)^2} \tag{3.80}$$

thereby giving

$$(\Delta A)^2(\Delta B)^2 \geqslant \tfrac{1}{4}\langle C\rangle^2$$

or, upon taking the positive square root,

$$\Delta A\Delta B \geqslant \tfrac{1}{2}|\langle C\rangle|$$

Substituting equation (3.77) into this result yields

$$\Delta A\Delta B \geqslant \tfrac{1}{2}|\langle[\hat{A}, \hat{B}]\rangle| \tag{3.81}$$

This general expression relates the uncertainties in the simultaneous measurements of A and B to the commutator of the corresponding operators $\hat{A}$ and $\hat{B}$ and is a general statement of the Heisenberg uncertainty principle.

Position–momentum uncertainty principle

We now consider the special case for which A is the variable x ($\hat{A} = x$) and B is the momentum p_x ($\hat{B} = -i\hbar\, d/dx$). The commutator $[\hat{A}, \hat{B}]$ may be evaluated by letting it operate on Ψ

$$[\hat{A}, \hat{B}]\Psi = -i\hbar\left(x\frac{d\Psi}{dx} - \frac{dx\Psi}{dx}\right) = i\hbar\Psi$$

so that $|\langle[\hat{A}, \hat{B}]\rangle| = \hbar$ and equation (3.81) gives

$$\Delta x\Delta p_x \geqslant \frac{\hbar}{2} \tag{3.82}$$

The Heisenberg position–momentum uncertainty principle (3.82) agrees with equation (2.26), which was derived by a different, but mathematically equivalent procedure. The relation (3.82) is consistent with (1.44), which is based on the Fourier transform properties of wave packets. The difference between the right-hand sides of (1.44) and (3.82) is due to the precise definition (3.75) of the uncertainties in equation (3.82).

Similar applications of equation (3.81) using the position–momentum pairs y, $\hat{p}_y$ and z, $\hat{p}_z$ yield

$$\Delta y \Delta p_y \geqslant \frac{\hbar}{2}, \quad \Delta z \Delta p_z \geqslant \frac{\hbar}{2}$$

Since x commutes with the operators $\hat{p}_y$ and $\hat{p}_z$, y commutes with $\hat{p}_x$ and $\hat{p}_z$, and z commutes with $\hat{p}_x$ and $\hat{p}_y$, the relation (3.81) gives

$$\Delta q_i \Delta p_j = 0, \quad i \neq j$$

where $q_1 = x$, $q_2 = y$, $q_3 = z$, $p_1 = p_x$, $p_2 = p_y$, $p_3 = p_z$. Thus, the position coordinate q_i and the momentum component p_j for $i \neq j$ may be precisely determined simultaneously.

Minimum uncertainty wave packet

The minimum value of the product $\Delta A \Delta B$ occurs for a particular state Ψ for which the relation (3.81) becomes an *equality*, i.e., when

$$\Delta A \Delta B = \tfrac{1}{2} |\langle [\hat{A}, \hat{B}] \rangle| \tag{3.83}$$

According to equation (3.79), this equality applies when

$$[\hat{A} - \langle A \rangle + i\lambda(\hat{B} - \langle B \rangle)]\Psi = 0 \tag{3.84}$$

where λ is given by (3.80). For the position–momentum example where $\hat{A} = x$ and $\hat{B} = -i\hbar \, d/dx$, equation (3.84) takes the form

$$\left(-i\hbar \frac{d}{dx} - \langle p_x \rangle \right) \Psi = \frac{i}{\lambda}(x - \langle x \rangle)\Psi$$

for which the solution is

$$\Psi = c e^{-(x - \langle x \rangle)^2 / 2\lambda\hbar} e^{i\langle p_x \rangle x / \hbar} \tag{3.85}$$

where c is a constant of integration and may be used to normalize Ψ. The real constant λ may be shown from equation (3.80) to be

$$\lambda = \frac{\hbar}{2(\Delta p_x)^2} = \frac{2(\Delta x)^2}{\hbar}$$

where the relation $\Delta x \Delta p_x = \hbar/2$ has been used, and is observed to be positive. Thus, the state function Ψ in equation (3.85) for a particle with minimum position–momentum uncertainty is a wave packet in the form of a plane wave $\exp[i\langle p_x \rangle x/\hbar]$ with wave number $k_0 = \langle p_x \rangle/\hbar$ multiplied by a gaussian modulating function centered at $\langle x \rangle$. Wave packets are discussed in Section

1.2. Only the spatial dependence of Ψ has been derived in equation (3.85). The state function Ψ may also depend on the time through the possible time dependence of the parameters c, λ, $\langle x \rangle$, and $\langle p_x \rangle$.

Energy–time uncertainty principle

We now wish to derive the energy–time uncertainty principle, which is discussed in Section 1.5 and expressed in equation (1.45). We show in Section 1.5 that for a wave packet associated with a free particle moving in the x-direction the product $\Delta E \Delta t$ is equal to the product $\Delta x \Delta p_x$ if ΔE and Δt are defined appropriately. However, this derivation does not apply to a particle in a potential field.

The position, momentum, and energy are all dynamical quantities and consequently possess quantum-mechanical operators from which expectation values at any given time may be determined. Time, on the other hand, has a unique role in non-relativistic quantum theory as an independent variable; dynamical quantities are functions of time. Thus, the 'uncertainty' in time cannot be related to a range of expectation values.

To obtain the energy-time uncertainty principle for a particle in a time-independent potential field, we set $\hat{A}$ equal to $\hat{H}$ in equation (3.81)

$$(\Delta E)(\Delta B) \geq \tfrac{1}{2} |\langle [\hat{H}, \hat{B}] \rangle|$$

where ΔE is the uncertainty in the energy as defined by (3.75) with $\hat{A} = \hat{H}$. Substitution of equation (3.72) into this expression gives

$$(\Delta E)(\Delta B) \geq \frac{\hbar}{2} \left| \frac{\mathrm{d}\langle B \rangle}{\mathrm{d}t} \right| \qquad (3.86)$$

In a short period of time Δt, the change in the expectation value of B is given by

$$\Delta B = \frac{\mathrm{d}\langle B \rangle}{\mathrm{d}t} \Delta t$$

When this expression is combined with equation (3.86), we obtain the desired result

$$(\Delta E)(\Delta t) \geq \frac{\hbar}{2} \qquad (3.87)$$

We see that the energy and time obey an uncertainty relation when Δt is defined as the period of time required for the expectation value of B to change by one standard deviation. This definition depends on the choice of the dynamical variable B so that Δt is relatively larger or smaller depending on that choice. If $\mathrm{d}\langle B \rangle / \mathrm{d}t$ is small so that B changes slowly with time, then the period Δt will be long and the uncertainty in the energy will be small.

Conversely, if B changes rapidly with time, then the period Δt for B to change by one standard deviation will be short and the uncertainty in the energy of the system will be large.

Problems

3.1 Which of the following operators are linear?

(a) $\sqrt{}$ (b) $\sin$ (c) $x\hat{D}_x$ (d) $\hat{D}_x x$

3.2 Demonstrate the validity of the relationships (3.4a) and (3.4b).

3.3 Show that

$$[\hat{A}, [\hat{B}, \hat{C}]] + [\hat{B}, [\hat{C}, \hat{A}]] + [\hat{C}, [\hat{A}, \hat{B}]] = 0$$

where $\hat{A}$, $\hat{B}$, and $\hat{C}$ are arbitrary linear operators.

3.4 Show that $(\hat{D}_x + x)(\hat{D}_x - x) = \hat{D}_x^2 - x^2 - 1$.

3.5 Show that xe^{-x^2} is an eigenfunction of the linear operator $(\hat{D}_x^2 - 4x^2)$. What is the eigenvalue?

3.6 Show that the operator $\hat{D}_x^2$ is hermitian. Is the operator $i\hat{D}_x^2$ hermitian?

3.7 Show that if the linear operators $\hat{A}$ and $\hat{B}$ do not commute, the operators $(\hat{A}\hat{B} + \hat{B}\hat{A})$ and $i[\hat{A}, \hat{B}]$ are hermitian.

3.8 If the real normalized functions $f(\mathbf{r})$ and $g(\mathbf{r})$ are not orthogonal, show that their sum $f(\mathbf{r}) + g(\mathbf{r})$ and their difference $f(\mathbf{r}) - g(\mathbf{r})$ are orthogonal.

3.9 Consider the set of functions $\psi_1 = e^{-x/2}$, $\psi_2 = xe^{-x/2}$, $\psi_3 = x^2e^{-x/2}$, $\psi_4 = x^3e^{-x/2}$, defined over the range $0 \leqslant x \leqslant \infty$. Use the Schmidt orthogonalization procedure to construct from the set ψ_i an orthogonal set of functions with $w(x) = 1$.

3.10 Evaluate the following commutators:

(a) $[x, \hat{p}_x]$ (b) $[x, \hat{p}_x^2]$ (c) $[x, \hat{H}]$ (d) $[\hat{p}_x, \hat{H}]$

3.11 Evaluate $[x, \hat{p}_x^3]$ and $[x^2, \hat{p}_x^2]$ using equations (3.4).

3.12 Using equation (3.4b), show by iteration that

$$[x^n, \hat{p}_x] = i\hbar n x^{n-1}$$

where n is a positive integer greater than zero.

3.13 Show that

$$[f(x), \hat{p}_x] = i\hbar \frac{df(x)}{dx}$$

3.14 Calculate the expectation values of x, x^2, $\hat{p}$, and $\hat{p}^2$ for a particle in a one-dimensional box in state ψ_n (see Section 2.5).

3.15 Calculate the expectation value of $\hat{p}^4$ for a particle in a one-dimensional box in state ψ_n.

3.16 A hermitian operator $\hat{A}$ has only three normalized eigenfunctions ψ_1, ψ_2, ψ_3, with corresponding eigenvalues $a_1 = 1$, $a_2 = 2$, $a_3 = 3$, respectively. For a particular state ϕ of the system, there is a 50% chance that a measure of A produces a_1 and equal chances for either a_2 or a_3.

(a) Calculate $\langle A \rangle$.

(b) Express the normalized wave function ϕ of the system in terms of the eigenfunctions of $\hat{A}$.

3.17 The wave function $\Psi(x)$ for a particle in a one-dimensional box of length a is

$$\Psi(x) = C \sin^7 \left(\frac{\pi x}{a} \right); \quad 0 \leqslant x \leqslant a$$

where C is a constant. What are the possible observed values for the energy and their respective probabilities?

3.18 If $|\psi\rangle$ is an eigenfunction of $\hat{H}$ with eigenvalue E, show that for any operator $\hat{A}$ the expectation value of $[\hat{H}, \hat{A}]$ vanishes, i.e.,

$$\langle \psi | [\hat{H}, \hat{A}] | \psi \rangle = 0$$

3.19 Derive both of the Ehrenfest theorems using equation (3.72).

3.20 Show that

$$\Delta H \Delta x \geqslant \frac{\hbar}{2m} \langle \hat{p}_x \rangle$$

4

Harmonic oscillator

In this chapter we treat in detail the quantum behavior of the harmonic oscillator. This physical system serves as an excellent example for illustrating the basic principles of quantum mechanics that are presented in Chapter 3. The Schrödinger equation for the harmonic oscillator can be solved rigorously and exactly for the energy eigenvalues and eigenstates. The mathematical process for the solution is neither trivial, as is the case for the particle in a box, nor excessively complicated. Moreover, we have the opportunity to introduce the *ladder operator* technique for solving the eigenvalue problem.

The harmonic oscillator is an important system in the study of physical phenomena in both classical and quantum mechanics. Classically, the harmonic oscillator describes the mechanical behavior of a spring and, by analogy, other phenomena such as the oscillations of charge flow in an electric circuit, the vibrations of sound-wave and light-wave generators, and oscillatory chemical reactions. The quantum-mechanical treatment of the harmonic oscillator may be applied to the vibrations of molecular bonds and has many other applications in quantum physics and field theory.

4.1 Classical treatment

The harmonic oscillator is an idealized one-dimensional physical system in which a single particle of mass m is attracted to the origin by a force F proportional to the displacement of the particle from the origin

$$F = -kx \tag{4.1}$$

The proportionality constant k is known as the *force constant*. The minus sign in equation (4.1) indicates that the force is in the opposite direction to the direction of the displacement. The typical experimental representation of the oscillator consists of a spring with one end stationary and with a mass m

attached to the other end. The spring is assumed to obey *Hooke's law*, that is to say, equation (4.1). The constant k is then often called the *spring constant*.

In classical mechanics the particle obeys Newton's second law of motion

$$F = ma = m\frac{d^2x}{dt^2} \tag{4.2}$$

where a is the acceleration of the particle and t is the time. The combination of equations (4.1) and (4.2) gives the differential equation

$$\frac{d^2x}{dt^2} = -\frac{k}{m}x$$

for which the solution is

$$x = A\sin(2\pi vt + b) = A\sin(\omega t + b) \tag{4.3}$$

where the *amplitude A* of the vibration and the *phase b* are the two constants of integration and where the frequency v and the angular frequency ω of vibration are related to k and m by

$$\omega = 2\pi v = \sqrt{\frac{k}{m}} \tag{4.4}$$

According to equation (4.3), the particle oscillates sinusoidally about the origin with frequency v and maximum displacement $\pm A$.

The potential energy V of a particle is related to the force F acting on it by the expression

$$F = -\frac{dV}{dx}$$

Thus, from equations (4.1) and (4.4), we see that for a harmonic oscillator the potential energy is given by

$$V = \tfrac{1}{2}kx^2 = \tfrac{1}{2}m\omega^2x^2 \tag{4.5}$$

The total energy E of the particle undergoing harmonic motion is given by

$$E = \tfrac{1}{2}mv^2 + V = \tfrac{1}{2}mv^2 + \tfrac{1}{2}m\omega^2x^2 \tag{4.6}$$

where v is the instantaneous velocity. If the oscillator is undisturbed by outside forces, the energy E remains fixed at a constant value. When the particle is at maximum displacement from the origin so that $x = \pm A$, the velocity v is zero and the potential energy is a maximum. As $|x|$ decreases, the potential decreases and the velocity increases keeping E constant. As the particle crosses the origin ($x = 0$), the velocity attains its maximum value $v = \sqrt{2E/m}$.

To relate the maximum displacement A to the constant energy E, we note that when $x = \pm A$, equation (4.6) becomes

$$E = \tfrac{1}{2}m\omega^2A^2$$

so that

$$A = \frac{1}{\omega}\sqrt{\frac{2E}{m}}$$

Thus, equation (4.3) takes the form

$$x = \frac{1}{\omega}\sqrt{\frac{2E}{m}}\,\sin(\omega t + b) \qquad (4.7)$$

By defining a reduced distance y as

$$y \equiv \omega\sqrt{\frac{m}{2E}}\,x \qquad (4.8)$$

so that the particle oscillates between $y = -1$ and $y = 1$, we may express the equation of motion (4.7) in a universal form that is independent of the total energy E

$$y(t) = \sin(\omega t + b) \qquad (4.9)$$

As the particle oscillates back and forth between $y = -1$ and $y = 1$, the probability that it will be observed between some value y and $y + dy$ is $P(y)\,dy$, where $P(y)$ is the *probability density*. Since the probability of finding the particle within the range $-1 \leqslant y \leqslant 1$ is unity (the particle must be somewhere in that range), the probability density is normalized

$$\int_{-1}^{1} P(y)\,dy = 1$$

The probability of finding the particle within the interval dy at a given distance y is proportional to the time dt spent in that interval

$$P(y)\,dy = c\,dt = c\frac{dt}{dy}\,dy$$

so that

$$P(y) = c\frac{dt}{dy}$$

where c is the proportionality constant. To find $P(y)$, we solve equation (4.9) for t

$$t(y) = \frac{1}{\omega}[\sin^{-1}(y) - b]$$

and then take the derivative to give

$$P(y) = \frac{c}{\omega}(1 - y^2)^{-1/2} = \frac{1}{\pi}(1 - y^2)^{-1/2} \qquad (4.10)$$

where c was determined by the normalization requirement. The probability density $P(y)$ for the oscillating particle is shown in Figure 4.1.

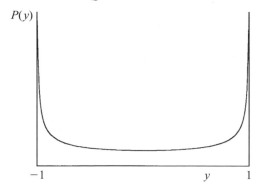

Figure 4.1 Classical probability density for an oscillating particle.

4.2 Quantum treatment

The classical Hamiltonian $H(x,\, p)$ for the harmonic oscillator is

$$H(x,\, p) = \frac{p^2}{2m} + V(x) = \frac{p^2}{2m} + \tfrac{1}{2}m\omega^2 x^2 \tag{4.11}$$

The Hamiltonian operator $\hat{H}(x,\, \hat{p})$ is obtained by replacing the momentum p in equation (4.11) with the momentum operator $\hat{p} = -i\hbar\, d/dx$

$$\hat{H} = \frac{\hat{p}^2}{2m} + \tfrac{1}{2}m\omega^2 x^2 = -\frac{\hbar^2}{2m}\frac{d^2}{dx^2} + \tfrac{1}{2}m\omega^2 x^2 \tag{4.12}$$

The Schrödinger equation is, then

$$-\frac{\hbar^2}{2m}\frac{d^2\psi(x)}{dx^2} + \tfrac{1}{2}m\omega^2 x^2 \psi(x) = E\psi(x) \tag{4.13}$$

It is convenient to introduce the dimensionless variable ξ by the definition

$$\xi = \left(\frac{m\omega}{\hbar}\right)^{1/2} x \tag{4.14}$$

so that the Hamiltonian operator becomes

$$\hat{H} = \frac{\hbar\omega}{2}\left(\xi^2 - \frac{d^2}{d\xi^2}\right) \tag{4.15}$$

Since the Hamiltonian operator is written in terms of the variable ξ rather than x, we should express the eigenstates in terms of ξ as well. Accordingly, we define the functions $\phi(\xi)$ by the relation

$$\phi(\xi) = \left(\frac{\hbar}{m\omega}\right)^{1/4}\psi(x) \tag{4.16}$$

If the functions $\psi(x)$ are normalized with respect to integration over x

$$\int_{-\infty}^{\infty} |\psi(x)|^2 \, dx = 1$$

then from equations (4.14) and (4.16) we see that the functions $\phi(\xi)$ are normalized with respect to integration over ξ

$$\int_{-\infty}^{\infty} |\phi(\xi)|^2 \, d\xi = 1$$

The Schrödinger equation (4.13) then takes the form

$$-\frac{d^2\phi(\xi)}{d\xi^2} + \xi^2\phi(\xi) = \frac{2E}{\hbar\omega}\phi(\xi) \tag{4.17}$$

Since the Hamiltonian operator is hermitian, the energy eigenvalues E are real.

There are two procedures available for solving this differential equation. The older procedure is the Frobenius or series solution method. The solution of equation (4.17) by this method is presented in Appendix G. In this chapter we use the more modern ladder operator procedure. Both methods give exactly the same results.

Ladder operators

We now solve the Schrödinger eigenvalue equation for the harmonic oscillator by the so-called factoring method using *ladder operators*. We introduce the two ladder operators $\hat{a}$ and $\hat{a}^\dagger$ by the definitions

$$\hat{a} \equiv \left(\frac{m\omega}{2\hbar}\right)^{1/2}\left(x + \frac{i\hat{p}}{m\omega}\right) = \frac{1}{\sqrt{2}}\left(\xi + \frac{d}{d\xi}\right) \tag{4.18a}$$

$$\hat{a}^\dagger \equiv \left(\frac{m\omega}{2\hbar}\right)^{1/2}\left(x - \frac{i\hat{p}}{m\omega}\right) = \frac{1}{\sqrt{2}}\left(\xi - \frac{d}{d\xi}\right) \tag{4.18b}$$

Application of equation (3.33) reveals that the operator $\hat{a}^\dagger$ is the adjoint of $\hat{a}$, which explains the notation. Since the operator $\hat{a}$ is not equal to its adjoint $\hat{a}^\dagger$, neither $\hat{a}$ nor $\hat{a}^\dagger$ is hermitian. (We follow here the common practice of using a lower case letter for the harmonic-oscillator ladder operators rather than our usual convention of using capital letters for operators.) We readily observe that

$$\hat{a}\hat{a}^\dagger = \frac{1}{2}\left(\xi^2 - \frac{d^2}{d\xi^2} + 1\right) = \frac{\hat{H}}{\hbar\omega} + \frac{1}{2} \tag{4.19a}$$

$$\hat{a}^\dagger\hat{a} = \frac{1}{2}\left(\xi^2 - \frac{d^2}{d\xi^2} - 1\right) = \frac{\hat{H}}{\hbar\omega} - \frac{1}{2} \tag{4.19b}$$

from which it follows that the commutator of $\hat{a}$ and $\hat{a}^\dagger$ is unity

$$[\hat{a}, \hat{a}^\dagger] = \hat{a}\hat{a}^\dagger - \hat{a}^\dagger\hat{a} = 1 \tag{4.20}$$

We next define the number operator $\hat{N}$ as the product $\hat{a}^\dagger\hat{a}$

$$\hat{N} \equiv \hat{a}^\dagger \hat{a} \tag{4.21}$$

The adjoint of $\hat{N}$ may be obtained as follows

$$\hat{N}^\dagger = (\hat{a}^\dagger \hat{a})^\dagger = \hat{a}^\dagger (\hat{a}^\dagger)^\dagger = \hat{a}^\dagger \hat{a} = \hat{N}$$

where the relations (3.40) and (3.37) have been used. We note that $\hat{N}$ is self-adjoint, making it hermitian and therefore having real eigenvalues. Equation (4.19b) may now be written in the form

$$\hat{H} = \hbar\omega(\hat{N} + \tfrac{1}{2}) \tag{4.22}$$

Since $\hat{H}$ and $\hat{N}$ differ only by the factor $\hbar\omega$ and an additive constant, they commute and, therefore, have the same eigenfunctions.

If the eigenvalues of $\hat{N}$ are represented by the parameter λ and the corresponding orthonormal eigenfunctions by $\phi_{\lambda i}(\xi)$ or, using Dirac notation, by $|\lambda i\rangle$, then we have

$$\hat{N}|\lambda i\rangle = \lambda|\lambda i\rangle \tag{4.23}$$

and

$$\hat{H}|\lambda i\rangle = \hbar\omega(\hat{N} + \tfrac{1}{2})|\lambda i\rangle = \hbar\omega(\lambda + \tfrac{1}{2})|\lambda i\rangle = E_\lambda|\lambda i\rangle \tag{4.24}$$

Thus, the energy eigenvalues E_λ are related to the eigenvalues of $\hat{N}$ by

$$E_\lambda = (\lambda + \tfrac{1}{2})\hbar\omega \tag{4.25}$$

The index i in $|\lambda i\rangle$ takes on integer values from 1 to g_λ, where g_λ is the degeneracy of the eigenvalue λ. We shall find shortly that each eigenvalue of $\hat{N}$ is non-degenerate, but in arriving at a general solution of the eigenvalue equation, we must initially allow for degeneracy.

From equations (4.20) and (4.21), we note that the product of $\hat{N}$ and either $\hat{a}$ or $\hat{a}^\dagger$ may be expressed as follows

$$\hat{N}\hat{a} = \hat{a}^\dagger \hat{a}\hat{a} = (\hat{a}\hat{a}^\dagger - 1)\hat{a} = \hat{a}(\hat{a}^\dagger \hat{a} - 1) = \hat{a}(\hat{N} - 1) \tag{4.26a}$$

$$\hat{N}\hat{a}^\dagger = \hat{a}^\dagger \hat{a}\hat{a}^\dagger = \hat{a}^\dagger(\hat{a}^\dagger \hat{a} + 1) = \hat{a}^\dagger(\hat{N} + 1) \tag{4.26b}$$

These identities are useful in the following discussion.

If we let the operator $\hat{N}$ act on the function $\hat{a}|\lambda i\rangle$, we obtain

$$\hat{N}\hat{a}|\lambda i\rangle = \hat{a}(\hat{N} - 1)|\lambda i\rangle = \hat{a}(\lambda - 1)|\lambda i\rangle = (\lambda - 1)\hat{a}|\lambda i\rangle \tag{4.27}$$

where equations (4.23) and (4.26a) have been introduced. Thus, we see that $\hat{a}|\lambda i\rangle$ is an eigenfunction of $\hat{N}$ with eigenvalue $\lambda - 1$. The operator $\hat{a}$ alters the eigenstate $|\lambda i\rangle$ to an eigenstate of $\hat{N}$ corresponding to a lower value for the eigenvalue, namely $\lambda - 1$. The energy of the oscillator is thereby reduced, according to (4.25), by $\hbar\omega$. As a consequence, the operator $\hat{a}$ is called a *lowering operator* or a *destruction operator*.

Letting $\hat{N}$ operate on the function $\hat{a}^\dagger|\lambda i\rangle$ gives

$$\hat{N}\hat{a}^\dagger|\lambda i\rangle = \hat{a}^\dagger(\hat{N}+1)|\lambda i\rangle = \hat{a}^\dagger(\lambda+1)|\lambda i\rangle = (\lambda+1)\hat{a}^\dagger|\lambda i\rangle \qquad (4.28)$$

where equations (4.23) and (4.26b) have been used. In this case we see that $\hat{a}^\dagger|\lambda i\rangle$ is an eigenfunction of $\hat{N}$ with eigenvalue $\lambda+1$. The operator $\hat{a}^\dagger$ changes the eigenstate $|\lambda i\rangle$ to an eigenstate of $\hat{N}$ with a higher value, $\lambda+1$, of the eigenvalue. The energy of the oscillator is increased by $\hbar\omega$. Thus, the operator $\hat{a}^\dagger$ is called a *raising operator* or a *creation operator*.

Quantization of the energy

In the determination of the energy eigenvalues, we first show that the eigenvalues λ of $\hat{N}$ are positive ($\lambda \geqslant 0$). Since the expectation value of the operator $\hat{N}$ for an oscillator in state $|\lambda i\rangle$ is λ, we have

$$\langle\lambda i|\hat{N}|\lambda i\rangle = \lambda\langle\lambda i|\lambda i\rangle = \lambda$$

The integral $\langle\lambda i|\hat{N}|\lambda i\rangle$ may also be transformed in the following manner

$$\langle\lambda i|\hat{N}|\lambda i\rangle = \langle\lambda i|\hat{a}^\dagger\hat{a}|\lambda i\rangle = \int(\hat{a}\phi_{\lambda i}^*)(\hat{a}\phi_{\lambda i})\,\mathrm{d}\tau = \int|\hat{a}\phi_{\lambda i}|^2\,\mathrm{d}\tau$$

The integral on the right must be positive, so that λ is positive and the eigenvalues of $\hat{N}$ and $\hat{H}$ cannot be negative.

For the condition $\lambda = 0$, we have

$$\int|\hat{a}\phi_{\lambda i}|^2\,\mathrm{d}\tau = 0$$

which requires that

$$\hat{a}|(\lambda=0)i\rangle = 0 \qquad (4.29)$$

For eigenvalues λ greater than zero, the quantity $\hat{a}|\lambda i\rangle$ is non-vanishing.

To find further restrictions on the values of λ, we select a suitably large, but otherwise arbitrary value of λ, say η, and continually apply the lowering operator $\hat{a}$ to the eigenstate $|\eta i\rangle$, thereby forming a succession of eigenvectors

$$\hat{a}|\eta i\rangle, \quad \hat{a}^2|\eta i\rangle, \quad \hat{a}^3|\eta i\rangle, \quad \ldots$$

with respective eigenvalues $\eta-1$, $\eta-2$, $\eta-3$, $\ldots$ We have already shown that if $|\eta i\rangle$ is an eigenfunction of $\hat{N}$, then $\hat{a}|\eta i\rangle$ is also an eigenfunction. By iteration, if $\hat{a}|\eta i\rangle$ is an eigenfunction of $\hat{N}$, then $\hat{a}^2|\eta i\rangle$ is an eigenfunction, and so forth, so that the members of the sequence are all eigenfunctions. Eventually this procedure gives an eigenfunction $\hat{a}^k|\eta i\rangle$ with eigenvalue $(\eta-k)$, k being a positive integer, such that $0 \leqslant (\eta-k) < 1$. The next step in the sequence would yield the eigenfunction $\hat{a}^{k+1}|\eta i\rangle$ with eigenvalue $\lambda = (\eta-k-1) < 0$, which is not allowed. Thus, the sequence must terminate by the condition

$$\hat{a}^{k+1}|\eta i\rangle = \hat{a}[\hat{a}^k|\eta i\rangle] = 0$$

The only circumstance in which $\hat{a}$ operating on an eigenvector yields the value zero is when the eigenvector corresponds to the eigenvalue $\lambda = 0$, as shown in equation (4.29). Since the eigenvalue of $\hat{a}^k|\eta i\rangle$ is $\eta - k$ and this eigenvalue equals zero, we have $\eta - k = 0$ and η must be an integer. The minimum value of $\lambda = \eta - k$ is, then, zero.

Beginning with $\lambda = 0$, we can apply the operator $\hat{a}^\dagger$ successively to $|0i\rangle$ to form a series of eigenvectors

$$\hat{a}^\dagger|0i\rangle, \quad \hat{a}^{\dagger 2}|0i\rangle, \quad \hat{a}^{\dagger 3}|0i\rangle, \quad \ldots$$

with respective eigenvalues 0, 1, 2, ... Thus, the eigenvalues of the operator $\hat{N}$ are the set of positive integers, so that $\lambda = 0, 1, 2, \ldots$ Since the value η was chosen arbitrarily and was shown to be an integer, this sequence generates all the eigenfunctions of $\hat{N}$. There are no eigenfunctions corresponding to non-integral values of λ. Since λ is now known to be an integer n, we replace λ by n in the remainder of this discussion of the harmonic oscillator.

The energy eigenvalues as related to λ in equation (4.25) are now expressed in terms of n by

$$E_n = (n + \tfrac{1}{2})\hbar\omega, \quad n = 0, 1, 2, \ldots \tag{4.30}$$

so that the energy is quantized in units of $\hbar\omega$. The lowest value of the energy or zero-point energy is $\hbar\omega/2$. Classically, the lowest energy for an oscillator is zero.

Non-degeneracy of the energy levels
To determine the degeneracy of the energy levels or, equivalently, of the eigenvalues of the number operator $\hat{N}$, we must first obtain the eigenvectors $|0i\rangle$ for the ground state. These eigenvectors are determined by equation (4.29). When equation (4.18a) is substituted for $\hat{a}$, equation (4.29) takes the form

$$\left(\frac{\mathrm{d}}{\mathrm{d}\xi} + \xi\right)|0i\rangle = \left(\frac{\mathrm{d}}{\mathrm{d}\xi} + \xi\right)\phi_{0i}(\xi) = 0$$

or

$$\frac{\mathrm{d}\phi_{0i}}{\phi_{0i}} = -\xi\,\mathrm{d}\xi$$

This differential equation may be integrated to give

$$\phi_{0i}(\xi) = c\mathrm{e}^{-\xi^2/2} = \mathrm{e}^{i\alpha}\pi^{-1/4}\mathrm{e}^{-\xi^2/2}$$

where the constant of integration c is determined by the requirement that the functions $\phi_{ni}(\xi)$ be normalized and $\mathrm{e}^{i\alpha}$ is a phase factor. We have used the standard integral (A.5) to evaluate c. We observe that all the solutions for the ground-state eigenfunction are proportional to one another. Thus, there exists only one independent solution and the ground state is non-degenerate. If we arbitrarily set α equal to zero so that $\phi_0(\xi)$ is real, then the ground-state eigenvector is

$$|0\rangle = \pi^{-1/4} e^{-\xi^2/2} \qquad (4.31)$$

We next show that if the eigenvalue n of the number operator $\hat{N}$ is non-degenerate, then the eigenvalue $n + 1$ is also non-degenerate. We begin with the assumption that there is only one eigenvector with the property that

$$\hat{N}|n\rangle = n|n\rangle$$

and consider the eigenvector $|(n + 1)i\rangle$, which satisfies

$$\hat{N}|(n + 1)i\rangle = (n + 1)|(n + 1)i\rangle$$

If we operate on $|(n + 1)i\rangle$ with the lowering operator $\hat{a}$, we obtain to within a multiplicative constant c the unique eigenfunction $|n\rangle$,

$$\hat{a}|(n + 1)i\rangle = c|n\rangle$$

We next operate on this expression with the adjoint of $\hat{a}$ to give

$$\hat{a}^\dagger \hat{a}|(n + 1)i\rangle = \hat{N}|(n + 1)i\rangle = (n + 1)|(n + 1)i\rangle = c\hat{a}^\dagger|n\rangle$$

from which it follows that

$$|(n + 1)i\rangle = \frac{c}{n + 1} \hat{a}^\dagger|n\rangle$$

Thus, all the eigenvectors $|(n + 1)i\rangle$ corresponding to the eigenvalue $n + 1$ are proportional to $\hat{a}^\dagger|n\rangle$ and are, therefore, not independent since they are proportional to each other. We conclude then that if the eigenvalue n is non-degenerate, then the eigenvalue $n + 1$ is non-degenerate.

Since we have shown that the ground state is non-degenerate, we see that the next higher eigenvalue $n = 1$ is also non-degenerate. But if the eigenvalue $n = 1$ is non-degenerate, then the eigenvalue $n = 2$ is non-degenerate. By iteration, all of the eigenvalues n of $\hat{N}$ are non-degenerate. From equation (4.30) we observe that all the energy levels E_n of the harmonic oscillator are non-degenerate.

4.3 Eigenfunctions

Lowering and raising operations

From equations (4.27) and (4.28) and the conclusions that the eigenvalues of $\hat{N}$ are non-degenerate and are positive integers, we see that $\hat{a}|n\rangle$ and $\hat{a}^\dagger|n\rangle$ are eigenfunctions of $\hat{N}$ with eigenvalues $n - 1$ and $n + 1$, respectively. Accordingly, we may write

$$\hat{a}|n\rangle = c_n|n - 1\rangle \qquad (4.32a)$$

and

$$\hat{a}^{\dagger}|n\rangle = c'_n|n+1\rangle \tag{4.32b}$$

where c_n and c'_n are proportionality constants, dependent on the value of n, and to be determined by the requirement that $|n-1\rangle$, $|n\rangle$, and $|n+1\rangle$ are normalized. To evaluate the numerical constants c_n and c'_n, we square both sides of equations (4.32a) and (4.32b) and integrate with respect to ξ to obtain

$$\int_{-\infty}^{\infty} |\hat{a}\phi_n|^2 \, d\xi = |c_n|^2 \int_{-\infty}^{\infty} \phi_{n-1}^* \phi_{n-1} \, d\xi \tag{4.33a}$$

and

$$\int_{-\infty}^{\infty} |\hat{a}^{\dagger}\phi_n|^2 \, d\xi = |c'_n|^2 \int_{-\infty}^{\infty} \phi_{n+1}^* \phi_{n+1} \, d\xi \tag{4.33b}$$

The integral on the left-hand side of equation (4.33a) may be evaluated as follows

$$\int_{-\infty}^{\infty} |\hat{a}\phi_n|^2 \, d\xi = \int_{-\infty}^{\infty} (\hat{a}\phi_n^*)(\hat{a}\phi_n) \, d\xi = \langle n|\hat{a}^{\dagger}\hat{a}|n\rangle = \langle n|\hat{N}|n\rangle = n$$

Similarly, the integral on the left-hand side of equation (4.33b) becomes

$$\int_{-\infty}^{\infty} |\hat{a}^{\dagger}\phi_n|^2 \, d\xi = \int_{-\infty}^{\infty} (\hat{a}^{\dagger}\phi_n^*)(\hat{a}^{\dagger}\phi_n) \, d\xi = \langle n|\hat{a}\hat{a}^{\dagger}|n\rangle = \langle n|\hat{N}+1|n\rangle = n+1$$

Since the eigenfunctions are normalized, we obtain

$$|c_n|^2 = n, \qquad |c'_n|^2 = n+1$$

Without loss of generality, we may let c_n and c'_n be real and positive, so that equations (4.32a) and (4.32b) become

$$\hat{a}|n\rangle = \sqrt{n}|n-1\rangle \tag{4.34a}$$

$$\hat{a}^{\dagger}|n\rangle = \sqrt{n+1}|n+1\rangle \tag{4.34b}$$

If the normalized eigenvector $|n\rangle$ is known, these relations may be used to obtain the eigenvectors $|n-1\rangle$ and $|n+1\rangle$, both of which will be normalized.

Excited-state eigenfunctions

We are now ready to obtain the set of simultaneous eigenfunctions for the commuting operators $\hat{N}$ and $\hat{H}$. The ground-state eigenfunction $|0\rangle$ has already been determined and is given by equation (4.31). The series of eigenfunctions $|1\rangle$, $|2\rangle$, $\ldots$ are obtained from equations (4.34b) and (4.18b), which give

$$|n+1\rangle = [2(n+1)]^{-1/2}\left(\xi - \frac{d}{d\xi}\right)|n\rangle \tag{4.35}$$

Thus, the eigenvector $|1\rangle$ is obtained from $|0\rangle$

$$|1\rangle = 2^{-1/2}\left(\xi - \frac{d}{d\xi}\right)(\pi^{-1/4}e^{-\xi^2/2}) = 2^{1/2}\pi^{-1/4}\xi e^{-\xi^2/2}$$

the eigenvector $|2\rangle$ from $|1\rangle$

$$|2\rangle = \tfrac{1}{2}\left(\xi - \frac{d}{d\xi}\right)(2^{1/2}\pi^{-1/4}\xi e^{-\xi^2/2}) = 2^{-1/2}\pi^{-1/4}(2\xi^2 - 1)e^{-\xi^2/2}$$

the eigenvector $|3\rangle$ from $|2\rangle$

$$|3\rangle = 6^{-1/2}\left(\xi - \frac{d}{d\xi}\right)(2^{-1/2}\pi^{-1/4}(2\xi^2 - 1)e^{-\xi^2/2})$$

$$= 3^{-1/2}\pi^{-1/4}(2\xi^3 - 3\xi)e^{-\xi^2/2}$$

and so forth, indefinitely. Each of the eigenfunctions obtained by this procedure is normalized.

When equation (4.18a) is combined with (4.34a), we have

$$|n - 1\rangle = (2n)^{-1/2}\left(\xi + \frac{d}{d\xi}\right)|n\rangle \tag{4.36}$$

Just as equation (4.35) allows one to go 'up the ladder' to obtain $|n + 1\rangle$ from $|n\rangle$, equation (4.36) allows one to go 'down the ladder' to obtain $|n - 1\rangle$ from $|n\rangle$. This lowering procedure maintains the normalization of each of the eigenvectors.

Another, but completely equivalent, way of determining the series of eigenfunctions may be obtained by first noting that equation (4.34b) may be written for the series $n = 0, 1, 2, \ldots$ as follows

$$|1\rangle = \hat{a}^\dagger|0\rangle$$

$$|2\rangle = 2^{-1/2}\hat{a}^\dagger|1\rangle = 2^{-1/2}(\hat{a}^\dagger)^2|0\rangle$$

$$|3\rangle = 3^{-1/2}\hat{a}^\dagger|2\rangle = (3!)^{-1/2}(\hat{a}^\dagger)^3|0\rangle$$

$$\vdots$$

Obviously, the expression for $|n\rangle$ is

$$|n\rangle = (n!)^{-1/2}(\hat{a}^\dagger)^n|0\rangle$$

Substitution of equation (4.18b) for $\hat{a}^\dagger$ and (4.31) for the ground-state eigenvector $|0\rangle$ gives

$$|n\rangle = (2^n n!)^{-1/2}\pi^{-1/4}\left(\xi - \frac{d}{d\xi}\right)^n e^{-\xi^2/2} \tag{4.37}$$

This equation may be somewhat simplified if we note that

$$\left(\xi - \frac{d}{d\xi}\right)e^{-\xi^2/2} = \left(\xi - \frac{d}{d\xi}\right)e^{\xi^2/2}e^{-\xi^2} = \xi e^{-\xi^2/2} - \xi e^{\xi^2/2}e^{-\xi^2} - e^{\xi^2/2}\frac{d}{d\xi}e^{-\xi^2}$$

$$= -e^{\xi^2/2}\frac{d}{d\xi}e^{-\xi^2}$$

so that

$$\left(\xi - \frac{d}{d\xi}\right)^n e^{-\xi^2/2} = (-1)^n e^{\xi^2/2}\frac{d^n}{d\xi^n}e^{-\xi^2} \tag{4.38}$$

Substitution of equation (4.38) into (4.37) gives

$$|n\rangle = (-1)^n(2^n n!)^{-1/2}\pi^{-1/4}e^{\xi^2/2}\frac{d^n}{d\xi^n}e^{-\xi^2} \tag{4.39}$$

which may be used to obtain the entire set of eigenfunctions of $\hat{N}$ and $\hat{\Pi}$.

Eigenfunctions in terms of Hermite polynomials

It is customary to express the eigenfunctions for the stationary states of the harmonic oscillator in terms of the Hermite polynomials. The infinite set of Hermite polynomials $H_n(\xi)$ is defined in Appendix D, which also derives many of the properties of those polynomials. In particular, equation (D.3) relates the Hermite polynomial of order n to the nth-order derivative which appears in equation (4.39)

$$H_n(\xi) = (-1)^n e^{\xi^2}\frac{d^n}{d\xi^n}e^{-\xi^2}$$

Therefore, we may express the eigenvector $|n\rangle$ in terms of the Hermite polynomial $H_n(\xi)$ by the relation

$$|n\rangle = \phi_n(\xi) = (2^n n!)^{-1/2}\pi^{-1/4}H_n(\xi)e^{-\xi^2/2} \tag{4.40}$$

The eigenstates $\psi_n(x)$ are related to the functions $\phi_n(\xi)$ by equation (4.16), so that we have

$$\psi_n(x) = (2^n n!)^{-1/2}\left(\frac{m\omega}{\pi\hbar}\right)^{1/4}H_n(\xi)e^{-\xi^2/2}$$

$$\tag{4.41}$$

$$\xi = \left(\frac{m\omega}{\hbar}\right)^{1/2}x$$

For reference, the Hermite polynomials for $n = 0$ to $n = 10$ are listed in Table 4.1. When needed, higher-order Hermite polynomials are most easily obtained from the recurrence relation (D.5). If only a single Hermite polynomial is wanted and the neighboring polynomials are not available, then equation (D.4) may be used.

Table 4.1. *Hermite polynomials*

n	$H_n(\xi)$
0	1
1	2ξ
2	$4\xi^2 - 2$
3	$8\xi^3 - 12\xi$
4	$16\xi^4 - 48\xi^2 + 12$
5	$32\xi^5 - 160\xi^3 + 120\xi$
6	$64\xi^6 - 480\xi^4 + 720\xi^2 - 120$
7	$128\xi^7 - 1344\xi^5 + 3360\xi^3 - 1680\xi$
8	$256\xi^8 - 3584\xi^6 + 13440\xi^4 - 13440\xi^2 + 1680$
9	$512\xi^9 - 9216\xi^7 + 48384\xi^5 - 80640\xi^3 + 30240\xi$
10	$1024\xi^{10} - 23040\xi^8 + 161280\xi^6 - 403200\xi^4 + 302400\xi^2 - 30240$

The functions $\phi_n(\xi)$ in equation (4.40) are identical to those defined by equation (D.15) and, therefore, form a complete set as shown in equation (D.19). Substituting equation (4.16) into (D.19) and applying the relation (C.5b), we see that the functions $\psi_n(x)$ in equation (4.41) form a complete set, so that

$$\sum_{n=0}^{\infty} \psi_n(x)\psi_n(x') = \delta(x - x') \qquad (4.42)$$

Physical interpretation

The first four eigenfunctions $\psi_n(x)$ for $n = 0, 1, 2, 3$ are plotted in Figure 4.2 and the corresponding functions $[\psi_n(x)]^2$ in Figure 4.3. These figures also show the outline of the potential energy $V(x)$ from equation (4.5) and the four corresponding energy levels from equation (4.30). The function $[\psi_n(x)]^2$ is the probability density as a function of x for the particle in the nth quantum state. The quantity $[\psi_n(x)]^2 \, dx$ at any point x gives the probability for finding the particle between x and $x + dx$.

We wish to compare the quantum probability distributions with those obtained from the classical treatment of the harmonic oscillator at the same energies. The classical probability density $P(y)$ as a function of the reduced distance y $(-1 \leqslant y \leqslant 1)$ is given by equation (4.10) and is shown in Figure 4.1. When equations (4.8), (4.14), and (4.30) are combined, we see that the maximum displacement in terms of ξ for a classical oscillator with energy $(n + \frac{1}{2})\hbar\omega$ is $\sqrt{2n + 1}$. For $\xi < -\sqrt{2n + 1}$ and $\xi > \sqrt{2n + 1}$, the classical

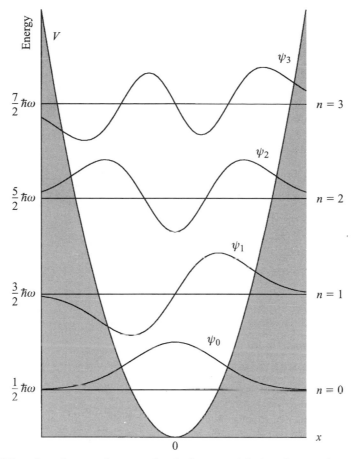

Figure 4.2 Wave functions and energy levels for a particle in a harmonic potential well. The outline of the potential energy is indicated by shading.

probability for finding the particle is equal to zero. These regions are shaded in Figures 4.2 and 4.3.

Each of the quantum probability distributions differs from the corresponding classical distribution in one very significant respect. In the quantum solution there is a non-vanishing probability of finding the particle outside the classically allowed region, i.e., in a region where the total energy is less than the potential energy. Since the Hermite polynomial $H_n(\xi)$ is of degree n, the wave function $\psi_n(x)$ has n nodes, a node being a point where a function touches or crosses the x-axis. The quantum probability density $[\psi_n(x)]^2$ is zero at a node. Within the classically allowed region, the wave function and the probability density oscillate with n nodes; outside that region the wave function and probability density rapidly approach zero with no nodes.

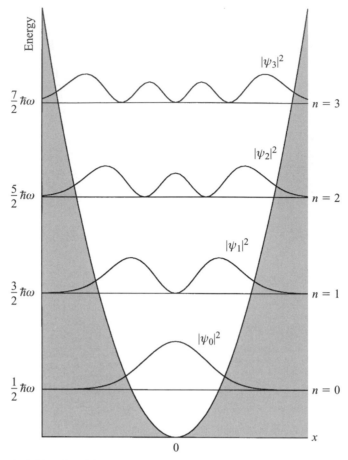

Figure 4.3 Probability densities and energy levels for a particle in a harmonic potential well. The outline of the potential energy is indicated by shading.

While the classical particle is most likely to be found near its maximum displacement, the probability density for the quantum particle in the ground state is largest at the origin. However, as the value of n increases, the quantum probability distribution begins to look more and more like the classical probability distribution. In Figure 4.4 the function $[\psi_{30}(x)]^2$ is plotted along with the classical result for an energy $30.5 \, \hbar\omega$. The average behavior of the rapidly oscillating quantum curve agrees well with the classical curve. This observation is an example of the Bohr correspondence principle, mentioned in Section 2.3. According to the correspondence principle, classical mechanics and quantum theory must give the same results in the limit of large quantum numbers.

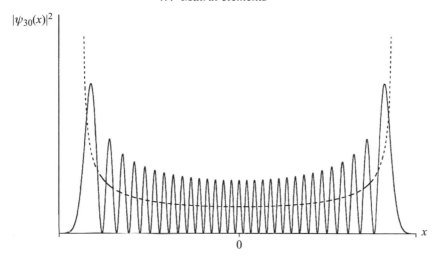

Figure 4.4 The probability density $|\psi_{30}(x)|^2$ for an oscillating particle in state $n = 30$. The dotted curve is the classical probability density for a particle with the same energy.

4.4 Matrix elements

In the application to an oscillator of some quantum-mechanical procedures, the matrix elements of x^n and $\hat{p}^n$ for a harmonic oscillator are needed. In this section we derive the matrix elements $\langle n'|x|n \rangle$, $\langle n'|x^2|n \rangle$, $\langle n'|\hat{p}|n \rangle$, and $\langle n'|\hat{p}^2|n \rangle$, and show how other matrix elements may be determined.

The ladder operators $\hat{a}$ and $\hat{a}^\dagger$ defined in equation (4.18) may be solved for x and for $\hat{p}$ to give

$$x = \left(\frac{\hbar}{2m\omega}\right)^{1/2}(\hat{a}^\dagger + \hat{a}) \tag{4.43a}$$

$$\hat{p} = i\left(\frac{m\hbar\omega}{2}\right)^{1/2}(\hat{a}^\dagger - \hat{a}) \tag{4.43b}$$

From equations (4.34) and the orthonormality of the harmonic oscillator eigenfunctions $|n\rangle$, we find that the matrix elements of $\hat{a}$ and $\hat{a}^\dagger$ are

$$\langle n'|\hat{a}|n \rangle = \sqrt{n}\langle n'|n - 1 \rangle = \sqrt{n}\delta_{n',n-1} \tag{4.44a}$$

$$\langle n'|\hat{a}^\dagger|n \rangle = \sqrt{n + 1}\langle n'|n + 1 \rangle = \sqrt{n + 1}\delta_{n',n+1} \tag{4.44b}$$

The set of equations (4.43) and (4.44) may be used to evaluate the matrix elements of any integral power of x and $\hat{p}$.

To find the matrix element $\langle n'|x|n \rangle$, we apply equations (4.43a) and (4.44) to obtain

$$\langle n'|x|n \rangle = \left(\frac{\hbar}{2m\omega}\right)^{1/2} (\langle n'|\hat{a}^\dagger|n \rangle + \langle n'|\hat{a}|n \rangle)$$

$$= \left(\frac{\hbar}{2m\omega}\right)^{1/2} (\sqrt{n+1}\delta_{n',n+1} + \sqrt{n}\delta_{n',n-1})$$

so that

$$\langle n+1|x|n \rangle = \sqrt{\frac{(n+1)\hbar}{2m\omega}} \qquad\qquad (4.45a)$$

$$\langle n-1|x|n \rangle = \sqrt{\frac{n\hbar}{2m\omega}} \qquad\qquad (4.45b)$$

$$\langle n'|x|n \rangle = 0 \quad \text{for} \quad n' \neq n+1, n-1 \qquad\qquad (4.45c)$$

If we replace n by $n-1$ in equation (4.45a), we obtain

$$\langle n|x|n-1 \rangle = \sqrt{\frac{n\hbar}{2m\omega}}$$

From equation (4.45b) we see that

$$\langle n-1|x|n \rangle = \langle n|x|n-1 \rangle$$

Likewise, we can show that

$$\langle n+1|x|n \rangle = \langle n|x|n+1 \rangle$$

In general, then, we have

$$\langle n'|x|n \rangle = \langle n|x|n' \rangle$$

To find the matrix element $\langle n'|\hat{p}|n \rangle$, we use equations (4.43b) and (4.44) to give

$$\langle n'|\hat{p}|n \rangle = i\left(\frac{m\hbar\omega}{2}\right)^{1/2} \langle n'|\hat{a}^\dagger - \hat{a}|n \rangle$$

$$= i\left(\frac{m\hbar\omega}{2}\right)^{1/2} (\sqrt{n+1}\delta_{n',n+1} - \sqrt{n}\delta_{n',n-1})$$

so that

$$\langle n+1|\hat{p}|n \rangle = i\sqrt{\frac{(n+1)m\hbar\omega}{2}} \qquad\qquad (4.46a)$$

$$\langle n-1|\hat{p}|n \rangle = -i\sqrt{\frac{nm\hbar\omega}{2}} \qquad\qquad (4.46b)$$

$$\langle n'|\hat{p}|n \rangle = 0 \quad \text{for} \quad n' \neq n+1, n-1 \qquad\qquad (4.46c)$$

We can easily show that

$$\langle n'|\hat{p}|n\rangle = -\langle n|\hat{p}|n'\rangle$$

The matrix element $\langle n'|x^2|n\rangle$ is

$$\langle n'|x^2|n\rangle = \frac{\hbar}{2m\omega}\langle n'|(\hat{a}^\dagger + \hat{a})^2|n\rangle = \frac{\hbar}{2m\omega}\langle n'|(\hat{a}^\dagger)^2 + \hat{a}^\dagger\hat{a} + \hat{a}\hat{a}^\dagger + \hat{a}^2|n\rangle$$

From equation (4.34) we have

$$
\begin{aligned}
(\hat{a}^\dagger)^2|n\rangle &= \sqrt{n+1}\,\hat{a}^\dagger|n+1\rangle = \sqrt{(n+1)(n+2)}|n+2\rangle \\
\hat{a}^\dagger\hat{a}|n\rangle &= \sqrt{n}\,\hat{a}^\dagger|n-1\rangle = n|n\rangle \\
\hat{a}\hat{a}^\dagger|n\rangle &= \sqrt{n+1}\,\hat{a}|n+1\rangle = (n+1)|n\rangle \\
\hat{a}^2|n\rangle &= \sqrt{n}\,\hat{a}|n-1\rangle = \sqrt{n(n-1)}|n-2\rangle
\end{aligned}
\tag{4.47}
$$

so that

$$\langle n'|x^2|n\rangle = \frac{\hbar}{2m\omega}[\sqrt{(n+1)(n+2)}\delta_{n',n+2} + (2n+1)\delta_{n'n}$$
$$+ \sqrt{n(n-1)}\delta_{n',n-2}]$$

We conclude that

$$\langle n+2|x^2|n\rangle = \langle n|x^2|n+2\rangle = \frac{\hbar}{2m\omega}\sqrt{(n+1)(n+2)} \tag{4.48a}$$

$$\langle n|x^2|n\rangle = \frac{\hbar}{m\omega}(n+\tfrac{1}{2}) \tag{4.48b}$$

$$\langle n-2|x^2|n\rangle = \langle n|x^2|n-2\rangle = \frac{\hbar}{2m\omega}\sqrt{n(n-1)} \tag{4.48c}$$

$$\langle n'|x^2|n\rangle = 0, \qquad n' \neq n+2,\ n,\ n-2 \tag{4.48d}$$

The matrix element $\langle n'|\hat{p}^2|n\rangle$ is obtained from equations (4.43b) and (4.47)

$$\langle n'|\hat{p}^2|n\rangle = -\left(\frac{m\hbar\omega}{2}\right)\langle n'|(\hat{a}^\dagger - \hat{a})^2|n\rangle = -\left(\frac{m\hbar\omega}{2}\right)\langle n'|(\hat{a}^\dagger)^2 - \hat{a}^\dagger\hat{a}$$
$$- \hat{a}\hat{a}^\dagger + \hat{a}^2|n\rangle$$
$$= -\left(\frac{m\hbar\omega}{2}\right)[\sqrt{(n+1)(n+2)}\delta_{n',n+2} - (2n+1)\delta_{n'n}$$
$$+ \sqrt{n(n-1)}\delta_{n',n-2}]$$

so that

$$\langle n+2|\hat{p}^2|n\rangle = \langle n|\hat{p}^2|n+2\rangle = -\left(\frac{m\hbar\omega}{2}\right)\sqrt{(n+1)(n+2)} \qquad (4.49a)$$

$$\langle n|\hat{p}^2|n\rangle = m\hbar\omega(n+\tfrac{1}{2}) \qquad (4.49b)$$

$$\langle n-2|\hat{p}^2|n\rangle = \langle n|\hat{p}^2|n-2\rangle = -\left(\frac{m\hbar\omega}{2}\right)\sqrt{n(n-1)} \qquad (4.49c)$$

$$\langle n'|\hat{p}^2|n\rangle = 0, \qquad n' \neq n+2, \, n, \, n-2 \qquad (4.49d)$$

Following this same procedure using the operators $(\hat{a}^\dagger \pm \hat{a})^k$, we can find the matrix elements of x^k and of $\hat{p}^k$ for any positive integral power k. In Chapters 9 and 10, we need the matrix elements of x^3 and x^4. The matrix elements $\langle n'|x^3|n\rangle$ are as follows:

$$\langle n+3|x^3|n\rangle = \langle n|x^3|n+3\rangle = \left(\frac{\hbar}{2m\omega}\right)^{3/2}\sqrt{(n+1)(n+2)(n+3)} \qquad (4.50a)$$

$$\langle n+1|x^3|n\rangle = \langle n|x^3|n+1\rangle = 3\left(\frac{(n+1)\hbar}{2m\omega}\right)^{3/2} \qquad (4.50b)$$

$$\langle n-1|x^3|n\rangle = \langle n|x^3|n-1\rangle = 3\left(\frac{n\hbar}{2m\omega}\right)^{3/2} \qquad (4.50c)$$

$$\langle n-3|x^3|n\rangle = \langle n|x^3|n-3\rangle = \left(\frac{\hbar}{2m\omega}\right)^{3/2}\sqrt{n(n-1)(n-2)} \qquad (4.50d)$$

$$\langle n'|x^3|n\rangle = 0, \qquad n' \neq n\pm1, \, n\pm3 \qquad (4.50e)$$

The matrix elements $\langle n'|x^4|n\rangle$ are as follows

$$\langle n+4|x^4|n\rangle = \langle n|x^4|n+4\rangle = \left(\frac{\hbar}{2m\omega}\right)^2\sqrt{(n+1)(n+2)(n+3)(n+4)}$$
$$(4.51a)$$

$$\langle n+2|x^4|n\rangle = \langle n|x^4|n+2\rangle = \frac{1}{2}\left(\frac{\hbar}{m\omega}\right)^2(2n+3)\sqrt{(n+1)(n+2)} \qquad (4.51b)$$

$$\langle n|x^4|n\rangle = \frac{3}{2}\left(\frac{\hbar}{m\omega}\right)^2\left(n^2+n+\tfrac{1}{2}\right) \qquad (4.51c)$$

$$\langle n-2|x^4|n\rangle = \langle n|x^4|n-2\rangle = \frac{1}{2}\left(\frac{\hbar}{m\omega}\right)^2(2n-1)\sqrt{n(n-1)} \qquad (4.51d)$$

$$\langle n-4|x^4|n\rangle = \langle n|x^4|n-4\rangle = \left(\frac{\hbar}{2m\omega}\right)^2\sqrt{n(n-1)(n-2)(n-3)} \qquad (4.51e)$$

$$\langle n'|x^4|n\rangle = 0, \qquad n' \neq n, \, n\pm2, \, n\pm4 \qquad (4.51f)$$

4.5 Heisenberg uncertainty relation

Using the results of Section 4.4, we may easily verify for the harmonic oscillator the Heisenberg uncertainty relation as discussed in Section 3.11. Specifically, we wish to show for the harmonic oscillator that

$$\Delta x \Delta p \geq \tfrac{1}{2}\hbar$$

where

$$(\Delta x)^2 = \langle (x - \langle x \rangle)^2 \rangle$$
$$(\Delta p)^2 = \langle (\hat{p} - \langle p \rangle)^2 \rangle$$

The expectation values of x and of $\hat{p}$ for a harmonic oscillator in eigenstate $|n\rangle$ are just the matrix elements $\langle n|x|n \rangle$ and $\langle n|\hat{p}|n \rangle$, respectively. These matrix elements are given in equations (4.45c) and (4.46c). We see that both vanish, so that $(\Delta x)^2$ reduces to the expectation value of x^2 or $\langle n|x^2|n \rangle$ and $(\Delta p)^2$ reduces to the expectation value of $\hat{p}^2$ or $\langle n|\hat{p}^2|n \rangle$. These matrix elements are given in equations (4.48b) and (4.49b). Therefore, we have

$$\Delta x = \left(\frac{\hbar}{m\omega} \right)^{1/2} (n + \tfrac{1}{2})^{1/2}$$

$$\Delta p = (m\hbar\omega)^{1/2}(n + \tfrac{1}{2})^{1/2}$$

and the product $\Delta x \Delta p$ is

$$\Delta x \Delta p = (n + \tfrac{1}{2})\hbar$$

For the ground state $(n = 0)$, we see that the product $\Delta x \Delta p$ equals the minimum allowed value $\hbar/2$. This result is consistent with the form (equation (3.85)) of the state function for minimum uncertainty. When the ground-state harmonic-oscillator values of $\langle x \rangle$, $\langle p \rangle$, and λ are substituted into equation (3.85), the ground-state eigenvector $|0\rangle$ in equation (4.31) is obtained. For excited states of the harmonic oscillator, the product $\Delta x \Delta p$ is greater than the minimum allowed value.

4.6 Three-dimensional harmonic oscillator

The harmonic oscillator may be generalized to three dimensions, in which case the particle is displaced from the origin in a general direction in cartesian space. The force constant is not necessarily the same in each of the three dimensions, so that the potential energy is

$$V = \tfrac{1}{2}k_x x^2 + \tfrac{1}{2}k_y y^2 + \tfrac{1}{2}k_z z^2 = \tfrac{1}{2}m(\omega_x^2 x^2 + \omega_y^2 y^2 + \omega_z^2 z^2)$$

where k_x, k_y, k_z are the respective force constants and ω_x, ω_y, ω_z are the respective classical angular frequencies of vibration.

The Schrödinger equation for this three-dimensional harmonic oscillator is

$$-\frac{\hbar^2}{2m}\left(\frac{\partial^2\psi}{\partial x^2} + \frac{\partial^2\psi}{\partial y^2} + \frac{\partial^2\psi}{\partial z^2}\right) + \tfrac{1}{2}m(\omega_x^2 x^2 + \omega_y^2 y^2 + \omega_z^2 z^2)\psi = E\psi$$

where $\psi(x, y, z)$ is the wave function. To solve this partial differential equation of three variables, we separate variables by making the substitution

$$\psi(x, y, z) = X(x)Y(y)Z(z) \tag{4.52}$$

where $X(x)$ is a function only of the variable x, $Y(y)$ only of y, and $Z(z)$ only of z. After division by $-\psi(x, y, z)$, the Schrödinger equation takes the form

$$\left(\frac{\hbar^2}{2mX}\frac{d^2X}{dx^2} - \tfrac{1}{2}m\omega_x^2 x^2\right) + \left(\frac{\hbar^2}{2mY}\frac{d^2Y}{dy^2} - \tfrac{1}{2}m\omega_y^2 y^2\right)$$

$$+ \left(\frac{\hbar^2}{2mZ}\frac{d^2Z}{dz^2} - \tfrac{1}{2}m\omega_z^2 z^2\right) = E$$

The first term on the left-hand side is a function only of the variable x and remains constant when y and z change but x does not. Similarly, the second term is a function only of y and does not change in value when x and z change but y does not. The third term depends only on z and keeps a constant value when only x and y change. However, the sum of these three terms is always equal to the constant energy E for all choices of x, y, z. Thus, each of the three independent terms must be equal to a constant

$$\frac{\hbar^2}{2mX}\frac{d^2X}{dx^2} - \tfrac{1}{2}m\omega_x^2 x^2 = E_x$$

$$\frac{\hbar^2}{2mY}\frac{d^2Y}{dy^2} - \tfrac{1}{2}m\omega_y^2 y^2 = E_y$$

$$\frac{\hbar^2}{2mZ}\frac{d^2Z}{dz^2} - \tfrac{1}{2}m\omega_z^2 z^2 = E_z$$

where the three separation constants E_x, E_y, E_z satisfy the relation

$$E_x + E_y + E_z = E \tag{4.53}$$

The differential equation for $X(x)$ is exactly of the form given by (4.13) for a one-dimensional harmonic oscillator. Thus, the eigenvalues E_x are given by equation (4.30)

$$E_{n_x} = (n_x + \tfrac{1}{2})\hbar\omega_x, \qquad n_x = 0, 1, 2, \ldots$$

and the eigenfunctions are given by (4.41)

$$X_{n_x}(x) = (2^{n_x} n_x!)^{-1/2} \left(\frac{m\omega_x}{\pi\hbar}\right)^{1/4} H_{n_x}(\xi)e^{-\xi^2/2}$$

$$\xi = \left(\frac{m\omega_x}{\hbar}\right)^{1/2} x$$

Similarly, the eigenvalues for the differential equations for $Y(y)$ and $Z(z)$ are, respectively

$$E_{n_y} = (n_y + \tfrac{1}{2})\hbar\omega_y, \qquad n_y = 0, 1, 2, \ldots$$

$$E_{n_z} = (n_z + \tfrac{1}{2})\hbar\omega_z, \qquad n_z = 0, 1, 2, \ldots$$

and the corresponding eigenfunctions are

$$Y_{n_y}(y) = (2^{n_y} n_y!)^{-1/2} \left(\frac{m\omega_y}{\pi\hbar}\right)^{1/4} H_{n_y}(\eta)e^{-\eta^2/2}$$

$$\eta = \left(\frac{m\omega_y}{\hbar}\right)^{1/2} y$$

$$Z_{n_z}(z) = (2^{n_z} n_z!)^{-1/2} \left(\frac{m\omega_z}{\pi\hbar}\right)^{1/4} H_{n_z}(\zeta)e^{-\zeta^2/2}$$

$$\zeta = \left(\frac{m\omega_z}{\hbar}\right)^{1/2} z$$

The energy levels for the three-dimensional harmonic oscillator are, then, given by the sum (equation (4.53))

$$E_{n_x,n_y,n_z} = (n_x + \tfrac{1}{2})\hbar\omega_x + (n_y + \tfrac{1}{2})\hbar\omega_y + (n_z + \tfrac{1}{2})\hbar\omega_z \qquad (4.54)$$

The total wave functions are given by equation (4.52)

$$\psi_{n_x,n_y,n_z}(x,\, y,\, z) = (2^{n_x+n_y+n_z} n_x!n_y!n_z!)^{-1/2} \left(\frac{m}{\pi\hbar}\right)^{3/4} (\omega_x\omega_y\omega_z)^{1/4}$$

$$\times H_{n_x}(\xi) H_{n_y}(\eta) H_{n_z}(\zeta)e^{-(\xi^2+\eta^2+\zeta^2)/2} \qquad (4.55)$$

An isotropic oscillator is one for which the restoring force is independent of the direction of the displacement and depends only on its magnitude. For such an oscillator, the directional force constants are equal to one another

$$k_x = k_y = k_z \equiv k$$

and, as a result, the angular frequencies are all the same

$$\omega_x = \omega_y = \omega_z \equiv \omega$$

In this case, the total energies are

$$E_{n_x,n_y,n_z} = (n_x + n_y + n_z + \tfrac{3}{2})\hbar\omega = (n + \tfrac{3}{2})\hbar\omega \qquad (4.56)$$

where n is called the total quantum number. All the energy levels for the isotropic three-dimensional harmonic oscillator, except for the lowest level, are degenerate. The degeneracy of the energy level E_n is $(n + 1)(n + 2)/2$.

Problems

4.1 Consider a classical particle of mass m in a parabolic potential well. At time t the displacement x of the particle from the origin is given by
$$x = a\sin(\omega t + b)$$
where a is a constant and ω is the angular frequency of the vibration. From this expression find the kinetic and potential energies as functions of time and show that the total energy remains constant throughout the motion.

4.2 Evaluate the constant c in equation (4.10). (To evaluate the integral, let $y = \cos\theta$.)

4.3 Show that $\hat{a}$ and $\hat{a}^\dagger$ in equations (4.18) are not hermitian and that $\hat{a}^\dagger$ is the adjoint of $\hat{a}$.

4.4 The operator $\hat{N} \equiv \hat{a}^\dagger\hat{a}$ is hermitian. Is the operator $\hat{a}\hat{a}^\dagger$ hermitian?

4.5 Evaluate the commutators $[\hat{H}, \hat{a}]$ and $[\hat{H}, \hat{a}^\dagger]$.

4.6 Calculate the expectation value of x^6 for the harmonic oscillator in the $n = 1$ state.

4.7 Consider a particle of mass m in a parabolic potential well. Calculate the probability of finding the particle in the classically allowed region when the particle is in its ground state.

4.8 Consider a particle of mass m in a one-dimensional potential well such that
$$V(x) = \tfrac{1}{2}m\omega^2 x^2, \qquad x \geq 0$$
$$= \infty, \qquad x < 0$$
What are the eigenfunctions and eigenvalues?

4.9 What is the probability density as a function of the momentum p of an oscillating particle in its ground state in a parabolic potential well? (First find the momentum-space wave function.)

4.10 Show that the wave functions $A_n(\gamma)$ in momentum space corresponding to $\phi_n(\xi)$ in equation (4.40) for a linear harmonic oscillator are
$$A_n(\gamma) = (2\pi)^{-1/2}\int_{-\infty}^{\infty}\phi_n(\xi)e^{-i\gamma\xi}\,d\xi$$
$$= i^{-n}(2^n n!\pi^{1/2})^{-1/2}e^{-\gamma^2/2}H_n(\gamma)$$
where $\xi \equiv (m\omega/\hbar)^{1/2}x$ and $\gamma \equiv (m\hbar\omega)^{-1/2}p$. (Use the generating function (D.1) to evaluate the Fourier integral.)

4.11 Using only equation (4.43b) and the fact that $\hat{a}^\dagger$ is the adjoint of $\hat{a}$, prove that
$$\langle n'|\hat{p}|n\rangle = -\langle n|\hat{p}|n'\rangle$$

4.12 Derive the relations (4.50) for the matrix elements $\langle n'|x^3|n\rangle$.

4.13 Derive the relations (4.51) for the matrix elements $\langle n'|x^4|n\rangle$.

4.14 Derive the result that the degeneracy of the energy level E_n for an isotropic three-dimensional harmonic oscillator is $(n+1)(n+2)/2$.

5

Angular momentum

Angular momentum plays an important role in both classical and quantum mechanics. In isolated classical systems the total angular momentum is a constant of motion. In quantum systems the angular momentum is important in studies of atomic, molecular, and nuclear structure and spectra and in studies of spin in elementary particles and in magnetism.

5.1 Orbital angular momentum

We first consider a particle of mass m moving according to the laws of classical mechanics. The angular momentum $\mathbf{L}$ of the particle with respect to the origin of the coordinate system is defined by the relation

$$\mathbf{L} \equiv \mathbf{r} \times \mathbf{p} \tag{5.1}$$

where $\mathbf{r}$ is the position vector given by equation (2.60) and $\mathbf{p}$ is the linear momentum given by equation (2.61). When expressed as a determinant, the angular momentum $\mathbf{L}$ is

$$\mathbf{L} = \begin{vmatrix} \mathbf{i} & \mathbf{j} & \mathbf{k} \\ x & y & z \\ p_x & p_y & p_z \end{vmatrix}$$

The components L_x, L_y, L_z of the vector $\mathbf{L}$ are

$$L_x = yp_z - zp_y$$

$$L_y = zp_x - xp_z \tag{5.2}$$

$$L_z = xp_y - yp_x$$

The square of the magnitude of the vector $\mathbf{L}$ is given in terms of these components by

$$L^2 = \mathbf{L} \cdot \mathbf{L} = L_x^2 + L_y^2 + L_z^2 \tag{5.3}$$

130

If a force $\mathbf{F}$ acts on the particle, then the torque $\mathbf{T}$ on the particle is defined as

$$\mathbf{T} = \mathbf{r} \times \mathbf{F} = \mathbf{r} \times \frac{d\mathbf{p}}{dt} \tag{5.4}$$

where Newton's second law that the force equals the rate of change of linear momentum, $\mathbf{F} = d\mathbf{p}/dt$, has been introduced. If we take the time derivative of equation (5.1), we obtain

$$\frac{d\mathbf{L}}{dt} = \left(\frac{d\mathbf{r}}{dt} \times \mathbf{p} \right) + \left(\mathbf{r} \times \frac{d\mathbf{p}}{dt} \right) = \mathbf{r} \times \frac{d\mathbf{p}}{dt} \tag{5.5}$$

since

$$\frac{d\mathbf{r}}{dt} \times \mathbf{p} = \frac{d\mathbf{r}}{dt} \times m \frac{d\mathbf{r}}{dt} = 0$$

Combining equations (5.4) and (5.5), we find that

$$\mathbf{T} = \frac{d\mathbf{L}}{dt} \tag{5.6}$$

If there is no force acting on the particle, the torque is zero. Consequently, the rate of change of the angular momentum is zero and the angular momentum is conserved.

The quantum-mechanical operators for the components of the orbital angular momentum are obtained by replacing p_x, p_y, p_z in the classical expressions (5.2) by their corresponding quantum operators,

$$\hat{L}_x = y\hat{p}_z - z\hat{p}_y = \frac{\hbar}{i} \left(y \frac{\partial}{\partial z} - z \frac{\partial}{\partial y} \right) \tag{5.7a}$$

$$\hat{L}_y = z\hat{p}_x - x\hat{p}_z = \frac{\hbar}{i} \left(z \frac{\partial}{\partial x} - x \frac{\partial}{\partial z} \right) \tag{5.7b}$$

$$\hat{L}_z = x\hat{p}_y - y\hat{p}_x = \frac{\hbar}{i} \left(x \frac{\partial}{\partial y} - y \frac{\partial}{\partial x} \right) \tag{5.7c}$$

Since y commutes with $\hat{p}_z$ and z commutes with $\hat{p}_y$, there is no ambiguity regarding the order of y and $\hat{p}_z$ and of z and $\hat{p}_y$ in constructing $\hat{L}_x$. Similar remarks apply to $\hat{L}_y$ and $\hat{L}_z$. The quantum-mechanical operator for $\mathbf{L}$ is

$$\hat{\mathbf{L}} = \mathbf{i}\hat{L}_x + \mathbf{j}\hat{L}_y + \mathbf{k}\hat{L}_z \tag{5.8}$$

and for L^2 is

$$\hat{L}^2 = \hat{\mathbf{L}} \cdot \hat{\mathbf{L}} = \hat{L}_x^2 + \hat{L}_y^2 + \hat{L}_z^2 \tag{5.9}$$

The operators $\hat{L}_x$, $\hat{L}_y$, $\hat{L}_z$ can easily be shown to be hermitian with respect to a set of functions of x, y, z that vanish at $\pm\infty$. As a consequence, $\hat{\mathbf{L}}$ and $\hat{L}^2$ are also hermitian.

Commutation relations

The commutator $[\hat{L}_x, \hat{L}_y]$ may be evaluated as follows

$$[\hat{L}_x, \hat{L}_y] = [y\hat{p}_z - z\hat{p}_y, z\hat{p}_x - x\hat{p}_z]$$

$$= [y\hat{p}_z, z\hat{p}_x] + [z\hat{p}_y, x\hat{p}_z] - [y\hat{p}_z, x\hat{p}_z] - [z\hat{p}_y, z\hat{p}_x]$$

The last two terms vanish because $y\hat{p}_z$ commutes with $x\hat{p}_z$ and because $z\hat{p}_y$ commutes with $z\hat{p}_x$. If we expand the remaining terms, we obtain

$$[\hat{L}_x, \hat{L}_y] = y\hat{p}_x\hat{p}_z z - y\hat{p}_x z\hat{p}_z + x\hat{p}_y z\hat{p}_z - x\hat{p}_y\hat{p}_z z = (x\hat{p}_y - y\hat{p}_x)[z, \hat{p}_z]$$

Introducing equations (3.44) and (5.7c), we have

$$[\hat{L}_x, \hat{L}_y] = i\hbar\hat{L}_z \tag{5.10a}$$

By a cyclic permutation of x, y, and z in equation (5.10a), we obtain the commutation relations for the other two pairs of operators

$$[\hat{L}_y, \hat{L}_z] = i\hbar\hat{L}_x \tag{5.10b}$$

$$[\hat{L}_z, \hat{L}_x] = i\hbar\hat{L}_y \tag{5.10c}$$

Equations (5.10) may be written in an equivalent form as

$$\hat{\mathbf{L}} \times \hat{\mathbf{L}} = i\hbar\hat{\mathbf{L}} \tag{5.11}$$

which may be demonstrated by expansion of the left-hand side.

5.2 Generalized angular momentum

In quantum mechanics we need to consider not only orbital angular momentum, but spin angular momentum as well. Whereas orbital angular momentum is expressed in terms of the x, y, z coordinates and their conjugate angular momenta, spin angular momentum is intrinsic to the particle and is not expressible in terms of a coordinate system. However, in quantum mechanics both types of angular momenta have common mathematical properties that are not dependent on a coordinate representation. For this reason we introduce generalized angular momentum and develop its mathematical properties according to the procedures of quantum theory.

Based on an analogy with orbital angular momentum, we define a generalized angular-momentum operator $\hat{\mathbf{J}}$ with components $\hat{J}_x, \hat{J}_y, \hat{J}_z$

$$\hat{\mathbf{J}} = \mathbf{i}\hat{J}_x + \mathbf{j}\hat{J}_y + \mathbf{k}\hat{J}_z$$

The operator $\hat{\mathbf{J}}$ is *any* hermitian operator which obeys the relation

$$\hat{\mathbf{J}} \times \hat{\mathbf{J}} = i\hbar\hat{\mathbf{J}} \tag{5.12}$$

or equivalently

$$[\hat{J}_x, \hat{J}_y] = i\hbar\hat{J}_z \tag{5.13a}$$

$$[\hat{J}_y, \hat{J}_z] = i\hbar\hat{J}_x \tag{5.13b}$$

$$[\hat{J}_z, \hat{J}_x] = i\hbar\hat{J}_y \tag{5.13c}$$

The square of the angular-momentum operator is defined by

$$\hat{J}^2 = \hat{\mathbf{J}} \cdot \hat{\mathbf{J}} = \hat{J}_x^2 + \hat{J}_y^2 + \hat{J}_z^2 \tag{5.14}$$

and is hermitian since $\hat{J}_x$, $\hat{J}_y$, and $\hat{J}_z$ are hermitian. The operator $\hat{J}^2$ commutes with each of the three operators $\hat{J}_x$, $\hat{J}_y$, $\hat{J}_z$. We first evaluate the commutator $[\hat{J}^2, \hat{J}_z]$

$$[\hat{J}^2, \hat{J}_z] = [\hat{J}_x^2, \hat{J}_z] + [\hat{J}_y^2, \hat{J}_z] + [\hat{J}_z^2, \hat{J}_z]$$

$$= \hat{J}_x[\hat{J}_x, \hat{J}_z] + [\hat{J}_x, \hat{J}_z]\hat{J}_x + \hat{J}_y[\hat{J}_y, \hat{J}_z] + [\hat{J}_y, \hat{J}_z]\hat{J}_y$$

$$= -i\hbar\hat{J}_x\hat{J}_y - i\hbar\hat{J}_y\hat{J}_x + i\hbar\hat{J}_y\hat{J}_x + i\hbar\hat{J}_x\hat{J}_y$$

$$= 0 \tag{5.15a}$$

where the fact that $\hat{J}_z$ commutes with itself and equations (3.4b) and (5.13) have been used. By similar expansions, we may also show that

$$[\hat{J}^2, \hat{J}_x] = 0 \tag{5.15b}$$

$$[\hat{J}^2, \hat{J}_y] = 0 \tag{5.15c}$$

Since the operator $\hat{J}^2$ commutes with each of the components $\hat{J}_x$, $\hat{J}_y$, $\hat{J}_z$ of $\hat{\mathbf{J}}$, but the three components do not commute with each other, we can obtain simultaneous eigenfunctions of $\hat{J}^2$ and one, but only one, of the three components of $\hat{\mathbf{J}}$. Following the usual convention, we arbitrarily select $\hat{J}_z$ and seek the simultaneous eigenfunctions of $\hat{J}^2$ and $\hat{J}_z$. Since angular momentum has the same dimensions as $\hbar$, we represent the eigenvalues of $\hat{J}^2$ by $\lambda\hbar^2$ and the eigenvalues of $\hat{J}_z$ by $m\hbar$, where λ and m are dimensionless and are real because $\hat{J}^2$ and $\hat{J}_z$ are hermitian. If the corresponding orthonormal eigenfunctions are denoted in Dirac notation by $|\lambda m\rangle$, then we have

$$\hat{J}^2|\lambda m\rangle = \lambda\hbar^2|\lambda m\rangle \tag{5.16a}$$

$$\hat{J}_z|\lambda m\rangle = m\hbar^2|\lambda m\rangle \tag{5.16b}$$

We implicitly assume that these eigenfunctions are uniquely determined by only the two parameters λ and m.

The expectation values of $\hat{J}^2$ and $\hat{J}_z^2$ are, according to (3.46), and (5.16)

$$\langle\hat{J}^2\rangle = \langle\lambda m|\hat{J}^2|\lambda m\rangle = \lambda\hbar^2$$

$$\langle\hat{J}_z^2\rangle = \langle\lambda m|\hat{J}_z^2|\lambda m\rangle = m^2\hbar^2$$

since the eigenfunctions $|\lambda m\rangle$ are normalized. Using equation (5.14) we may also write

$$\langle \hat{J}^2 \rangle = \langle \hat{J}_x^2 \rangle + \langle \hat{J}_y^2 \rangle + \langle \hat{J}_z^2 \rangle$$

Since $\hat{J}_x$ and $\hat{J}_y$ are hermitian, the expectation values of $\hat{J}_x^2$ and $\hat{J}_y^2$ are real and positive, so that

$$\langle \hat{J}^2 \rangle \geqslant \langle \hat{J}_z^2 \rangle$$

from which it follows that

$$\lambda \geqslant m^2 \geqslant 0 \qquad\qquad (5.17)$$

Ladder operators

We have already introduced the use of ladder operators in Chapter 4 to find the eigenvalues for the harmonic oscillator. We employ the same technique here to obtain the eigenvalues of $\hat{J}^2$ and $\hat{J}_z$. The requisite ladder operators $\hat{J}_+$ and $\hat{J}_-$ are defined by the relations

$$\hat{J}_+ \equiv \hat{J}_x + i\hat{J}_y \qquad\qquad (5.18a)$$
$$\hat{J}_- \equiv \hat{J}_x - i\hat{J}_y \qquad\qquad (5.18b)$$

Neither $\hat{J}_+$ nor $\hat{J}_-$ is hermitian. Application of equation (3.33) shows that they are adjoints of each other. Using the definitions (5.18) and (5.14) and the commutation relations (5.13) and (5.15), we can readily prove the following relationships

$$[\hat{J}_z, \hat{J}_+] = \hbar\hat{J}_+ \qquad\qquad (5.19a)$$

$$[\hat{J}_z, \hat{J}_-] = -\hbar\hat{J}_- \qquad\qquad (5.19b)$$

$$[\hat{J}^2, \hat{J}_+] = 0 \qquad\qquad (5.19c)$$

$$[\hat{J}^2, \hat{J}_-] = 0 \qquad\qquad (5.19d)$$

$$[\hat{J}_+, \hat{J}_-] = 2\hbar\hat{J}_z \qquad\qquad (5.19e)$$

$$\hat{J}_+\hat{J}_- = \hat{J}^2 - \hat{J}_z^2 + \hbar\hat{J}_z \qquad\qquad (5.19f)$$

$$\hat{J}_-\hat{J}_+ = \hat{J}^2 - \hat{J}_z^2 - \hbar\hat{J}_z \qquad\qquad (5.19g)$$

If we let the operator $\hat{J}^2$ act on the function $\hat{J}_+|\lambda m\rangle$ and observe that, according to equation (5.19c), $\hat{J}^2$ and $\hat{J}_+$ commute, we obtain

$$\hat{J}^2\hat{J}_+|\lambda m\rangle = \hat{J}_+\hat{J}^2|\lambda m\rangle = \lambda\hbar^2\hat{J}_+|\lambda m\rangle$$

where (5.16a) was also used. We note that $\hat{J}_+|\lambda m\rangle$ is an eigenfunction of $\hat{J}^2$ with eigenvalue $\lambda\hbar^2$. Thus, the operator $\hat{J}_+$ has no effect on the eigenvalues of

$\hat{J}^2$ because $\hat{J}^2$ and $\hat{J}_+$ commute. However, if the operator $\hat{J}_z$ acts on the function $\hat{J}_+|\lambda m\rangle$, we have

$$\hat{J}_z\hat{J}_+|\lambda m\rangle = \hat{J}_+\hat{J}_z|\lambda m\rangle + \hbar\hat{J}_+|\lambda m\rangle = m\hbar\hat{J}_+|\lambda m\rangle + \hbar\hat{J}_+|\lambda m\rangle$$

$$= (m+1)\hbar\hat{J}_+|\lambda m\rangle \qquad (5.20)$$

where equations (5.19a) and (5.16b) were used. Thus, the function $\hat{J}_+|\lambda m\rangle$ is an eigenfunction of $\hat{J}_z$ with eigenvalue $(m+1)\hbar$. Writing equation (5.16b) as

$$\hat{J}_z|\lambda,\, m+1\rangle = (m+1)\hbar|\lambda,\, m+1\rangle$$

we see from equation (5.20) that $\hat{J}_+|\lambda m\rangle$ is proportional to $|\lambda,\, m+1\rangle$

$$\hat{J}_+|\lambda m\rangle = c_+|\lambda,\, m+1\rangle \qquad (5.21)$$

where c_+ is the proportionality constant. The operator $\hat{J}_+$ is, therefore, a *raising operator*, which alters the eigenfunction $|\lambda m\rangle$ for the eigenvalue $m\hbar$ to the eigenfunction for $(m+1)\hbar$.

The proportionality constant c_+ in equation (5.21) may be evaluated by squaring both sides of equation (5.21) to give

$$\langle\lambda m|\hat{J}_-\hat{J}_+|\lambda m\rangle = |c_+|^2\langle\lambda,\, m+1|\lambda,\, m+1\rangle$$

since the bra $\langle\lambda m|\hat{J}_-$ is the adjoint of the ket $\hat{J}_+|\lambda m\rangle$. Using equations (5.16) and (5.19g) and the normality of the eigenfunctions, we have

$$|c_+|^2 = \langle\lambda m|\hat{J}^2 \quad \hat{J}_z^2 - \hbar\hat{J}_z|\lambda m\rangle = (\lambda - m^2 - m)\hbar^2$$

and equation (5.21) becomes

$$\hat{J}_+|\lambda m\rangle = \sqrt{\lambda - m(m+1)}\,\hbar|\lambda,\, m+1\rangle \qquad (5.22)$$

In equation (5.22) we have arbitrarily taken c_+ to be real and positive.

We next let the operators $\hat{J}^2$ and $\hat{J}_z$ act on the function $\hat{J}_-|\lambda m\rangle$ to give

$$\hat{J}^2\hat{J}_-|\lambda m\rangle - \hat{J}_-\hat{J}^2|\lambda m\rangle - \lambda\hbar^2\hat{J}_-|\lambda m\rangle$$

$$\hat{J}_z\hat{J}_-|\lambda m\rangle - \hat{J}_-\hat{J}_z|\lambda m\rangle - \hbar\hat{J}_-|\lambda m\rangle = (m-1)\hbar J_-|\lambda m\rangle$$

where we have used equations (5.16), (5.19b), and (5.19d). The function $\hat{J}_-|\lambda m\rangle$ is a simultaneous eigenfunction of $\hat{J}^2$ and $\hat{J}_z$ with eigenvalues $\lambda\hbar^2$ and $(m-1)\hbar$, respectively. Accordingly, the function $\hat{J}_-|\lambda m\rangle$ is proportional to $|\lambda,\, m-1\rangle$

$$\hat{J}_-|\lambda m\rangle = c_-|\lambda,\, m-1\rangle \qquad (5.23)$$

where c_- is the proportionality constant. The operator $\hat{J}_-$ changes the eigenfunction $|\lambda m\rangle$ to the eigenfunction $|\lambda,\, m-1\rangle$ for a lower value of the eigenvalue of $\hat{J}_z$ and is, therefore, a *lowering operator*.

To evaluate the proportionality constant c_- in equation (5.23), we square both sides of (5.23) and note that the bra $\langle\lambda m|\hat{J}_+$ is the adjoint of the ket $\hat{J}_-|\lambda m\rangle$, giving

$$|c_-|^2 = \langle \lambda m | \hat{J}_+ \hat{J}_- | \lambda m \rangle = \langle \lambda m | \hat{J}^2 - \hat{J}_z^2 + \hbar \hat{J}_z | \lambda m \rangle = (\lambda - m^2 + m)\hbar^2$$

where equation (5.19f) was also used. Equation (5.23) then becomes

$$\hat{J}_- | \lambda m \rangle = \sqrt{\lambda - m(m-1)}\ \hbar | \lambda,\ m-1 \rangle \qquad (5.24)$$

where we have taken c_- to be real and positive. This choice is consistent with the selection above of c_+ as real and positive.

Determination of the eigenvalues

We now apply the raising and lowering operators to find the eigenvalues of $\hat{J}^2$ and $\hat{J}_z$. Equation (5.17) tells us that for a given value of λ, the parameter m has a maximum and a minimum value, the maximum value being positive and the minimum value being negative. For the special case in which λ equals zero, the parameter m must, of course, be zero as well.

We select arbitrary values for λ, say ξ, and for m, say η, where $0 \leq \eta^2 \leq \xi$ so that (5.17) is satisfied. Application of the raising operator $\hat{J}_+$ to the corresponding ket $|\xi \eta\rangle$ gives the ket $|\xi,\ \eta + 1\rangle$. Successive applications of $\hat{J}_+$ give $|\xi,\ \eta + 2\rangle$, $|\xi,\ \eta + 3\rangle$, etc. After k such applications, we obtain the ket $|\xi j\rangle$, where $j = \eta + k$ and $j^2 \leq \xi$. The value of j is such that an additional application of $\hat{J}_+$ produces the ket $|\xi,\ j + 1\rangle$ with $(j+1)^2 > \xi$ (that is to say, it produces a ket $|\lambda m\rangle$ with $m^2 > \lambda$), which is not possible. Accordingly, the sequence must terminate by the condition $\hat{J}_+ |\xi j\rangle = 0$. From equation (5.22), this condition is given by

$$\hat{J}_+ |\xi j\rangle = \sqrt{\xi - j(j+1)}\ \hbar |\xi,\ j+1\rangle = 0$$

which is valid only if the coefficient of $|\xi,\ j + 1\rangle$ vanishes, so that we have $\xi = j(j+1)$.

We now apply the lowering operator $\hat{J}_-$ to the ket $|\xi j\rangle$ successively to construct the series of kets $|\xi,\ j - 1\rangle$, $|\xi,\ j - 2\rangle$, etc. After a total of n applications of $\hat{J}_-$, we obtain the ket $|\xi j'\rangle$, where $j' = j - n$ is the minimum value of m allowed by equation (5.17). Therefore, this lowering sequence must terminate by the condition

$$\hat{J}_- |\xi j'\rangle = \sqrt{\xi - j'(j' - 1)}\ \hbar |\xi,\ j' - 1\rangle = 0$$

where equation (5.24) has been introduced. This condition is valid only if the coefficient of $|\xi,\ j' - 1\rangle$ vanishes, giving $\xi = j'(j' - 1)$.

The parameter ξ has two conditions imposed upon it

$$\xi = j(j+1)$$

$$\xi = j'(j' - 1)$$

giving the relation

$$j(j+1) = j'(j'-1)$$

The solution to this quadratic equation gives $j' = -j$. The other solution, $j' = j + 1$, is not physically meaningful because j' must be less than j. We have shown, therefore, that the parameter m ranges from $-j$ to j

$$-j \leqslant m \leqslant j$$

If we combine the conclusion that $j' = -j$ with the relation $j' = j - n$, we see that $j = n/2$, where $n = 0, 1, 2, \ldots$ Thus, the allowed values of j are the integers $0, 1, 2, \ldots$ (if n is even) and the half-integers $\frac{1}{2}, \frac{3}{2}, \frac{5}{2}, \ldots$ (if n is odd) and the allowed values of m are $-j, -j+1, \ldots, j-1, j$.

We began this analysis with an arbitrary value for λ, namely $\lambda = \xi$, and an arbitrary value for m, namely $m = \eta$. We showed that, in order to satisfy requirement (5.17), the parameter ξ must satisfy $\xi = j(j+1)$, where j is restricted to integral or half-integral values. Since the value ξ was chosen arbitrarily, we conclude that the only allowed values for λ are

$$\lambda = j(j+1) \tag{5.25}$$

The parameter η is related to j by $j = \eta + k$, where k is the number of successive applications of $\hat{J}_+$ until $|\xi\eta\rangle$ is transformed into $|\xi j\rangle$. Since k must be a positive integer, the parameter η must be restricted to integral or half-integral values. However, the value η was chosen arbitrarily, leading to the conclusion that the only allowed values of m are $m = -j, -j+1, \ldots, j-1, j$. Thus, we have found all of the allowed values for λ and for m and, therefore, all of the eigenvalues of $\hat{J}^2$ and $\hat{J}_z$.

In view of equation (5.25), we now denote the eigenkets $|\lambda m\rangle$ by $|jm\rangle$. Equations (5.16) may now be written as

$$\hat{J}^2|jm\rangle = j(j+1)\hbar^2|jm\rangle, \qquad j = 0, \tfrac{1}{2}, 1, \tfrac{3}{2}, 2, \ldots \tag{5.26a}$$

$$\hat{J}_z|jm\rangle = m\hbar|jm\rangle, \qquad m = -j, -j+1, \ldots, j-1, j \tag{5.26b}$$

Each eigenvalue of $\hat{J}^2$ is $(2j+1)$-fold degenerate, because there are $(2j+1)$ values of m for a given value of j. Equations (5.22) and (5.24) become

$$\hat{J}_+|jm\rangle = \sqrt{j(j+1) - m(m+1)}\,\hbar|j, m+1\rangle$$

$$= \sqrt{(j-m)(j+m+1)}\,\hbar|j, m+1\rangle \tag{5.27a}$$

$$\hat{J}_-|jm\rangle = \sqrt{j(j+1) - m(m-1)}\,\hbar|j, m-1\rangle$$

$$= \sqrt{(j+m)(j-m+1)}\,\hbar|j, m-1\rangle \tag{5.27b}$$

5.3 Application to orbital angular momentum

We now apply the results of the quantum-mechanical treatment of generalized angular momentum to the case of orbital angular momentum. The orbital angular momentum operator $\hat{\mathbf{L}}$, defined in Section 5.1, is identified with the operator $\hat{\mathbf{J}}$ of Section 5.2. Likewise, the operators $\hat{L}^2$, $\hat{L}_x$, $\hat{L}_y$, and $\hat{L}_z$ are identified with $\hat{J}^2$, $\hat{J}_x$, $\hat{J}_y$, and $\hat{J}_z$, respectively. The parameter j of Section 5.2 is denoted by l when applied to orbital angular momentum. The simultaneous eigenfunctions of $\hat{L}^2$ and $\hat{L}_z$ are denoted by $|lm\rangle$, so that we have

$$\hat{L}^2|lm\rangle = l(l+1)\hbar^2|lm\rangle \tag{5.28a}$$

$$\hat{L}_z|lm\rangle = m\hbar|lm\rangle, \qquad m = -l, -l+1, \ldots, l-1, l \tag{5.28b}$$

Our next objective is to find the analytical forms for these simultaneous eigenfunctions. For that purpose, it is more convenient to express the operators $\hat{L}_x$, $\hat{L}_y$, $\hat{L}_z$, and $\hat{L}^2$ in spherical polar coordinates r, θ, φ rather than in cartesian coordinates x, y, z. The relationships between r, θ, φ and x, y, z are shown in Figure 5.1. The transformation equations are

$$x = r\sin\theta\cos\varphi \tag{5.29a}$$

$$y = r\sin\theta\sin\varphi \tag{5.29b}$$

$$z = r\cos\theta \tag{5.29c}$$

$$r = (x^2 + y^2 + z^2)^{1/2} \tag{5.29d}$$

$$\theta = \cos^{-1}(z/(x^2 + y^2 + z^2)^{1/2}) \tag{5.29e}$$

$$\varphi = \tan^{-1}(y/x) \tag{5.29f}$$

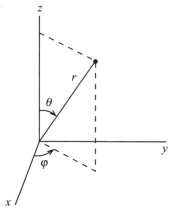

Figure 5.1 Spherical polar coordinate system.

These coordinates are defined over the following intervals

$$-\infty \leqslant x, y, z \leqslant \infty, \quad 0 \leqslant r \leqslant \infty, \quad 0 \leqslant \theta \leqslant \pi, \quad 0 \leqslant \varphi \leqslant 2\pi$$

The volume element $d\tau = dx\,dy\,dz$ becomes $d\tau = r^2 \sin\theta\,dr\,d\theta\,d\varphi$ in spherical polar coordinates.

To transform the partial derivatives $\partial/\partial x$, $\partial/\partial y$, $\partial/\partial z$, which appear in the operators $\hat{L}_x$, $\hat{L}_y$, $\hat{L}_z$ of equations (5.7), we use the expressions

$$\frac{\partial}{\partial x} = \left(\frac{\partial r}{\partial x}\right)_{y,z}\frac{\partial}{\partial r} + \left(\frac{\partial \theta}{\partial x}\right)_{y,z}\frac{\partial}{\partial \theta} + \left(\frac{\partial \varphi}{\partial x}\right)_{y,z}\frac{\partial}{\partial \varphi}$$

$$= \sin\theta\cos\varphi\,\frac{\partial}{\partial r} + \frac{\cos\theta\cos\varphi}{r}\frac{\partial}{\partial \theta} - \frac{\sin\varphi}{r\sin\theta}\frac{\partial}{\partial \varphi} \qquad (5.30a)$$

$$\frac{\partial}{\partial y} = \left(\frac{\partial r}{\partial y}\right)_{x,z}\frac{\partial}{\partial r} + \left(\frac{\partial \theta}{\partial y}\right)_{x,z}\frac{\partial}{\partial \theta} + \left(\frac{\partial \varphi}{\partial y}\right)_{x,z}\frac{\partial}{\partial \varphi}$$

$$= \sin\theta\sin\varphi\,\frac{\partial}{\partial r} + \frac{\cos\theta\sin\varphi}{r}\frac{\partial}{\partial \theta} + \frac{\cos\varphi}{r\sin\theta}\frac{\partial}{\partial \varphi} \qquad (5.30b)$$

$$\frac{\partial}{\partial z} = \left(\frac{\partial r}{\partial z}\right)_{x,y}\frac{\partial}{\partial r} + \left(\frac{\partial \theta}{\partial z}\right)_{x,y}\frac{\partial}{\partial \theta} + \left(\frac{\partial \varphi}{\partial z}\right)_{x,y}\frac{\partial}{\partial \varphi}$$

$$= \cos\theta\,\frac{\partial}{\partial r} - \frac{\sin\theta}{r}\frac{\partial}{\partial \theta} \qquad (5.30c)$$

Substitution of these three expressions into equations (5.7) gives

$$\hat{L}_x = \frac{\hbar}{i}\left(-\sin\varphi\,\frac{\partial}{\partial \theta} - \cot\theta\cos\varphi\,\frac{\partial}{\partial \varphi}\right) \qquad (5.31a)$$

$$\hat{L}_y = \frac{\hbar}{i}\left(\cos\varphi\,\frac{\partial}{\partial \theta} - \cot\theta\sin\varphi\,\frac{\partial}{\partial \varphi}\right) \qquad (5.31b)$$

$$\hat{L}_z = \frac{\hbar}{i}\frac{\partial}{\partial \varphi} \qquad (5.31c)$$

By squaring each of the operators $\hat{L}_x$, $\hat{L}_y$, $\hat{L}_z$ and adding, we find that $\hat{L}^2$ is given in spherical polar coordinates by

$$\hat{L}^2 = -\hbar^2\left[\frac{1}{\sin\theta}\frac{\partial}{\partial \theta}\left(\sin\theta\frac{\partial}{\partial \theta}\right) + \frac{1}{\sin^2\theta}\frac{\partial^2}{\partial \varphi^2}\right] \qquad (5.32)$$

Since the variable r does not appear in any of these operators, their eigenfunctions are independent of r and are functions only of the variables θ and φ. The simultaneous eigenfunctions $|lm\rangle$ of $\hat{L}^2$ and $\hat{L}_z$ will now be denoted by the function $Y_{lm}(\theta, \varphi)$ so as to acknowledge explicitly their dependence on the angles θ and φ.

The eigenvalue equation for $\hat{L}_z$ is

$$\hat{L}_z Y_{lm}(\theta, \varphi) = \frac{\hbar}{i} \frac{\partial}{\partial \varphi} Y_{lm}(\theta, \varphi) = m\hbar Y_{lm}(\theta, \varphi) \tag{5.33}$$

where equations (5.28b) and (5.31c) have been combined. Equation (5.33) may be written in the form

$$\frac{dY_{lm}(\theta, \varphi)}{Y_{lm}(\theta, \varphi)} = im \, d\varphi \quad (\theta \text{ held constant})$$

the solution of which is

$$Y_{lm}(\theta, \varphi) = \Theta_{lm}(\theta) e^{im\varphi} \tag{5.34}$$

where $\Theta_{lm}(\theta)$ is the 'constant of integration' and is a function only of the variable θ. Thus, we have shown that $Y_{lm}(\theta, \varphi)$ is the product of two functions, one a function only of θ, the other a function only of φ

$$Y_{lm}(\theta, \varphi) = \Theta_{lm}(\theta)\Phi_m(\varphi) \tag{5.35}$$

We have also shown that the function $\Phi_m(\varphi)$ involves only the parameter m and not the parameter l.

The function $\Phi_m(\varphi)$ must be single-valued and continuous at all points in space in order for $Y_{lm}(\theta, \varphi)$ to be an eigenfunction of $\hat{L}^2$ and $\hat{L}_z$. If $\Phi_m(\varphi)$ and hence $Y_{lm}(\theta, \varphi)$ are not single-valued and continuous at some point φ_0, then the derivative of $Y_{lm}(\theta, \varphi)$ with respect to φ would produce a delta function at the point φ_0 and equation (5.33) would not be satisfied. Accordingly, we require that

$$\Phi_m(\varphi) = \Phi_m(\varphi + 2\pi)$$

or

$$e^{im\varphi} = e^{im(\varphi + 2\pi)}$$

so that

$$e^{2im\pi} = 1$$

This equation is valid only if m is an integer, positive or negative

$$m = 0, \pm 1, \pm 2, \ldots$$

We showed in Section 5.2 that the parameter m for generalized angular momentum can equal either an integer or a half-integer. However, in the case of orbital angular momentum, the parameter m can only be an integer; the half-integer values for m are not allowed. Since the permitted values of m are $-l$, $-l + 1, \ldots, l - 1, l$, the parameter l can have only integer values in the case of orbital angular momentum; half-integer values for l are also not allowed.

Ladder operators

The ladder operators for orbital angular momentum are

$$\hat{L}_+ \equiv \hat{L}_x + i\hat{L}_y$$
$$\hat{L}_- \equiv \hat{L}_x - i\hat{L}_y \qquad (5.36)$$

and are identified with the ladder operators $\hat{J}_+$ and $\hat{J}_-$ of Section 5.2. Substitution of (5.31a) and (5.31b) into (5.36) yields

$$\hat{L}_+ = \hbar e^{i\varphi}\left(\frac{\partial}{\partial\theta} + i\cot\theta\,\frac{\partial}{\partial\varphi}\right) \qquad (5.37a)$$

$$\hat{L}_- = \hbar e^{-i\varphi}\left(-\frac{\partial}{\partial\theta} + i\cot\theta\,\frac{\partial}{\partial\varphi}\right) \qquad (5.37b)$$

where equation (A.31) has been used. When applied to orbital angular momentum, equations (5.27) take the form

$$\hat{L}_+ Y_{lm}(\theta,\,\varphi) = \sqrt{(l-m)(l+m+1)}\,\hbar Y_{l,m+1}(\theta,\,\varphi) \qquad (5.38a)$$
$$\hat{L}_- Y_{lm}(\theta,\,\varphi) = \sqrt{(l+m)(l-m+1)}\,\hbar Y_{l,m-1}(\theta,\,\varphi) \qquad (5.38b)$$

For the case where m is equal to its minimum value, $m = -l$, equation (5.38b) becomes

$$\hat{L}_- Y_{l,-l}(\theta,\,\varphi) = 0$$

or

$$\left(-\frac{\partial}{\partial\theta} + i\cot\theta\,\frac{\partial}{\partial\varphi}\right)Y_{l,-l}(\theta,\,\varphi) = 0$$

when equation (5.37b) is introduced. Substitution of $Y_{l,-l}(\theta,\,\varphi)$ from equation (5.34) gives

$$\left(-\frac{\partial}{\partial\theta} + i\cot\theta\,\frac{\partial}{\partial\varphi}\right)\Theta_{l,-l}(\theta)e^{-il\varphi} = \left(-\frac{\partial}{\partial\theta} + l\cot\theta\right)\Theta_{l,-l}(\theta)e^{-il\varphi} = 0$$

Dividing by $e^{-il\varphi}$, we obtain the differential equation

$$d\ln\Theta_{l,-l}(\theta) = l\cot\theta\,d\theta = \frac{l\cos\theta}{\sin\theta}\,d\theta = \frac{l}{\sin\theta}\,d\sin\theta = l\,d\ln\sin\theta$$

which has the solution

$$\Theta_{l,-l}(\theta) = A_l\sin^l\theta \qquad (5.39)$$

where A_l is the constant of integration.

Normalization of $Y_{l,-l}(\theta,\,\varphi)$
Following the usual custom, we require that the eigenfunctions $Y_{lm}(\theta,\,\varphi)$ be normalized, so that

$$\int_0^{2\pi} \int_0^{\pi} Y_{lm}^*(\theta, \varphi) Y_{lm}(\theta, \phi) \sin\theta \, d\theta \, d\varphi$$

$$= \int_0^{\pi} \Theta_{lm}^*(\theta) \Theta_{lm}(\theta) \sin\theta \, d\theta \int_0^{2\pi} \Phi_m^*(\varphi) \Phi_m(\varphi) \, d\varphi = 1$$

where the θ- and φ-dependent parts of the volume element $d\tau$ are included in the integration. For convenience, we require that each of the two factors $\Theta_{lm}(\theta)$ and $\Phi_m(\varphi)$ be normalized. Writing $\Phi_m(\varphi)$ as

$$\Phi_m(\varphi) = A e^{im\varphi}$$

we find that

$$\int_0^{2\pi} (A e^{im\varphi})^* (A e^{im\varphi}) \, d\varphi = |A|^2 \int_0^{2\pi} d\varphi = 1$$

$$A = e^{i\alpha}/\sqrt{2\pi}$$

giving

$$\Phi_m(\varphi) = \frac{1}{\sqrt{2\pi}} e^{im\varphi} \tag{5.40}$$

where we have arbitrarily set α equal to zero in the phase factor $e^{i\alpha}$ associated with the normalization constant.

The function $\Theta_{l,-l}(\theta)$ is given by equation (5.39) and the value of the constant of integration A_l is determined by the normalization condition

$$\int_0^{\pi} [\Theta_{l,-l}(\theta)]^* \Theta_{l,-l}(\theta) \sin\theta \, d\theta = |A_l|^2 \int_0^{\pi} \sin^{2l+1}\theta \, d\theta = 1 \tag{5.41}$$

We need to evaluate the integral I_l

$$I_l \equiv \int_0^{\pi} \sin^{2l+1}\theta \, d\theta = -\int_1^{-1} (1 - \mu^2)^l \, d\mu = \int_{-1}^1 (1 - \mu^2)^l \, d\mu$$

where we have defined the variable μ by the relation

$$\mu \equiv \cos\theta \tag{5.42}$$

so that

$$1 - \mu^2 = \sin^2\theta, \qquad d\mu = -\sin\theta \, d\theta$$

The integral I_l may be transformed as follows

$$I_l = \int_{-1}^1 (1 - \mu^2)^{l-1} \, d\mu - \int_{-1}^1 (1 - \mu^2)^{l-1} \mu^2 \, d\mu = I_{l-1} + \int_{-1}^1 \frac{\mu}{2l} \, d(1 - \mu^2)^l$$

$$= I_{l-1} - \int_{-1}^1 \frac{(1 - \mu^2)^l}{2l} \, d\mu = I_{l-1} - \frac{1}{2l} I_l$$

where we have integrated by parts and noted that the integrated term vanishes. Solving for I_l, we obtain a recurrence relation for the integral

$$I_l = \frac{2l}{2l+1} I_{l-1} \tag{5.43}$$

Since I_0 is given by

$$I_0 = \int_{-1}^{1} d\mu = 2$$

we can obtain I_l by repeated application of equation (5.43) starting with I_0

$$I_l = \frac{(2l)(2l-2)(2l-4)\cdots 2}{(2l+1)(2l-1)(2l-3)\cdots 3} I_0 = \frac{2^{2l+1}(l!)^2}{(2l+1)!}$$

where we have noted that

$$(2l)(2l-2)\cdots 2 = 2^l l!$$

$$(2l+1)(2l-1)(2l-3)\cdots 3$$

$$= \frac{(2l+1)(2l)(2l-1)(2l-2)(2l-3)\cdots 3\times 2\times 1}{(2l)(2l-2)\cdots 2} = \frac{(2l+1)!}{2^l l!}$$

Substituting this result into equation (5.41), we find that

$$|A_l| = \frac{1}{2^l l!} \sqrt{\frac{(2l+1)!}{2}}$$

It is customary to let α equal zero in the phase factor $e^{i\alpha}$ for $\Theta_{l,-l}(\theta)$, so that

$$\Theta_{l,-l}(\theta) = \frac{1}{2^l l!} \sqrt{\frac{(2l+1)!}{2}} \sin^l \theta \tag{5.44}$$

Combining equations (5.35), (5.40) and (5.44), we obtain the normalized eigenfunction

$$Y_{l,-l}(\theta, \varphi) = \frac{1}{2^l l!} \sqrt{\frac{(2l+1)!}{4\pi}} \sin^l \theta \, e^{il\varphi} \tag{5.45}$$

Spherical harmonics

The functions $Y_{lm}(\theta, \varphi)$ are known as *spherical harmonics* and may be obtained from $Y_{l,-l}(\theta, \varphi)$ by repeated application of the raising operator $\hat{L}_+$ according to (5.38a). By this procedure, the spherical harmonics $Y_{l,-l+1}(\theta, \varphi)$, $Y_{l,-l+2}(\theta, \varphi)$, ..., $Y_{l,-1}(\theta, \varphi)$, $Y_{l0}(\theta, \varphi)$, $Y_{l1}(\theta, \varphi)$, ..., $Y_{ll}(\theta, \varphi)$ may be determined. Since the starting function $Y_{l,-l}(\theta, \varphi)$ is normalized, each of the spherical harmonics generated from equation (5.38a) will also be normalized.

We may readily derive a general expression for the spherical harmonic $Y_{lm}(\theta, \varphi)$ which results from the repeated application of $\hat{L}_+$ to $Y_{l,-l}(\theta, \varphi)$. We begin with equation (5.38a) with m set equal to $-l$

$$Y_{l,-l+1} = \frac{1}{\sqrt{2l}\hbar} \hat{L}_+ Y_{l,-l} \tag{5.46}$$

For m equal to $-l+1$, equation (5.38a) gives

$$Y_{l,-l+2} = \frac{1}{\sqrt{2(2l-1)}\ \hbar}\ \hat{L}_+ Y_{l,-l+1} = \frac{1}{\sqrt{2(2l)(2l-1)}\ \hbar^2}\ \hat{L}_+^2 Y_{l,-l}$$

where equation (5.46) has been introduced in the last term. If we continue in the same pattern, we find

$$Y_{l,-l+3} = \frac{1}{\sqrt{3(2l-2)}\ \hbar}\ \hat{L}_+ Y_{l,-l+2} = \frac{1}{\sqrt{2\cdot 3(2l)(2l-1)(2l-2)}\ \hbar^3}\ \hat{L}_+^3 Y_{l,-l}$$

$$\vdots$$

$$Y_{l,-l+k} = \sqrt{\frac{(2l-k)!}{k!(2l)!}}\ \frac{1}{\hbar^k}\ \hat{L}_+^k Y_{l,-l}$$

where k is the number of steps in this sequence. We now set $k = l + m$ in the last expression to obtain

$$Y_{lm} = \sqrt{\frac{(l-m)!}{(l+m)!(2l)!}}\frac{1}{\hbar^{l+m}}\ \hat{L}_+^{l+m} Y_{l,-l} \tag{5.47}$$

If the number of steps k is less than the value of l, then the integer m is negative; if k equals l, then m is zero; if k is greater than l, then m is positive; and finally if k equals $2l$, then m equals its largest value of l.

The next step in this derivation is the evaluation of $\hat{L}_+^{l+m} Y_{l,-l}$ using equation (5.37a). If the operator $\hat{L}_+$ in (5.37a) acts on $Y_{l,-l}(\theta, \varphi)$ as given in (5.45), we have

$$\hat{L}_+ Y_{l,-l} = c_l \hbar e^{i\varphi}\left(\frac{\partial}{\partial\theta} + i\cot\theta\,\frac{\partial}{\partial\phi}\right)\sin^l\theta\, e^{-il\varphi}$$

$$= c_l \hbar e^{-i(l-1)\varphi}\left(\frac{d}{d\theta} + l\cot\theta\right)\sin^l\theta$$

$$= c_l \hbar e^{-i(l-1)\varphi}\left(\frac{d}{d\theta} + l\cot\theta\right)\frac{\sin^{2l}\theta}{\sin^l\theta}$$

$$= c_l \hbar e^{-i(l-1)\varphi}\left[\frac{1}{\sin^l\theta}\frac{d}{d\theta}\sin^{2l}\theta - l\sin^{l-1}\theta\cos\theta + l\sin^{l-1}\theta\cos\theta\right]$$

$$= -c_l \hbar e^{-i(l-1)\varphi}\frac{1}{\sin^{l-1}\theta}\frac{d}{d(\cos\theta)}\sin^{2l}\theta$$

where for brevity we have defined c_l as

$$c_l = \frac{1}{2^l l!}\sqrt{\frac{(2l+1)!}{4\pi}} \tag{5.48}$$

We then operate on this result with $\hat{L}_+$ to obtain

$$\hat{L}_+^2 Y_{l,-l} = -c_l \hbar^2 e^{i\varphi} \left(\frac{\partial}{\partial\theta} + i\cot\theta \frac{\partial}{\partial\varphi} \right) \left(\frac{e^{-i(l-1)\varphi}}{\sin^{l-1}\theta} \frac{d}{d(\cos\theta)} \sin^{2l}\theta \right)$$

$$= -c_l \hbar^2 e^{-i(l-2)\varphi} \left(\frac{d}{d\theta} + (l-1)\cot\theta \right) \left(\frac{1}{\sin^{l-1}\theta} \frac{d}{d(\cos\theta)} \sin^{2l}\theta \right)$$

$$= c_l \hbar^2 e^{-i(l-2)\varphi} \frac{1}{\sin^{l-2}\theta} \frac{d^2}{d(\cos\theta)^2} \sin^{2l}\theta$$

After k such applications of $\hat{L}_+$ to the function $Y_{l,-l}(\theta, \varphi)$, we have

$$\hat{L}_+^k Y_{l,-l} = (-\hbar)^k c_l e^{-i(l-k)\varphi} \frac{1}{\sin^{l-k}\theta} \frac{d^k}{d(\cos\theta)^k} \sin^{2l}\theta$$

If we set $k = l + m$ in this expression, we obtain the desired result

$$\hat{L}_+^{l+m} Y_{l,-l} = (-\hbar)^{l+m} c_l e^{im\varphi} \sin^m\theta \frac{d^{l+m}}{d(\cos\theta)^{l+m}} \sin^{2l}\theta \qquad (5.49)$$

The general expression for $Y_{lm}(\theta, \varphi)$ is obtained by substituting equation (5.49) into (5.47) with c_l given by equation (5.48)

$$Y_{lm}(\theta, \varphi) = \frac{(-1)^{l+m}}{2^l l!} \sqrt{\frac{(2l+1)}{4\pi} \frac{(l-m)!}{(l+m)!}} e^{im\varphi} \sin^m\theta \frac{d^{l+m}}{d(\cos\theta)^{l+m}} \sin^{2l}\theta$$

$$(5.50)$$

When $Y_{lm}(\theta, \varphi)$ is decomposed into its two normalized factors according to equations (5.35) and (5.40), we have

$$\Theta_{lm}(\theta) = \frac{(-1)^{l\,|\,m}}{2^l l!} \sqrt{\frac{(2l+1)}{2} \frac{(l-m)!}{(l+m)!}} \sin^m\theta \frac{d^{l+m}}{d(\cos\theta)^{l+m}} \sin^{2l}\theta \qquad (5.51)$$

$$\Phi_m(\varphi) = \frac{1}{\sqrt{2\pi}} e^{im\varphi} \qquad (5.52)$$

The spherical harmonics for $l = 0, 1, 2, 3$ are listed in Table 5.1. We note that the function $\Theta_{l,-m}(\theta)$ is related to $\Theta_{lm}(\theta)$ by

$$\Theta_{l,-m}(\theta) = (-1)^m \Theta_{lm}(\theta) \qquad (5.53)$$

and that the complex conjugate $Y_{lm}^*(\theta, \varphi)$ is related to $Y_{lm}(\theta, \varphi)$ by

$$Y_{lm}^*(\theta, \varphi) = (-1)^m Y_{l,-m} \qquad (5.54)$$

Because both $\hat{L}^2$ and $\hat{L}_z$ are hermitian, the spherical harmonics $Y_{lm}(\theta, \varphi)$ form an orthogonal set, so that

$$\int_0^{2\pi} \int_0^{\pi} Y_{l'm'}^*(\theta, \varphi) Y_{lm}(\theta, \varphi) \sin\theta \, d\theta \, d\varphi = \delta_{ll'} \delta_{mm'} \qquad (5.55)$$

Table 5.1. *Spherical harmonics* $Y_{lm}(\theta, \varphi)$ *for* $l = 0, 1, 2, 3$

$Y_{00} = \left(\dfrac{1}{4\pi}\right)^{1/2}$

$Y_{30} = \left(\dfrac{7}{16\pi}\right)^{1/2} (5\cos^3\theta - 3\cos\theta)$

$Y_{10} = \left(\dfrac{3}{4\pi}\right)^{1/2} \cos\theta$

$Y_{3,\pm1} = \mp\left(\dfrac{21}{64\pi}\right)^{1/2} \sin\theta(5\cos^2\theta - 1)e^{\pm i\varphi}$

$Y_{1,\pm1} = \mp\left(\dfrac{3}{8\pi}\right)^{1/2} \sin\theta\, e^{\pm i\varphi}$

$Y_{3,\pm2} = \left(\dfrac{105}{32\pi}\right)^{1/2} \sin^2\theta\cos\theta\, e^{\pm2i\varphi}$

$Y_{20} = \left(\dfrac{5}{16\pi}\right)^{1/2} (3\cos^2\theta - 1)$

$Y_{3,\pm3} = \mp\left(\dfrac{35}{64\pi}\right)^{1/2} \sin^3\theta\, e^{\pm3i\varphi}$

$Y_{2,\pm1} = \mp\left(\dfrac{15}{8\pi}\right)^{1/2} \sin\theta\cos\theta\, e^{\pm i\varphi}$

$Y_{2,\pm2} = \left(\dfrac{15}{32\pi}\right)^{1/2} \sin^2\theta\, e^{\pm2i\varphi}$

If we introduce equation (5.35) into (5.55), we have

$$\int_0^\pi \Theta^*_{l'm'}(\theta)\Theta_{lm}(\theta)\sin\theta\, d\theta \int_0^{2\pi} \Phi^*_{m'}(\varphi)\Phi_m(\varphi)\, d\varphi = \delta_{ll'}\delta_{mm'}$$

The integral over the angle φ is

$$\int_0^{2\pi} \Phi^*_{m'}(\varphi)\Phi_m(\varphi)\, d\varphi = \frac{1}{2\pi}\int_0^{2\pi} e^{-im'\varphi}e^{im\varphi}\, d\varphi = \frac{1}{2\pi}\int_0^{2\pi} e^{i(m-m')\varphi}\, d\varphi$$

where equation (5.52) has been introduced. Since m and m' are integers, this integral vanishes unless $m = m'$, so that

$$\int_0^{2\pi} \Phi^*_{m'}(\varphi)\Phi_m(\varphi)\, d\varphi = \delta_{mm'} \tag{5.56}$$

from which it follows that

$$\int_0^\pi \Theta^*_{l'm}(\theta)\Theta_{lm}(\theta)\sin\theta\, d\theta = \delta_{ll'} \tag{5.57}$$

Note that in equation (5.57) the same value for m appears in both $\Theta^*_{l'm}(\theta)$ and $\Theta_{lm}(\theta)$. Thus, the functions $\Theta_{lm}(\theta)$ and $\Theta_{l'm}(\theta)$ for $l \neq l'$ are orthogonal, but the functions $\Theta_{lm}(\theta)$ and $\Theta_{l'm'}(\theta)$ are not orthogonal. However, for $m \neq m'$, the spherical harmonics $Y_{lm}(\theta, \varphi)$ and $Y_{l'm'}(\theta, \varphi)$ are orthogonal because of equation (5.56).

Relationship of spherical harmonics to associated Legendre polynomials

The functions $\Theta_{lm}(\theta)$ and consequently the spherical harmonics $Y_{lm}(\theta, \varphi)$ are related to the associated Legendre polynomials, whose definition and properties are presented in Appendix E. To show this relationship, we make the substitution of equation (5.42) for $\cos\theta$ in equation (5.51) and obtain

$$\Theta_{lm} = \frac{(-1)^m}{2^l l!} \sqrt{\frac{(2l+1)}{2} \frac{(l-m)!}{(l+m)!}} (1-\mu^2)^{m/2} \frac{d^{l+m}}{d\mu^{l+m}} (\mu^2-1)^l \qquad (5.58)$$

Equation (E.13) relates the associated Legendre polynomial $P_l^m(\mu)$ to the $(l+m)$th-order derivative in equation (5.58)

$$P_l^m(\mu) = \frac{1}{2^l l!} (1-\mu^2)^{m/2} \frac{d^{l+m}}{d\mu^{l+m}} (\mu^2-1)^l$$

where l and m are *positive* integers ($l, m \geq 0$) such that $m \leq l$. Thus, for positive m we have the relation

$$\Theta_{lm}(\theta) = (-1)^m \sqrt{\frac{(2l+1)}{2} \frac{(l-m)!}{(l+m)!}} P_l^m(\cos\theta), \qquad m \geq 0$$

For negative m, we may write $m = -|m|$ and note that equation (5.53) states

$$\Theta_{l,-|m|}(\theta) = (-1)^m \Theta_{l,|m|}(\theta)$$

so that we have

$$\Theta_{l,-|m|}(\theta) = \sqrt{\frac{(2l+1)}{2} \frac{(l-|m|)!}{(l+|m|)!}} P_l^{|m|}(\cos\theta)$$

These two results may be combined as

$$\Theta_{lm}(\theta) = \varepsilon \sqrt{\frac{(2l+1)}{2} \frac{(l-|m|)!}{(l+|m|)!}} P_l^{|m|}(\cos\theta)$$

where $\varepsilon = (-1)^m$ for $m > 0$ and $\varepsilon = 1$ for $m \leq 0$. Accordingly, the spherical harmonics $Y_{lm}(\theta, \varphi)$ are related to the associated Legendre polynomials by

$$Y_{lm}(\theta, \varphi) = \varepsilon \sqrt{\frac{(2l+1)}{4\pi} \frac{(l-|m|)!}{(l+|m|)!}} P_l^{|m|}(\cos\theta) e^{im\varphi}$$

$$\varepsilon = (-1)^m, \qquad m > 0 \qquad (5.59)$$

$$= 1, \qquad m \leq 0$$

The eigenvalues and eigenfunctions of the orbital angular momentum operator $\hat{L}^2$ may also be obtained by solving the differential equation $\hat{L}^2 \psi = \lambda \hbar^2 \psi$ using the Frobenius or series solution method. The application of this method is presented in Appendix G and, of course, gives the same results as the procedure using ladder operators. However, the Frobenius method may not be used to obtain the eigenvalues and eigenfunctions of the generalized angular momentum operators $\hat{\mathbf{J}}^2$ and $\hat{J}_z$ because their eigenfunctions do not have a spatial representation.

5.4 The rigid rotor

The motion of a rigid diatomic molecule serves as an application of the quantum-mechanical treatment of angular momentum to a chemical system. A rigid diatomic molecule consists of two particles of masses m_1 and m_2 which rotate about their center of mass while keeping the distance between them fixed at a value R. Although a diatomic molecule also undergoes vibrational motion in which the interparticle distance oscillates about some equilibrium value, that type of motion is neglected in the model being considered here; the interparticle distance is frozen at its equilibrium value R. Such a rotating system is called a *rigid rotor*.

We begin with a consideration of a classical particle i with mass m_i rotating in a plane at a constant distance r_i from a fixed center as shown in Figure 5.2. The time τ for the particle to make a complete revolution on its circular path is equal to the distance traveled divided by its linear velocity v_i

$$\tau = \frac{2\pi r_i}{v_i} \tag{5.60}$$

The reciprocal of τ gives the number of cycles per unit time, which is the frequency v of the rotation. The velocity v_i may then be expressed as

$$v_i = \frac{2\pi r_i}{\tau} = 2\pi v r_i = \omega r_i \tag{5.61}$$

where $\omega = 2\pi v$ is the *angular velocity*. According to equation (5.1), the angular momentum $\mathbf{L}_i$ of particle i is

$$\mathbf{L}_i = \mathbf{r}_i \times \mathbf{p}_i = m_i(\mathbf{r}_i \times \mathbf{v}_i) \tag{5.62}$$

Since the linear velocity vector $\mathbf{v}_i$ is perpendicular to the radius vector $\mathbf{r}_i$, the magnitude L_i of the angular momentum is

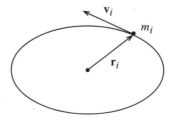

Figure 5.2 Motion of a rotating particle.

$$L_i = m_i r_i v_i \sin(\pi/2) = m_i r_i v_i = \omega m_i r_i^2 \tag{5.63}$$

where equation (5.61) has been introduced.

We next apply these classical relationships to the rigid diatomic molecule. Since the molecule is rotating freely about its center of mass, the potential energy is zero and the classical-mechanical Hamiltonian function H is just the kinetic energy of the two particles,

$$H = \frac{p_1^2}{2m_1} + \frac{p_2^2}{2m_2} = \tfrac{1}{2}m_1 v_1^2 + \tfrac{1}{2}m_2 v_2^2 \tag{5.64}$$

If we substitute equation (5.61) for each particle into (5.64) while noting that the angular velocity ω must be the same for both particles, we obtain

$$H = \tfrac{1}{2}\omega^2(m_1 r_1^2 + m_2 r_2^2) = \tfrac{1}{2}I\omega^2 \tag{5.65}$$

where we have defined the *moment of inertia* I by

$$I = m_1 r_1^2 + m_2 r_2^2 \tag{5.66}$$

In general, moments of inertia are determined relative to an axis of rotation. In this case the axis is perpendicular to the interparticle distance R and passes through the center of mass. Thus, we have

$$r_1 + r_2 = R$$

and

$$m_1 r_1 = m_2 r_2$$

or, upon inversion

$$r_1 = \frac{m_2}{m_1 + m_2} R$$

$$r_2 = \frac{m_1}{m_1 + m_2} R \tag{5.67}$$

Substitution of equations (5.67) into (5.66) gives

$$I = \mu R^2 \tag{5.68}$$

where the *reduced mass* μ is defined by

$$\mu = \frac{m_1 m_2}{m_1 + m_2} \tag{5.69}$$

The total angular momentum L for the two-particle system is given by

$$L = L_1 + L_2 = \omega(m_1 r_1^2 + m_2 r_2^2) = I\omega \tag{5.70}$$

where equations (5.63) and (5.66) are used. A comparison of equations (5.65) and (5.70) shows that

$$H = \frac{L^2}{2I} \tag{5.71}$$

Accordingly, the quantum-mechanical Hamiltonian operator $\hat{H}$ for this system is proportional to the square of the angular momentum operator $\hat{L}^2$

$$\hat{H} = \frac{1}{2I}\hat{L}^2 \qquad (5.72)$$

Thus, the operators $\hat{H}$ and $\hat{L}^2$ have the same eigenfunctions, namely, the spherical harmonics $Y_{Jm}(\theta, \varphi)$ as given in equation (5.50). It is customary in discussions of the rigid rotor to replace the quantum number l by the index J in the eigenfunctions and eigenvalues.

The eigenvalues of $\hat{H}$ are obtained by noting that

$$\hat{H}Y_{Jm}(\theta, \varphi) = \frac{1}{2I}\hat{L}^2 Y_{Jm}(\theta, \varphi) = \frac{J(J+1)\hbar^2}{2I}Y_{Jm}(\theta, \varphi) \qquad (5.73)$$

where l is replaced by J in equation (5.28a). Thus, the energy levels E_J for the rigid rotor are given by

$$E_J = J(J+1)\frac{\hbar^2}{2I} = J(J+1)B, \qquad J = 0, 1, 2, \ldots \qquad (5.74)$$

where $B = \hbar^2/2I$ is the *rotational constant* for the diatomic molecule. The energy levels E_J are shown in Figure 5.3. We observe that as J increases, the difference between successive levels also increases.

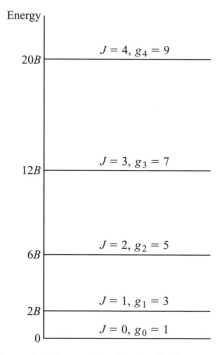

Figure 5.3 Energy levels of a rigid rotor.

To find the degeneracy of the eigenvalue E_J, we note that for a given value of J, the quantum number m has values $m = 0, \pm 1, \pm 2, \ldots, \pm J$. Accordingly, there are $(2J + 1)$ spherical harmonics for each value of J and the energy level E_J is $(2J + 1)$-fold degenerate. The ground-state energy level E_0 is non-degenerate.

5.5 Magnetic moment

Atoms are observed to have magnetic moments. To understand how an electron circulating about a nuclear core can give rise to a magnetic moment, we may apply classical theory. We consider an electron of mass m_e and charge $-e$ bound to a fixed nucleus of charge Ze by a central coulombic force $F(r)$ with potential $V(r)$

$$F(r) = -\frac{dV(r)}{dr} = \frac{-Ze^2}{4\pi\varepsilon_0 r^2} \tag{5.75}$$

$$V(r) = \frac{-Ze^2}{4\pi\varepsilon_0 r} \tag{5.76}$$

Equation (5.75) is Coulomb's law for the force between two charged particles separated by a distance r. In SI units, the charge e is expressed in coulombs (C), while ε_0 is the *permittivity of free space* with the value

$$\varepsilon_0 = 8.854\,19 \times 10^{-12} \text{ J}^{-1} \text{ C}^2 \text{ m}^{-1}$$

According to classical mechanics, a stable circular orbit of radius r and angular velocity ω is established for the electron if the centrifugal force $m_e r\omega^2$ balances the attractive coulombic force

$$m_e r\omega^2 = \frac{Ze^2}{4\pi\varepsilon_0 r^2}$$

This assumption is the basis of the Bohr model for the hydrogen-like atom. When solved for ω, this balancing equation is

$$\omega = \left(\frac{Ze^2}{4\pi\varepsilon_0 m_e r^3}\right)^{1/2} \tag{5.77}$$

An electron in a circular orbit with an angular velocity ω passes each point in the orbit $\omega/2\pi$ times per second. This electronic motion constitutes an electric current I, defined as the amount of charge passing a given point per second, so that

$$I = \frac{e\omega}{2\pi} \tag{5.78}$$

From the definition of the magnetic moment in electrodynamics, a circulating current I enclosing a small area A gives rise to a magnetic moment **M** of magnitude M given by

$$M = IA \qquad (5.79)$$

The area A enclosed by the circular electronic orbit of radius r is πr^2. From equation (5.63) we have the relation $L = m_e \omega r^2$. Thus, the magnitude of the magnetic moment is related to the magnitude L of the angular momentum by

$$M = \frac{eL}{2m_e} \qquad (5.80)$$

The direction of the vector **L** is determined by equation (5.62). By convention, the direction of the current I is opposite to the direction of rotation of the negatively charged electron, i.e., opposite to the direction of the vector **v**. Consequently, the vector **M** points in the opposite direction from **L** (see Figure 5.4) and equation (5.80) in vector form is

$$\mathbf{M} = \frac{-e}{2m_e}\mathbf{L} = \frac{-\mu_B}{\hbar}\mathbf{L} \qquad (5.81)$$

Since the units of **L** are those of $\hbar$, we have defined in equation (5.81) the *Bohr magneton* μ_B as

$$\mu_B \equiv \frac{e\hbar}{2m_e} = 9.274\,02 \times 10^{-24} \ \mathrm{JT}^{-1} \qquad (5.82)$$

The relationship (equation (5.81)) between **M** and **L** depends only on fundamental constants, the electronic mass and charge, and does not depend on any of the variables used in the derivation. Although this equation was obtained by applying classical theory to a circular orbit, it is more generally valid. It applies to elliptical orbits as well as to classical motion with attractive forces other than r^{-2} dependence. For any orbit in any central force field, the angular

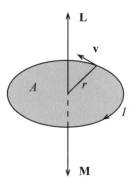

Figure 5.4 The magnetic moment **M** and the orbital angular momentum **L** of an electron in a circular orbit.

momentum is conserved and, since equation (5.81) applies, the magnetic moment is constant in both magnitude and direction. Moreover, equation (5.81) is also valid for orbital motion in quantum mechanics.

Interaction with a magnetic field

The potential energy V of an atom with a magnetic moment $\mathbf{M}$ in a magnetic field $\mathbf{B}$ is

$$V = -\mathbf{M} \cdot \mathbf{B} = -MB \cos \theta \qquad (5.83)$$

where θ is the angle between $\mathbf{M}$ and $\mathbf{B}$. The force $\mathbf{F}$ acting on the atom due to the magnetic field is

$$\mathbf{F} = -\nabla V$$

or

$$
\begin{aligned}
F_x &= -\mathbf{M} \cdot \frac{\partial \mathbf{B}}{\partial x} = -M \cos \theta \frac{\partial B}{\partial x} \\
F_y &= -\mathbf{M} \cdot \frac{\partial \mathbf{B}}{\partial y} = -M \cos \theta \frac{\partial B}{\partial y} \\
F_z &= -\mathbf{M} \cdot \frac{\partial \mathbf{B}}{\partial z} = -M \cos \theta \frac{\partial B}{\partial z}
\end{aligned}
\qquad (5.84)
$$

If the magnetic field is uniform, then the partial derivatives of B vanish and the force on the atom is zero.

According to electrodynamics, the force $\mathbf{F}$ for a non-uniform magnetic field produces on the atom a torque $\mathbf{T}$ given by

$$\mathbf{T} = \mathbf{M} \times \mathbf{B} = -\frac{\mu_B}{\hbar} \mathbf{L} \times \mathbf{B} \qquad (5.85)$$

where equation (5.81) has been introduced as well. From the relation $\mathbf{T} = d\mathbf{L}/dt$ in equation (5.6), we have

$$\frac{d\mathbf{L}}{dt} = -\frac{\mu_B}{\hbar} \mathbf{L} \times \mathbf{B} \qquad (5.86)$$

Thus, the torque changes the direction of the angular momentum vector $\mathbf{L}$ and the vector $d\mathbf{L}/dt$ is perpendicular to both $\mathbf{L}$ and $\mathbf{B}$, as shown in Figure 5.5. As a result of this torque, the vector $\mathbf{L}$ precesses around the direction of the magnetic field $\mathbf{B}$ with a constant angular velocity ω_L. This motion is known as *Larmor precession* and the angular velocity ω_L is called the *Larmor frequency*. Since the magnetic moment $\mathbf{M}$ is antiparallel to the angular moment $\mathbf{L}$, it also precesses about the magnetic field vector $\mathbf{B}$.

From equation (5.61), the Larmor angular frequency or velocity ω_L is equal to the velocity of the end of the vector $\mathbf{L}$ divided by the radius of the circular path shown in Figure 5.5

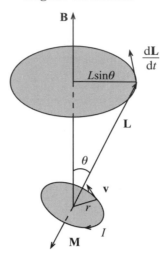

Figure 5.5 The motion in a magnetic field **B** of the orbital angular momentum vector **L**.

$$\omega_{\rm L} = \frac{|{\rm d}{\bf L}/{\rm d}t|}{L \sin \theta}$$

The magnitude of the vector ${\rm d}{\bf L}/{\rm d}t$ is obtained from equation (5.86) as

$$\left| \frac{{\rm d}{\bf L}}{{\rm d}t} \right| = \frac{\mu_{\rm B}}{\hbar} LB \sin \theta$$

so that

$$\omega_{\rm L} = \frac{\mu_{\rm B} B}{\hbar} \tag{5.87}$$

If we take the z-axis of the coordinate system parallel to the magnetic field vector **B**, then the projection of **L** on **B** is L_z and $\cos \theta$ in equation (5.83) is

$$\cos \theta = \frac{L_z}{L}$$

In quantum mechanics, the only allowed values of L are $\sqrt{l(l+1)}\,\hbar$ with $l = 0, 1, \ldots$ and the only allowed values of L_z are $m\hbar$ with $m = 0, \pm 1, \ldots,$ $\pm l$. Accordingly, the angle θ is quantized, being restricted to values for which

$$\cos \theta = \frac{m}{\sqrt{l(l+1)}}, \quad l = 0, 1, 2, \ldots, \quad m = 0, \pm 1, \ldots, \pm l \tag{5.88}$$

The possible orientations of **L** with respect to **B** for the case $l = 3$ are illustrated in Figure 5.6. Classically, all values between 0 and π are allowed for the angle θ. When equations (5.81) and (5.88) are substituted into (5.83), we find that the potential energy V is also quantized

$$V = m\mu_{\rm B} B, \quad m = 0, \pm 1, \ldots, \pm l \tag{5.89}$$

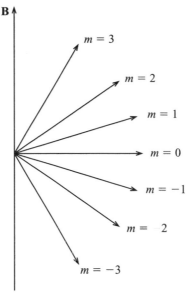

B

$m = 3$

$m = 2$

$m = 1$

$m = 0$

$m = -1$

$m = \ 2$

$m = -3$

Figure 5.6 Possible orientations in a magnetic field **B** of the orbital angular momentum vector **L** for the case $l = 3$

Problems

5.1 Show that each of the operators $\hat{L}_x$, $\hat{L}_y$, $\hat{L}_z$ is hermitian.

5.2 Evaluate the following commutators:

(a) $[\hat{L}_x, x]$ (b) $[\hat{L}_x, \hat{p}_x]$ (c) $[\hat{L}_x, y]$ (d) $[\hat{L}_x, \hat{p}_y]$

5.3 Using the commutation relation (5.10b), find the expectation value of $\hat{L}_x$ for a system in state $|lm\rangle$.

5.4 Apply the uncertainty principle to the operators $\hat{L}_x$ and $\hat{L}_y$ to obtain an expression for $\wedge\hat{l}_x\wedge\hat{l}_y$. Evaluate the expression for a system in state $|lm\rangle$.

5.5 Show that the operator $\hat{J}^2$ commutes with $\hat{J}_x$ and with $\hat{J}_y$.

5.6 Show that $\hat{J}_+$ and $\hat{J}_-$ as defined by equations (5.18) are adjoints of each other.

5.7 Prove the relationships (5.19a)–(5.19g).

5.8 Show that the choice for c_- in equation (5.24) is consistent with c_+ in equation (5.22).

5.9 Using the raising and lowering operators $\hat{J}_+$ and $\hat{J}_-$, show that

$$\langle jm|\hat{J}_x|jm\rangle = \langle jm|\hat{J}_y|jm\rangle = 0$$

5.10 Show that

$$\langle jm|\hat{J}_x^2|jm\rangle = \langle jm|\hat{J}_y^2|jm\rangle = \tfrac{1}{2}[j(j+1) - m^2]\hbar^2$$

5.11 Show that $|j, m\rangle$ are eigenfunctions of $[\hat{J}_x, \hat{J}_+]$ and of $[\hat{J}_y, \hat{J}_+]$. Find the eigenvalues of each of these commutators.

6

The hydrogen atom

A theoretical understanding of the structure and behavior of the hydrogen atom is essential to the fields of physics and chemistry. As the simplest atomic system, hydrogen must be understood before one can proceed to the treatment of more complex atoms, molecules, and atomic and molecular aggregates. The hydrogen atom is one of the few examples for which the Schrödinger equation can be solved exactly to obtain its wave functions and energy levels. The resulting agreement between theoretically derived and experimental quantities serves as confirmation of the applicability of quantum mechanics to a real chemical system. Further, the results of the quantum-mechanical treatment of atomic hydrogen are often used as the basis for approximate treatments of more complex atoms and molecules, for which the Schrödinger equation cannot be solved.

The study of the hydrogen atom also played an important role in the development of quantum theory. The Lyman, Balmer, and Paschen series of spectral lines observed in incandescent atomic hydrogen were found to obey the empirical equation

$$\nu = \mathsf{R}c\left(\frac{1}{n_1^2} - \frac{1}{n_2^2}\right), \qquad n_2 > n_1$$

where ν is the frequency of a spectral line, c is the speed of light, $n_1 = 1, 2, 3$ for the Lyman, Balmer, and Paschen series, respectively, n_2 is an integer determining the various lines in a given series, and R is the so-called Rydberg constant, which has the same value for each of the series. Neither the existence of these spectral lines nor the formula which describes them could be explained by classical theory. In 1913, N. Bohr postulated that the electron in a hydrogen atom revolves about the nucleus in a circular orbit with an angular momentum that is quantized. He then applied Newtonian mechanics to the electronic motion and obtained quantized energy levels and quantized orbital radii. From the Planck relation $\Delta E = E_{n_1} - E_{n_2} = h\nu$, Bohr was able to reproduce the experimental spectral lines and obtain a theoretical value for the Rydberg constant that agrees exactly with the experimentally determined value. Further investigations, however, showed that the Bohr model is not an accurate representation of the hydrogen-atom structure, even though it gives the correct formula for the energy levels, and led eventually to Schrödinger's wave mechanics. Schrödinger also used the hydrogen atom to illustrate his new theory.

6.1 Two-particle problem

In order to apply quantum-mechanical theory to the hydrogen atom, we first need to find the appropriate Hamiltonian operator and Schrödinger equation. As preparation for establishing the Hamiltonian operator, we consider a classical system of two interacting point particles with masses m_1 and m_2 and instantaneous positions $\mathbf{r}_1$ and $\mathbf{r}_2$ as shown in Figure 6.1. In terms of their cartesian components, these position vectors are

$$\mathbf{r}_1 = \mathbf{i}x_1 + \mathbf{j}y_1 + \mathbf{k}z_1$$

$$\mathbf{r}_2 = \mathbf{i}x_2 + \mathbf{j}y_2 + \mathbf{k}z_2$$

The vector distance between the particles is designated by $\mathbf{r}$

$$\mathbf{r} \equiv \mathbf{r}_2 - \mathbf{r}_1 = \mathbf{i}x + \mathbf{j}y + \mathbf{k}z \qquad (6.1)$$

where

$$x = x_2 - x_1, \qquad y = y_2 - y_1, \qquad z = z_2 - z_1$$

The center of mass of the two-particle system is located by the vector $\mathbf{R}$ with cartesian components, X, Y, Z

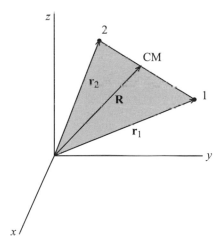

Figure 6.1 The center of mass (CM) of a two-particle system.

$$\mathbf{R} = \mathbf{i}X + \mathbf{j}Y + \mathbf{k}Z$$

By definition, the center of mass is related to $\mathbf{r}_1$ and $\mathbf{r}_2$ by

$$\mathbf{R} = \frac{m_1\mathbf{r}_1 + m_2\mathbf{r}_2}{M} \tag{6.2}$$

where $M = m_1 + m_2$ is the total mass of the system. We may express $\mathbf{r}_1$ and $\mathbf{r}_2$ in terms of $\mathbf{R}$ and $\mathbf{r}$ using equations (6.1) and (6.2)

$$\mathbf{r}_1 = \mathbf{R} - \frac{m_2}{M}\mathbf{r}$$

$$\mathbf{r}_2 = \mathbf{R} + \frac{m_1}{M}\mathbf{r} \tag{6.3}$$

If we restrict our interest to systems for which the potential energy V is a function only of the relative position vector $\mathbf{r}$, then the classical Hamiltonian function H is given by

$$H = \frac{|\mathbf{p}_1|^2}{2m_1} + \frac{|\mathbf{p}_2|^2}{2m_2} + V(\mathbf{r}) \tag{6.4}$$

where the momenta $\mathbf{p}_1$ and $\mathbf{p}_2$ for the two particles are

$$\mathbf{p}_1 = m_1\frac{d\mathbf{r}_1}{dt}, \qquad \mathbf{p}_2 = m_2\frac{d\mathbf{r}_2}{dt}$$

These momenta may be expressed in terms of the time derivatives of $\mathbf{R}$ and $\mathbf{r}$ by substitution of equation (6.3)

$$\mathbf{p}_1 = m_1\left(\frac{d\mathbf{R}}{dt} - \frac{m_2}{M}\frac{d\mathbf{r}}{dt}\right)$$

$$\mathbf{p}_2 = m_2\left(\frac{d\mathbf{R}}{dt} + \frac{m_1}{M}\frac{d\mathbf{r}}{dt}\right) \tag{6.5}$$

Substitution of equation (6.5) into (6.4) yields

$$H = \tfrac{1}{2}M\left|\frac{d\mathbf{R}}{dt}\right|^2 + \tfrac{1}{2}\mu\left|\frac{d\mathbf{r}}{dt}\right|^2 + V(\mathbf{r}) \tag{6.6}$$

where the cross terms have canceled out and we have defined the *reduced mass* μ by

$$\mu \equiv \frac{m_1 m_2}{m_1 + m_2} = \frac{m_1 m_2}{M} \tag{6.7}$$

The momenta $\mathbf{p}_R$ and $\mathbf{p}_r$, corresponding to the center of mass position $\mathbf{R}$ and the relative position variable $\mathbf{r}$, respectively, may be defined as

$$\mathbf{p}_R \equiv M\frac{d\mathbf{R}}{dt}, \qquad \mathbf{p}_r \equiv \mu\frac{d\mathbf{r}}{dt}$$

In terms of these momenta, the classical Hamiltonian becomes

$$H = \frac{|\mathbf{p}_R|^2}{2M} + \frac{|\mathbf{p}_r|^2}{2\mu} + V(\mathbf{r}) \tag{6.8}$$

We see that the kinetic energy contribution to the Hamiltonian is the sum of two parts, the kinetic energy due to the translational motion of the center of mass of the system as a whole and the kinetic energy due to the relative motion of the two particles. Since the potential energy $V(\mathbf{r})$ is assumed to be a function only of the relative position coordinate $\mathbf{r}$, the motion of the center of mass of the system is unaffected by the potential energy.

The quantum-mechanical Hamiltonian operator $\hat{H}$ is obtained by replacing $|\mathbf{p}_R|^2$ and $|\mathbf{p}_r|^2$ in equation (6.8) by the operators $-\hbar^2\nabla_R^2$ and $-\hbar^2\nabla_r^2$, respectively, where

$$\nabla_R^2 = \frac{\partial^2}{\partial X^2} + \frac{\partial^2}{\partial Y^2} + \frac{\partial^2}{\partial Z^2} \tag{6.9a}$$

$$\nabla_r^2 = \frac{\partial^2}{\partial x^2} + \frac{\partial^2}{\partial y^2} + \frac{\partial^2}{\partial z^2} \tag{6.9b}$$

The resulting Schrödinger equation is, then,

$$\left[-\frac{\hbar^2}{2M}\nabla_R^2 - \frac{\hbar^2}{2\mu}\nabla_r^2 + V(\mathbf{r}) \right] \Psi(\mathbf{R}, \mathbf{r}) = E\Psi(\mathbf{R}, \mathbf{r}) \tag{6.10}$$

This partial differential equation may be readily separated by writing the wave function $\Psi(\mathbf{R}, \mathbf{r})$ as the product of two functions, one a function only of the center of mass variables X, Y, Z and the other a function only of the relative coordinates x, y, z

$$\Psi(\mathbf{R}, \mathbf{r}) = \chi(X, Y, Z)\psi(x, y, z) = \chi(\mathbf{R})\psi(\mathbf{r})$$

With this substitution, equation (6.10) separates into two independent partial differential equations

$$-\frac{\hbar^2}{2M}\nabla_R^2\chi(\mathbf{R}) = E_R\chi(\mathbf{R}) \tag{6.11}$$

$$-\frac{\hbar^2}{2\mu}\nabla_r^2\psi(\mathbf{r}) + V(\mathbf{r})\psi(\mathbf{r}) = E_r\psi(\mathbf{r}) \tag{6.12}$$

where

$$E = E_R + E_r$$

Equation (6.11) is the Schrödinger equation for the translational motion of a free particle of mass M, while equation (6.12) is the Schrödinger equation for a hypothetical particle of mass μ moving in a potential field $V(\mathbf{r})$. Since the energy E_R of the translational motion is a positive constant ($E_R \geqslant 0$), the solutions of equation (6.11) are not relevant to the structure of the two-particle system and we do not consider this equation any further.

6.2 The hydrogen-like atom

The Schrödinger equation (6.12) for the relative motion of a two-particle system is applicable to the hydrogen-like atom, which consists of a nucleus of charge $+Ze$ and an electron of charge $-e$. The differential equation applies to H for $Z = 1$, He^+ for $Z = 2$, Li^{2+} for $Z = 3$, and so forth. The potential energy $V(r)$ of the interaction between the nucleus and the electron is a function of their separation distance $r = |\mathbf{r}| = (x^2 + y^2 + z^2)^{1/2}$ and is given by Coulomb's law (equation (5.76)), which in SI units is

$$V(r) = -\frac{Ze^2}{4\pi\varepsilon_0 r}$$

where *meter* is the unit of length, *joule* is the unit of energy, *coulomb* is the unit of charge, and ε_0 is the *permittivity of free space*. Another system of units, used often in the older literature and occasionally in recent literature, is the CGS gaussian system, in which Coulomb's law is written as

$$V(r) = -\frac{Ze^2}{r}$$

In this system, *centimeter* is the unit of length, *erg* is the unit of energy, and *statcoulomb* (also called the *electrostatic unit* or *esu*) is the unit of charge. In this book we accommodate both systems of units and write Coulomb's law in the form

$$V(r) = -\frac{Ze'^2}{r} \tag{6.13}$$

where $e' = e$ for CGS units or $e' = e/(4\pi\varepsilon_0)^{1/2}$ for SI units.

Equation (6.12) cannot be solved analytically when expressed in the cartesian coordinates x, y, z, but can be solved when expressed in spherical polar coordinates r, θ, φ, by means of the transformation equations (5.29). The laplacian operator ∇_r^2 in spherical polar coordinates is given by equation (A.61) and may be obtained by substituting equations (5.30) into (6.9b) to yield

$$\nabla_r^2 = \frac{1}{r^2}\frac{\partial}{\partial r}\left(r^2\frac{\partial}{\partial r}\right) + \frac{1}{r^2}\left[\frac{1}{\sin\theta}\frac{\partial}{\partial\theta}\left(\sin\theta\frac{\partial}{\partial\theta}\right) + \frac{1}{\sin^2\theta}\frac{\partial^2}{\partial\varphi^2}\right]$$

If this expression is compared with equation (5.32), we see that

$$\nabla_r^2 = \frac{1}{r^2}\frac{\partial}{\partial r}\left(r^2\frac{\partial}{\partial r}\right) - \frac{1}{\hbar^2 r^2}\hat{L}^2$$

where $\hat{L}^2$ is the square of the orbital angular momentum operator. With the laplacian operator ∇_r^2 expressed in spherical polar coordinates, the Schrödinger equation (6.12) becomes

$$\hat{H}\psi(r,\ \theta,\ \varphi) = E\psi(r,\ \theta,\ \varphi)$$

with

$$\hat{H} = -\frac{\hbar^2}{2\mu r^2}\frac{\partial}{\partial r}\left(r^2\frac{\partial}{\partial r}\right) + \frac{1}{2\mu r^2}\hat{L}^2 + V(r) \tag{6.14}$$

The operator $\hat{L}^2$ in equation (5.32) commutes with the Hamiltonian operator $\hat{H}$ in (6.14) because $\hat{L}^2$ commutes with itself and does not involve the variable r. Likewise, the operator $\hat{L}_z$ in equation (5.31c) commutes with $\hat{H}$ because it commutes with $\hat{L}^2$ as shown in (5.15a) and also does not involve the variable r. Thus, we have

$$[\hat{H},\ \hat{L}^2] = 0, \qquad [\hat{H},\ \hat{L}_z] = 0, \qquad [\hat{L}^2,\ \hat{L}_z] = 0$$

and the operators $\hat{H}$, $\hat{L}^2$, and $\hat{L}_z$ have simultaneous eigenfunctions,

$$\hat{H}\psi(r,\ \theta,\ \varphi) = E\psi(r,\ \theta,\ \varphi) \tag{6.15a}$$

$$\hat{L}^2\psi(r,\ \theta,\ \varphi) = l(l+1)\hbar^2\psi(r,\ \theta,\ \varphi), \qquad l = 0, 1, 2, \ldots \tag{6.15b}$$

$$\hat{L}_z\psi(r,\ \theta,\ \varphi) = m\hbar\psi(r,\ \theta,\ \varphi), \qquad m = -l, -l+1, \ldots, l-1, l \tag{6.15c}$$

The simultaneous eigenfunctions of $\hat{L}^2$ and $\hat{L}_z$ are the spherical harmonics $Y_{lm}(\theta,\ \varphi)$ given by equations (5.50) and (5.59). Since neither $\hat{L}^2$ nor $\hat{L}_z$ involve the variable r, any specific spherical harmonic may be multiplied by an arbitrary function of r and the result is still an eigenfunction. Thus, we may write $\psi(r,\ \theta,\ \varphi)$ as

$$\psi(r,\ \theta,\ \varphi) = R(r)Y_{lm}(\theta,\ \varphi) \tag{6.16}$$

Substitution of equations (6.13), (6.14), (6.15b), and (6.16) into (6.15a) gives

$$\hat{H}_l R(r) = ER(r) \tag{6.17}$$

where

$$\hat{H}_l = -\frac{\hbar^2}{2\mu r^2}\left[\frac{d}{dr}\left(r^2\frac{d}{dr}\right) - l(l+1)\right] - \frac{Ze'^2}{r} \tag{6.18}$$

and where the common factor $Y_{lm}(\theta,\ \varphi)$ has been divided out.

6.3 The radial equation

Our next task is to solve the radial equation (6.17) to obtain the radial function $R(r)$ and the energy E. The many solutions of the differential equation (6.17) depend not only on the value of l, but also on the value of E. Therefore, the solutions are designated as $R_{El}(r)$. Since the potential energy $-Ze'^2/r$ is always negative, we are interested in solutions with negative total energy, i.e., where $E \leq 0$. It is customary to require that the functions $R_{El}(r)$ be normalized. Since the radial part of the volume element in spherical coordinates is $r^2 \, dr$, the normalization criterion is

$$\int_0^\infty [R_{El}(r)]^2 r^2 \, dr = 1 \tag{6.19}$$

Through an explicit integration by parts, we can show that

$$\int_0^\infty R_{El}(r)[\hat{H}_l R_{E'l}(r)] r^2 \, dr = \int_0^\infty R_{E'l}(r)[\hat{H}_l R_{El}(r)] r^2 \, dr$$

Thus, the operator $\hat{H}_l$ is hermitian and the radial functions $R_{El}(r)$ constitute an orthonormal set with a weighting function $w(r)$ equal to r^2

$$\int_0^\infty R_{El}(r) R_{E'l}(r) r^2 \, dr = \delta_{EE'} \tag{6.20}$$

where $\delta_{EE'}$ is the Kronecker delta and equation (6.19) has been included.

We next make the following conventional change of variables

$$\lambda = \frac{\mu Z e'^2}{\hbar(-2\mu E)^{1/2}} \tag{6.21}$$

$$\rho = \frac{2(-2\mu E)^{1/2} r}{\hbar} = \frac{2\mu Z e'^2 r}{\lambda \hbar^2} = \frac{2Zr}{\lambda a_\mu} \tag{6.22}$$

where $a_\mu = \hbar^2/\mu e'^2$. We also make the substitution

$$R_{El}(r) = \left(\frac{2Z}{\lambda a_\mu}\right)^{3/2} S_{\lambda l}(\rho) \tag{6.23}$$

Equations (6.17) and (6.18) now take the form

$$\left(\rho^2 \frac{d^2}{d\rho^2} + 2\rho \frac{d}{d\rho} + \lambda\rho - \frac{\rho^2}{4}\right) S_{\lambda l} = l(l+1) S_{\lambda l} \tag{6.24}$$

where the first term has been expanded and the entire expression has been multiplied by ρ^2.

To be a suitable wave function, $S_{\lambda l}(\rho)$ must be well-behaved, i.e., it must be continuous, single-valued, and quadratically integrable. Thus, $\rho S_{\lambda l}$ vanishes when $\rho \to \infty$ because $S_{\lambda l}$ must vanish sufficiently fast. Since $S_{\lambda l}$ is finite everywhere, $\rho S_{\lambda l}$ also vanishes at $\rho = 0$. Substitution of equations (6.22) and (6.23) into (6.19) shows that $S_{\lambda l}(\rho)$ is normalized with a weighting function $w(\rho)$ equal to ρ^2

$$\int_0^\infty [S_{\lambda l}(\rho)]^2 \rho^2 \, d\rho = 1 \tag{6.25}$$

Equation (6.24) may be solved by the Frobenius or series solution method as presented in Appendix G. However, in this chapter we employ the newer procedure using ladder operators.

Ladder operators

We now solve equation (6.24) by means of ladder operators, analogous to the method used in Chapter 4 for the harmonic oscillator and in Chapter 5 for the angular momentum.[1] We define the operators $\hat{A}_\lambda$ and $\hat{B}_\lambda$ as

$$\hat{A}_\lambda \equiv -\rho \frac{d}{d\rho} - \frac{\rho}{2} + \lambda - 1 \qquad (6.26a)$$

$$\hat{B}_\lambda \equiv \rho \frac{d}{d\rho} - \frac{\rho}{2} + \lambda \qquad (6.26b)$$

We now show that the operator $\hat{A}_\lambda$ is the adjoint of $\hat{B}_\lambda$ and vice versa. Thus, neither $\hat{A}_\lambda$ nor $\hat{B}_\lambda$ is hermitian. For any arbitrary well-behaved functions $f(\rho)$ and $g(\rho)$, we consider the integral

$$\int_0^\infty f(\rho)[\hat{A}_\lambda g(\rho)]\,d\rho = -\int_0^\infty f\rho \frac{dg}{d\rho}\,d\rho + \int_0^\infty f\left(-\frac{\rho}{2} + \lambda - 1\right) g\,d\rho$$

where (6.26a) has been used. Integration by parts of the first term on the right-hand side with the realization that the integrated part vanishes yields

$$\int_0^\infty f\hat{A}_\lambda g\,d\rho = \int_0^\infty g\frac{d}{d\rho}(\rho f)\,d\rho + \int_0^\infty f\left(-\frac{\rho}{2} + \lambda - 1\right) g\,d\rho$$

$$= \int_0^\infty g\left(\rho\frac{d}{d\rho} - \frac{\rho}{2} + \lambda\right) f\,d\rho$$

Substitution of (6.26b) gives

$$\int_0^\infty f(\rho)[\hat{A}_\lambda g(\rho)]\,d\rho = \int_0^\infty g(\rho)[\hat{B}_\lambda f(\rho)]\,d\rho \qquad (6.27)$$

showing that, according to equation (3.33)

$$\hat{A}_\lambda^\dagger = \hat{B}_\lambda, \qquad \hat{B}_\lambda^\dagger = \hat{A}_\lambda$$

We readily observe from (6.26a) and (6.26b) that

$$\hat{B}_\lambda\hat{A}_\lambda = -\rho^2 \frac{d^2}{d\rho^2} - 2\rho\frac{d}{d\rho} - \lambda\rho + \frac{\rho^2}{4} + \lambda(\lambda - 1) \qquad (6.28a)$$

$$\hat{A}_\lambda\hat{B}_\lambda = -\rho^2 \frac{d^2}{d\rho^2} - 2\rho\frac{d}{d\rho} - (\lambda - 1)\rho + \frac{\rho^2}{4} + \lambda(\lambda - 1) \qquad (6.28b)$$

Equation (6.24) can then be written in the form

$$\hat{B}_\lambda\hat{A}_\lambda S_{\lambda l} = [\lambda(\lambda - 1) - l(l + 1)]S_{\lambda l} \qquad (6.29)$$

showing that the functions $S_{\lambda l}(\rho)$ are also eigenfunctions of $\hat{B}_\lambda\hat{A}_\lambda$. From equation (6.28b) we obtain

---

[1] We follow here the treatment by D. D. Fitts (1995) *J. Chem. Educ.* **72**, 1066. However, the definitions of the lowering operator and the constants $a_{\lambda l}$ and $b_{\lambda l}$ have been changed.

$$\hat{A}_\lambda \hat{B}_\lambda S_{\lambda-1,l} = [\lambda(\lambda - 1) - l(l + 1)]S_{\lambda-1,l} \tag{6.30}$$

when λ is replaced by $\lambda - 1$ in equation (6.24).

If we operate on both sides of equation (6.29) with the operator $\hat{A}_\lambda$, we obtain

$$\hat{A}_\lambda \hat{B}_\lambda \hat{A}_\lambda S_{\lambda l} = [\lambda(\lambda - 1) - l(l + 1)]\hat{A}_\lambda S_{\lambda l} \tag{6.31}$$

Comparison of this result with equation (6.30) leads to the conclusion that $\hat{A}_\lambda S_{\lambda l}$ and $S_{\lambda-1,l}$ are, except for a multiplicative constant, the same function. We implicitly assume here that $S_{\lambda l}$ is uniquely determined by only two parameters, λ and l. Accordingly, we may write

$$\hat{A}_\lambda S_{\lambda l} = a_{\lambda l} S_{\lambda-1,l} \tag{6.32}$$

where $a_{\lambda l}$ is a numerical constant, dependent in general on the values of λ and l, to be determined by the requirement that $S_{\lambda l}$ and $S_{\lambda-1,l}$ be normalized. Without loss of generality, we can take $a_{\lambda l}$ to be real. The function $\hat{A}_\lambda S_{\lambda l}$ is an eigenfunction of the operator in equation (6.24) with eigenvalue decreased by one. Thus, the operator $\hat{A}_\lambda$ transforms the eigenfunction $S_{\lambda l}$ determined by λ, l into the eigenfunction $S_{\lambda-1,l}$ determined by $\lambda - 1$, l. For this reason the operator $\hat{A}_\lambda$ is a lowering ladder operator.

Following an analogous procedure, we now operate on both sides of equation (6.30) with the operator $\hat{B}_\lambda$ to obtain

$$\hat{B}_\lambda \hat{A}_\lambda \hat{B}_\lambda S_{\lambda-1,l} = [\lambda(\lambda - 1) - l(l + 1)]\hat{B}_\lambda S_{\lambda-1,l} \tag{6.33}$$

Comparing equations (6.29) and (6.33) shows that $\hat{B}_\lambda S_{\lambda-1,l}$ and $S_{\lambda l}$ are proportional to each other

$$\hat{B}_\lambda S_{\lambda-1,l} = b_{\lambda l} S_{\lambda l} \tag{6.34}$$

where $b_{\lambda l}$ is the proportionality constant, assumed real, to be determined by the requirement that $S_{\lambda-1,l}$ and $S_{\lambda l}$ be normalized. The operator $\hat{B}_\lambda$ transforms the eigenfunction $S_{\lambda-1,l}$ into the eigenfunction $S_{\lambda l}$ with eigenvalue λ increased by one. Accordingly, the operator $\hat{B}_\lambda$ is a raising ladder operator.

The next step is to evaluate the numerical constants $a_{\lambda l}$ and $b_{\lambda l}$. In order to accomplish these evaluations, we must first investigate some mathematical properties of the eigenfunctions $S_{\lambda l}(\rho)$.

Orthonormal properties of $S_{\lambda l}(\rho)$

Although the functions $R_{nl}(r)$ according to equation (6.20) form an orthogonal set with $w(r) = r^2$, the orthogonal relationships do not apply to the set of functions $S_{\lambda l}(\rho)$ with $w(\rho) = \rho^2$. Since the variable ρ introduced in equation (6.22) depends not only on r, but also on the eigenvalue E, or equivalently on λ, the situation is more complex. To determine the proper orthogonal relationships for $S_{\lambda l}(\rho)$, we express equation (6.24) in the form

$$\hat{H}'_l S_{\lambda l} = -\lambda S_{\lambda l} \tag{6.35}$$

where $\hat{H}'_l$ is defined by

$$\hat{H}'_l = \rho \frac{d^2}{d\rho^2} + 2\frac{d}{d\rho} - \frac{\rho}{4} - \frac{l(l+1)}{\rho} \tag{6.36}$$

By means of integration by parts, we can readily show that this operator $\hat{H}'_l$ is hermitian for a weighting function $w(\rho)$ equal to ρ, thereby implying the orthogonal relationships

$$\int_0^\infty S_{\lambda l}(\rho) S_{\lambda' l}(\rho) \rho \, d\rho = 0 \quad \text{for} \quad \lambda \neq \lambda' \tag{6.37}$$

In order to complete the characterization of integrals of $S_{\lambda l}(\rho)$, we need to consider the case where $\lambda = \lambda'$ for $w(\rho) = \rho$. Recall that the functions $S_{\lambda l}(\rho)$ are normalized for $w(\rho) = \rho^2$ as expressed in equation (6.25). The same result does not apply for $w(\rho) = \rho$. We begin by expressing the desired integral in a slightly different form

$$\int_0^\infty [S_{\lambda l}(\rho)]^2 \rho \, d\rho = \frac{1}{2}\int_0^\infty [S_{\lambda l}(\rho)]^2 \, d(\rho^2)$$

Integration of the right-hand side by parts gives

$$\int_0^\infty [S_{\lambda l}(\rho)]^2 \rho \, d\rho = \frac{1}{2}\left[\rho^2 [S_{\lambda l}(\rho)]^2\right]_0^\infty - \int_0^\infty \rho^2 S_{\lambda l}\left[\frac{d}{d\rho} S_{\lambda l}\right] d\rho$$

If $S_{\lambda l}(\rho)$ is well-behaved, the integrated term vanishes. From equation (6.26a) we may write

$$\rho \frac{d}{d\rho} = -\hat{A}_\lambda - \frac{\rho}{2} + \lambda - 1$$

so that

$$\rho \frac{d}{d\rho} S_{\lambda l} = -\hat{A}_\lambda S_{\lambda l} - \tfrac{1}{2}\rho S_{\lambda l} + (\lambda - 1)S_{\lambda l}$$

$$= -a_{\lambda l} S_{\lambda-1,l} - \tfrac{1}{2}\rho S_{\lambda l} + (\lambda - 1)S_{\lambda l}$$

where equation (6.32) has been introduced. The integral then takes the form

$$\int_0^\infty [S_{\lambda l}(\rho)]^2 \rho \, d\rho = a_{\lambda l}\int_0^\infty S_{\lambda l} S_{\lambda-1,l}\rho \, d\rho + \tfrac{1}{2}\int_0^\infty [S_{\lambda l}]^2 \rho^2 \, d\rho$$

$$- (\lambda - 1)\int_0^\infty [S_{\lambda l}]^2 \rho \, d\rho$$

Since the first integral on the right-hand side vanishes according to equation (6.37) and the second integral equals unity according to (6.25), the result is

$$\int_0^\infty [S_{\lambda l}(\rho)]^2 \rho \, d\rho = \frac{1}{2\lambda} \tag{6.38}$$

Combining equation (6.38) with (6.37), we obtain

$$\int_0^\infty S_{\lambda l}(\rho) S_{\lambda' l}(\rho) \rho \, d\rho = \frac{1}{2\lambda} \delta_{\lambda\lambda'} \tag{6.39}$$

Evaluation of the constants $a_{\lambda l}$ and $b_{\lambda l}$

To evaluate the numerical constant $a_{\lambda l}$, which is defined in equation (6.32), we square both sides of (6.32), multiply through by ρ, and integrate with respect to ρ to obtain

$$\int_0^\infty \rho(\hat{A}_\lambda S_{\lambda l})(\hat{A}_\lambda S_{\lambda l}) \, d\rho = a_{\lambda l}^2 \int_0^\infty (S_{\lambda-1,l})^2 \rho \, d\rho \tag{6.40}$$

Application of equation (6.27) with $f = \rho \hat{A}_\lambda S_{\lambda l}$ and $g = S_{\lambda l}$ to the left-hand side and substitution of equation (6.38) on the right-hand side give

$$\int_0^\infty S_{\lambda l} \hat{B}_\lambda(\rho \hat{A}_\lambda S_{\lambda l}) \, d\rho = a_{\lambda l}^2/2(\lambda - 1) \tag{6.41}$$

The expression $\hat{B}_\lambda(\rho \hat{A}_\lambda S_{\lambda l})$ may be simplified as follows

$$\hat{B}_\lambda(\rho \hat{A}_\lambda S_{\lambda l}) = \rho \frac{d}{d\rho}(\rho \hat{A}_\lambda S_{\lambda l}) + \left(-\frac{\rho}{2} + \lambda\right)\rho \hat{A}_\lambda S_{\lambda l}$$

$$= \rho \hat{A}_\lambda S_{\lambda l} + \rho^2 \frac{d}{d\rho}(\hat{A}_\lambda S_{\lambda l}) + \rho\left(-\frac{\rho}{2} + \lambda\right)\hat{A}_\lambda S_{\lambda l}$$

$$= \rho \hat{A}_\lambda S_{\lambda l} + \rho \hat{B}_\lambda \hat{A}_\lambda S_{\lambda l}$$

$$= \rho a_{\lambda l} S_{\lambda-1,l} + [\lambda(\lambda - 1) - l(l + 1)]\rho S_{\lambda l}$$

where equations (6.26b), (6.32), and (6.29) have been used. When this result is substituted back into (6.41), we have

$$a_{\lambda l}\int_0^\infty S_{\lambda l} S_{\lambda-1,l}\rho \, d\rho + [\lambda(\lambda - 1) - l(l + 1)]\int_0^\infty S_{\lambda l}^2 \rho \, d\rho = a_{\lambda l}^2/2(\lambda - 1) \tag{6.42}$$

According to equation (6.39), the first integral vanishes and the second integral equals $(2\lambda)^{-1}$, giving the result

$$a_{\lambda l}^2 = \left(\frac{\lambda - 1}{\lambda}\right)[\lambda(\lambda - 1) - l(l + 1)]$$

$$= \left(\frac{\lambda - 1}{\lambda}\right)(\lambda + l)(\lambda - l - 1) \tag{6.43}$$

Substitution into (6.32) gives

$$\hat{A}_\lambda S_{\lambda l} = \left[\left(\frac{\lambda - 1}{\lambda}\right)(\lambda + l)(\lambda - l - 1)\right]^{1/2} S_{\lambda - 1, l} \qquad (6.44)$$

where we have arbitrarily taken the positive square root.

The numerical constant $b_{\lambda l}$, defined in equation (6.34), may be determined by an analogous procedure, beginning with the square of both sides of equation (6.34) and using equations (6.27), (6.26a), (6.34), (6.30), and (6.39). We obtain

$$b_{\lambda l}^2 = \left(\frac{\lambda}{\lambda - 1}\right)[\lambda(\lambda - 1) - l(l + 1)] = \left(\frac{\lambda}{\lambda - 1}\right)(\lambda + l)(\lambda - l - 1) \qquad (6.45)$$

so that equation (6.34) becomes

$$\hat{B}_\lambda S_{\lambda - 1, l} = \left[\left(\frac{\lambda}{\lambda - 1}\right)(\lambda + l)(\lambda - l - 1)\right]^{1/2} S_{\lambda, l} \qquad (6.46)$$

Taking the positive square root here will turn out to be consistent with the choice in equation (6.44).

Quantization of the energy

The parameter λ is positive, since otherwise the radial variable ρ, which is inversely proportional to λ, would be negative. Furthermore, the parameter λ cannot be zero if the transformations in equations (6.21), (6.22), and (6.23) are to remain valid. To find further restrictions on λ we must consider separately the cases where $l = 0$ and where $l \geqslant 1$.

For $l = 0$, equation (6.44) takes the form

$$\hat{A}_\lambda S_{\lambda 0} - (\lambda - 1)S_{\lambda - 1, 0} \qquad (6.47)$$

Suppose we begin with a suitably large value of λ, say ξ, and continually apply the lowering operator to both sides of equation (6.47) with $\lambda = \xi$

$$\hat{A}_{\xi - 1}\hat{A}_\xi S_{\xi 0} = (\xi - 1)(\xi - 2)S_{\xi - 2, 0}$$

$$\hat{A}_{\xi - 2}\hat{A}_{\xi - 1}\hat{A}_\xi S_{\xi 0} = (\xi - 1)(\xi - 2)(\xi - 3)S_{\xi - 3, 0}$$

$$\vdots$$

Eventually this procedure produces an eigenfunction $S_{\xi - k, 0}$, k being a positive integer, such that $0 < (\xi - k) \leqslant 1$. The next step in the sequence would give a function $S_{\xi - k - 1, 0}$ or $S_{\lambda 0}$ with $\lambda = (\xi - k - 1) \leqslant 0$, which is not allowed. Thus, the sequence must terminate with the condition

$$\hat{A}_{\xi - k}S_{\xi - k, 0} = (\xi - k - 1)S_{\xi - k - 1, 0} = 0$$

which can only occur if $(\xi - k) = 1$. Thus, ξ must be an integer and the minimum value of λ for $l = 0$ is $\lambda = 1$.

For the situations in which $l \geqslant 1$, we note that the quantities $a_{\lambda l}^2$ in equation

(6.43) and $b_{\lambda l}^2$ in equation (6.45), being squares of real numbers, must be positive. Consequently, the factor $(\lambda - l - 1)$ must be positive, so that $\lambda \geq (l + 1)$.

We now select some appropriately large value ξ of the parameter λ in equation (6.44) and continually apply the lowering operator to both sides of the equation in the same manner as in the $l = 0$ case. Eventually we obtain $S_{\xi-k,l}$ such that $(l + 1) \leq (\xi - k) < (l + 2)$. The next step in the sequence would give $S_{\xi-k-1,l}$ or $S_{\lambda l}$ with $\lambda = (\xi - k - 1) < (l + 1)$, which is not allowed, so that the sequence must be terminated according to

$$\hat{A}_{\xi-k} S_{\xi-k,l} = a_{\xi-k,l} S_{\xi-k-1,l}$$

$$= \left[\left(\frac{\xi - k - 1}{\xi - k} \right) (\xi - k + l)(\xi - k - l - 1) \right]^{1/2} S_{\xi-k-1,l}$$

$$= 0$$

for some value of k. Thus, ξ must be an integer for $a_{\xi-k,l}$ to vanish. As k increases during the sequence, the constant $a_{\xi-k,l}$ vanishes when $k = (\xi - l - 1)$ or $(\xi - k) = (l + 1)$. The minimum value of λ is then $l + 1$.

Combining the conclusions of both cases, we see that the minimum value of λ is $l + 1$ for $l = 0, 1, 2, \ldots$ Beginning with the value $\lambda = l + 1$, we can apply equation (6.46) to yield an infinite progression of eigenfunctions $S_{nl}(\rho)$ for each value of l ($l = 0, 1, 2, \ldots$), where λ can take on only integral values, $\lambda = n = l + 1, l + 2, l + 3, \ldots$ Since ξ in both cases was chosen arbitrarily and was shown to be an integer, equation (6.46) generates all of the eigenfunctions $S_{\lambda l}(\rho)$ for each value of l. There are no eigenfunctions corresponding to non-integral values of λ. Since λ is now shown to be an integer n, in the remainder of this presentation we replace λ by n.

Solving equation (6.21) for the energy E and replacing λ by n, we obtain the quantized energy levels for the hydrogen-like atom

$$E_n = -\frac{\mu Z^2 e'^4}{2\hbar^2 n^2} = -\frac{Z^2 e'^2}{2a_\mu n^2}, \qquad n = 1, 2, 3, \ldots \qquad (6.48)$$

These energy levels agree with the values obtained in the earlier Bohr theory.

Electronic energies are often expressed in the unit *electron volt* (eV). An electron volt is defined as the kinetic energy of an electron accelerated through a potential difference of 1 volt. Thus, we have

$$1 \text{ eV} = (1.602\,177 \times 10^{-19} \text{ C}) \times (1.000\,000 \text{ V}) = 1.602\,177 \times 10^{-19} \text{ J}$$

The ground-state energy E_1 of a hydrogen atom ($Z = 1$) as given by equation (6.48) is

$$E_1 = -2.178\,68 \times 10^{-18} \text{ J} = -13.598 \text{ eV}$$

This is the energy required to remove the electron from the ground state of a hydrogen atom to a state of zero kinetic energy at infinity and is also known as the *ionization potential* of the hydrogen atom.

Determination of the eigenfunctions

Equation (6.47) may be used to obtain the ground state ($n = 1$, $l = 0$) eigenfunction $S_{10}(\rho)$. Introducing the definition of $\hat{A}_n$ in equation (6.26a), we have

$$\hat{A}_1 S_{10} = -\left(\rho\frac{d}{d\rho} + \frac{\rho}{2}\right)S_{10} = 0$$

or

$$\frac{dS_{10}}{d\rho} = -\frac{S_{10}}{2}$$

from which it follows that

$$S_{10} = ce^{-\rho/2} = 2^{-1/2}e^{-\rho/2}$$

where the constant c of integration was evaluated by applying equations (6.25), (A.26), and (A.28).

The series of eigenfunctions S_{20}, S_{30}, ... are readily obtained from equations (6.46) and (6.26b) with $\lambda = n$, $l = 0$

$$\hat{B}_n S_{n-1,0} = \left(\rho\frac{d}{d\rho} - \frac{\rho}{2} + n\right)S_{n-1,0} = nS_{n0}$$

Thus, S_{20} is

$$S_{20} = \frac{1}{2}\left(\rho\frac{d}{d\rho} - \frac{\rho}{2} + 2\right)2^{-1/2}e^{-\rho/2}$$

$$= \frac{1}{2\sqrt{2}}(2 - \rho)e^{-\rho/2}$$

and S_{30} is

$$S_{30} = \frac{1}{3}\left(\rho\frac{d}{d\rho} - \frac{\rho}{2} + 3\right)\frac{1}{2\sqrt{2}}(2 - \rho)e^{-\rho/2}$$

$$= \frac{1}{6\sqrt{2}}(6 - 6\rho + \rho^2)e^{-\rho/2}$$

and so forth *ad infinitum*. Each eigenfunction is normalized.

The eigenfunctions for $l > 0$ are determined in a similar manner. A general formula for the eigenfunction $S_{l+1,l}$, which is the starting function for evaluating the series S_{nl} with fixed l, is obtained from equations (6.44) and (6.26a) with $l = n = l + 1$

$$\hat{A}_{l+1}S_{l+1,l} = -\left(\rho\frac{d}{d\rho} + \frac{\rho}{2} + l\right)S_{l+1,l} = 0$$

or

$$\rho\frac{dS_{l+1,l}}{d\rho} = \left(l - \frac{\rho}{2}\right)S_{l+1,l}$$

Integration gives

$$S_{l+1,l} = [(2l+2)!]^{-1/2}\rho^l e^{-\rho/2} \qquad (6.49)$$

where the integration constant was evaluated using equations (6.25), (A.26), and (A.28).

The eigenfunction S_{21} from equation (6.49) is

$$S_{21} = \frac{1}{2\sqrt{6}}\rho e^{-\rho/2}$$

and equations (6.46) and (6.26b) for $l = 1$ give

$$S_{31} = \frac{1}{\sqrt{6}}\left(\rho\frac{d}{d\rho} - \frac{\rho}{2} + 3\right)\frac{1}{2\sqrt{6}}\rho e^{-\rho/2}$$

$$= \frac{1}{12}(4-\rho)\rho e^{-\rho/2}$$

$$S_{41} = \sqrt{\frac{3}{40}}\left(\rho\frac{d}{d\rho} - \frac{\rho}{2} + 4\right)\frac{1}{12}(4-\rho)\rho e^{-\rho/2}$$

$$= \frac{1}{8\sqrt{30}}(20 - 10\rho + \rho^2)\rho e^{-\rho/2}$$

$$\vdots$$

The functions S_{31}, S_{41}, ... are automatically normalized as specified by equation (6.25). The normalized eigenfunctions $S_{nl}(\rho)$ for $l = 2, 3, 4, \ldots$ with $n \geqslant (l+1)$ are obtained by the same procedure.

A general formula for S_{nl} involves the repeated application of $\hat{B}_k$ for $k = l+2, l+3, \ldots, n-1, n$ to $S_{l+1,l}$ in equation (6.49). The raising operator must be applied $(n-l-1)$ times. The result is

$$S_{nl} = (b_{nl})^{-1}(b_{n-1,l})^{-1}\cdots(b_{l+2,l})^{-1}\hat{B}_n\hat{B}_{n-1}\cdots\hat{B}_{l+2}S_{l+1,l}$$

$$= \left[\frac{(l+1)(2l+1)!}{n(n+l)!(n-l-1)!(2l+2)!}\right]^{1/2}\left(\rho\frac{d}{d\rho} - \frac{\rho}{2} + n\right)$$

$$\times \left(\rho\frac{d}{d\rho} - \frac{\rho}{2} + n - 1\right)\cdots\left(\rho\frac{d}{d\rho} - \frac{\rho}{2} + l + 2\right)\rho^l e^{-\rho/2} \qquad (6.50)$$

Just as equation (6.46) can be used to go 'up the ladder' to obtain $S_{n,l}$ from

$S_{n-1,l}$, equation (6.44) allows one to go 'down the ladder' and obtain $S_{n-1,l}$ from S_{nl}. Taking the positive square root in going from equation (6.43) to (6.44) is consistent with taking the positive square root in going from equation (6.45) to (6.46); the signs of the functions S_{nl} are maintained in the raising and lowering operations. In all cases the ladder operators yield normalized eigenfunctions if the starting eigenfunction is normalized.

The radial factors of the hydrogen-like atom total wave functions $\psi(r, \theta, \varphi)$ are related to the functions $S_{nl}(\rho)$ by equation (6.23). Thus, we have

$$R_{10} = 2\left(\frac{Z}{a_\mu}\right)^{3/2} e^{-\rho/2}$$

$$R_{20} = \frac{1}{2\sqrt{2}}\left(\frac{Z}{a_\mu}\right)^{3/2}(2 - \rho)e^{-\rho/2}$$

$$R_{30} = \frac{1}{9\sqrt{3}}\left(\frac{Z}{a_\mu}\right)^{3/2}(6 - 6\rho + \rho^2)e^{-\rho/2}$$

$$\vdots$$

$$R_{21} = \frac{1}{2\sqrt{6}}\left(\frac{Z}{a_\mu}\right)^{3/2}\rho e^{-\rho/2}$$

$$R_{31} = \frac{1}{9\sqrt{6}}\left(\frac{Z}{a_\mu}\right)^{3/2}(4 - \rho)\rho e^{-\rho/2}$$

$$R_{41} = \frac{1}{32\sqrt{15}}\left(\frac{Z}{a_\mu}\right)^{3/2}(20 - 10\rho + \rho^2)\rho e^{-\rho/2}$$

$$\vdots$$

and so forth.

A more extensive listing appears in Table 6.1.

Radial functions in terms of associated Laguerre polynomials

The radial functions $S_{nl}(\rho)$ and $R_{nl}(r)$ may be expressed in terms of the associated Laguerre polynomials $L_k^j(\rho)$, whose definition and mathematical properties are discussed in Appendix F. One method for establishing the relationship between $S_{nl}(\rho)$ and $L_k^j(\rho)$ is to relate $S_{nl}(\rho)$ in equation (6.50) to the polynomial $L_k^j(\rho)$ in equation (F.15). That process, however, is long and tedious. Instead, we show that both quantities are solutions of the same differential equation.

Table 6.1. *Radial functions R_{nl} for the hydrogen-like atom for*
$n = 1$ to 6. The variable ρ is given by $\rho = 2Zr/na_\mu$

$$R_{10} = 2(Z/a_\mu)^{3/2} e^{-\rho/2}$$

$$R_{20} = \frac{(Z/a_\mu)^{3/2}}{2\sqrt{2}} (2 - \rho) e^{-\rho/2}$$

$$R_{21} = \frac{(Z/a_\mu)^{3/2}}{2\sqrt{6}} \rho e^{-\rho/2}$$

$$R_{30} = \frac{(Z/a_\mu)^{3/2}}{9\sqrt{3}} (6 - 6\rho + \rho^2) e^{-\rho/2}$$

$$R_{31} = \frac{(Z/a_\mu)^{3/2}}{9\sqrt{6}} (4 - \rho)\rho \, e^{-\rho/2}$$

$$R_{32} = \frac{(Z/a_\mu)^{3/2}}{9\sqrt{30}} \rho^2 \, e^{-\rho/2}$$

$$R_{40} = \frac{(Z/a_\mu)^{3/2}}{96} (24 - 36\rho + 12\rho^2 - \rho^3) e^{-\rho/2}$$

$$R_{41} = \frac{(Z/a_\mu)^{3/2}}{32\sqrt{15}} (20 - 10\rho + \rho^2)\rho \, e^{-\rho/2}$$

$$R_{42} = \frac{(Z/a_\mu)^{3/2}}{96\sqrt{5}} (6 - \rho)\rho^2 \, e^{-\rho/2}$$

$$R_{43} = \frac{(Z/a_\mu)^{3/2}}{96\sqrt{35}} \rho^3 \, e^{-\rho/2}$$

$$R_{50} = \frac{(Z/a_\mu)^{3/2}}{300\sqrt{5}} (120 - 240\rho + 120\rho^2 - 20\rho^3 + \rho^4) e^{-\rho/2}$$

$$R_{51} = \frac{(Z/a_\mu)^{3/2}}{150\sqrt{30}} (120 - 90\rho + 18\rho^2 - \rho^3)\rho \, e^{-\rho/2}$$

$$R_{52} = \frac{(Z/a_\mu)^{3/2}}{150\sqrt{70}} (42 - 14\rho + \rho^2)\rho^2 \, e^{-\rho/2}$$

$$R_{53} = \frac{(Z/a_\mu)^{3/2}}{300\sqrt{70}} (8 - \rho)\rho^3 \, e^{-\rho/2}$$

$$R_{54} = \frac{(Z/a_\mu)^{3/2}}{900\sqrt{70}} \rho^4 \, e^{-\rho/2}$$

$$R_{60} = \frac{(Z/a_\mu)^{3/2}}{2160\sqrt{6}} (720 - 1800\rho + 1200\rho^2 - 300\rho^3 + 30\rho^4 - \rho^5) e^{-\rho/2}$$

Table 6.1. (*cont.*)

$$R_{61} = \frac{(Z/a_\mu)^{3/2}}{432\sqrt{210}}(840 - 840\rho + 252\rho^2 - 28\rho^3 + \rho^4)\rho\,e^{-\rho/2}$$

$$R_{62} = \frac{(Z/a_\mu)^{3/2}}{864\sqrt{105}}(336 - 168\rho + 24\rho^2 - \rho^3)\rho^2\,e^{-\rho/2}$$

$$R_{63} = \frac{(Z/a_\mu)^{3/2}}{2592\sqrt{35}}(72 - 18\rho + \rho^2)\rho^3\,e^{-\rho/2}$$

$$R_{64} = \frac{(Z/a_\mu)^{3/2}}{12\,960\sqrt{7}}(10 - \rho)\rho^4\,e^{-\rho/2}$$

$$R_{65} = \frac{(Z/a_\mu)^{3/2}}{12\,960\sqrt{77}}\rho^5\,e^{-\rho/2}$$

We observe that the solutions $S_{nl}(\rho)$ of the differential equation (6.24) contain the factor $\rho^l e^{-\rho/2}$. Therefore, we define the function $F_{nl}(\rho)$ by

$$S_{nl}(\rho) = F_{nl}(\rho)\rho^l e^{-\rho/2}$$

and substitute this expression into equation (6.24) with $\lambda = n$ to obtain

$$\rho\frac{d^2 F_{nl}}{d\rho^2} + (2l + 2 - \rho)\frac{dF_{nl}}{d\rho} + (n - l - 1)F_{nl} = 0 \qquad (6.51)$$

where we have also divided the equation by the common factor ρ.

The differential equation satisfied by the associated Laguerre polynomials is given by equation (F.16) as

$$\rho\frac{d^2 L_k^j}{d\rho^2} + (j + 1 - \rho)\frac{dL_k^j}{d\rho} + (k - j)L_k^j = 0$$

If we let $k = n + l$ and $j = 2l + 1$, then this equation takes the form

$$\rho\frac{d^2 L_{n+l}^{2l+1}}{d\rho^2} + (2l + 2 - \rho)\frac{dL_{n+l}^{2l+1}}{d\rho} + (n - l - 1)L_{n+l}^{2l+1} = 0 \qquad (6.52)$$

We have already found that the set of functions $S_{nl}(\rho)$ contains all the solutions to (6.24). Therefore, a comparison of equations (6.51) and (6.52) shows that F_{nl} is proportional to L_{n+l}^{2l+1}. Thus, the function $S_{nl}(\rho)$ is related to the polynomial $L_{n+l}^{2l+1}(\rho)$ by

$$S_{nl}(\rho) = c_{nl}\rho^l e^{-\rho/2} L_{n+l}^{2l+1}(\rho) \qquad (6.53)$$

The proportionality constants c_{nl} in equation (6.53) are determined by the normalization condition (6.25). When equation (6.53) is substituted into (6.25), we have

$$c_{nl}^2 \int_0^\infty \rho^{2l+1} e^{-\rho} [L_{n+l}^{2l+1}(\rho)]^2 \, d\rho = 1$$

The value of the integral is given by equation (F.25) with $\alpha = n + l$ and $j = 2l + 1$, so that

$$c_{nl}^2 \frac{2n[(n+l)!]^3}{(n-l-1)!} = 1$$

and $S_{nl}(\rho)$ in equation (6.53) becomes

$$S_{nl}(\rho) = -\left(\frac{(n-l-1)!}{2n[(n+l)!]^3}\right)^{1/2} \rho^l e^{-\rho/2} L_{n+l}^{2l+1}(\rho) \qquad (6.54)$$

Taking the negative square root maintains the sign of $S_{nl}(\rho)$.

Equations (6.39) and (F.22), with $S_{nl}(\rho)$ and $L_k^j(\rho)$ related by (6.54), are identical. From equations (F.23) and (F.24), we find

$$\int_0^\infty S_{nl}(\rho) S_{n\pm1,l}(\rho) \rho^2 \, d\rho = -\frac{1}{2} \sqrt{\frac{(n-l)(n+l+1)}{n(n+1)}}$$

$$\int_0^\infty S_{nl}(\rho) S_{n',l}(\rho) \rho^2 \, d\rho = 0, \qquad n' \neq n, n \pm 1$$

The normalized radial functions $R_{nl}(r)$ may be expressed in terms of the associated Laguerre polynomials by combining equations (6.22), (6.23), and (6.54)

$$R_{nl}(r) = -\sqrt{\frac{4(n-l-1)! Z^3}{n^4[(n+l)!]^3 a_\mu^3}} \left(\frac{2Zr}{na_\mu}\right)^l e^{-Zr/na_0} L_{n+l}^{2l+1}(2Zr/na_\mu) \qquad (6.55)$$

Solution for positive energies

There are also solutions to the radial differential equation (6.17) for positive values of the energy E, which correspond to the ionization of the hydrogen-like atom. In the limit $r \to \infty$, equations (6.17) and (6.18) for positive E become

$$\frac{d^2 R(r)}{dr^2} + \frac{2\mu E}{\hbar^2} R(r) = 0$$

for which the solution is

$$R(r) = c e^{\pm i(2\mu E)^{1/2} r/\hbar}$$

where c is a constant of integration. This solution has oscillatory behavior at infinity and leads to an acceptable, well-behaved eigenfunction of equation (6.17) for all positive eigenvalues E. Thus, the radial equation (6.17) has a continuous range of positive eigenvalues as well as the discrete set (equation (6.48)) of negative eigenvalues. The corresponding eigenfunctions represent unbound or scattering states and are useful in the study of electron–ion collisions and scattering phenomena. In view of the complexity of the analysis for obtaining the eigenfunctions and eigenvalues of equation (6.17) for positive E and the unimportance of these quantities in most problems of chemical interest, we do not consider this case any further.

Infinite nuclear mass

The energy levels E_n and the radial functions $R_{nl}(r)$ depend on the reduced mass μ of the two-particle system

$$\mu = \frac{m_N m_e}{m_N + m_e} = \frac{m_e}{1 + \dfrac{m_e}{m_N}}$$

where m_N is the nuclear mass and m_e is the electronic mass. The value of m_e is $9.109\,39 \times 10^{-31}$ kg. For hydrogen, the nuclear mass is the protonic mass, $1.672\,62 \times 10^{-27}$ kg, so that μ is 9.1044×10^{-31} kg. For heavier hydrogen-like atoms, the nuclear mass is, of course, greater than the protonic mass. In the limit $m_N \rightarrow \infty$, the reduced mass and the electronic mass are the same. In the classical two-particle problem of Section 6.1, this limit corresponds to the nucleus remaining at a fixed point in space.

In most applications, the reduced mass is sufficiently close in value to the electronic mass m_e that it is customary to replace μ in the expressions for the energy levels and wave functions by m_e. The parameter $a_\mu = \hbar^2/\mu e'^2$ is thereby replaced by $a_0 = \hbar^2/m_e e'^2$. The quantity a_0 is, according to the earlier Bohr theory, the radius of the circular orbit of the electron in the ground state of the hydrogen atom ($Z = 1$) with a stationary nucleus. Except in Section 6.5, where this substitution is not appropriate, we replace μ by m_e and a_μ by a_0 in the remainder of this book.

6.4 Atomic orbitals

We have shown that the simultaneous eigenfunctions $\psi(r, \theta, \varphi)$ of the operators $\hat{H}$, $\hat{L}^2$, and $\hat{l}_z$ have the form

$$\psi_{nlm}(r, \theta, \varphi) = |nlm\rangle = R_{nl}(r)Y_{lm}(\theta, \varphi) \tag{6.56}$$

where for convenience we have introduced the Dirac notation. The radial functions $R_{nl}(r)$ and the spherical harmonics $Y_{lm}(\theta, \varphi)$ are listed in Tables 6.1 and 5.1, respectively. These eigenfunctions depend on the three quantum numbers n, l, and m. The integer n is called the *principal* or *total quantum number* and determines the energy of the atom. The *azimuthal quantum number* l determines the total angular momentum of the electron, while the *magnetic quantum number m* determines the *z*-component of the angular momentum. We have found that the allowed values of n, l, and m are

$$m = 0, \pm 1, \pm 2, \ldots$$

$$l = |m|, |m| + 1, |m| + 2, \ldots$$

$$n = l + 1, l + 2, l + 3, \ldots$$

This set of relationships may be inverted to give

$$n = 1, 2, 3, \ldots$$

$$l = 0, 1, 2, \ldots, n - 1$$

$$m = -l, -l + 1, \ldots, -1, 0, 1, \ldots, l - 1, l$$

These eigenfunctions form an orthonormal set, so that

$$\langle n' l' m' | nlm \rangle = \delta_{nn'} \delta_{ll'} \delta_{mm'}$$

The energy levels of the hydrogen-like atom depend only on the principal quantum number n and are given by equation (6.48), with a_μ replaced by a_0, as

$$E_n = -\frac{Z^2 e'^2}{2 a_0 n^2}, \qquad n = 1, 2, 3, \ldots \qquad (6.57)$$

To find the degeneracy g_n of E_n, we note that for a specific value of n there are n different values of l. For each value of l, there are $(2l + 1)$ different values of m, giving $(2l + 1)$ eigenfunctions. Thus, the number of wave functions corresponding to n is given by

$$g_n = \sum_{l=0}^{n-1}(2l + 1) = 2 \sum_{l=0}^{n-1} l + \sum_{l=0}^{n-1} 1$$

The first summation on the right-hand side is the sum of integers from 0 to $(n - 1)$ and is equal to $n(n - 1)/2$ (n terms multiplied by the average value of each term). The second summation on the right-hand side has n terms, each equal to unity. Thus, we obtain

$$g_n = n(n - 1) + n = n^2$$

showing that each energy level is n^2-fold degenerate. The ground-state energy level E_1 is non-degenerate.

The wave functions $|nlm\rangle$ for the hydrogen-like atom are often called *atomic orbitals*. It is customary to indicate the values 0, 1, 2, 3, 4, 5, 6, 7, $\ldots$ of the azimuthal quantum number l by the letters s, p, d, f, g, h, i, k, $\ldots$, respectively. Thus, the ground-state wave function $|100\rangle$ is called the 1s atomic orbital, $|200\rangle$ is called the 2s orbital, $|210\rangle$, $|211\rangle$, and $|21 -1\rangle$ are called 2p orbitals, and so forth. The first four letters, standing for *sharp*, *principal*, *diffuse*, and *fundamental*, originate from an outdated description of spectral lines. The letters which follow are in alphabetical order with j omitted.

s orbitals
The 1s atomic orbital $|1s\rangle$ is

$$|1s\rangle = |100\rangle = R_{10}(r)Y_{00}(\theta,\,\varphi) = \frac{1}{\pi^{1/2}}\left(\frac{Z}{a_0}\right)^{3/2} e^{-Zr/a_0} \qquad (6.58)$$

where $R_{10}(r)$ and $Y_{00}(0,\,\varphi)$ are obtained from Tables 6.1 and 5.1. Likewise, the orbital $|2s\rangle$ is

$$|2s\rangle = |200\rangle = \frac{(Z/a_0)^{3/2}}{4\sqrt{2\pi}}\left(2 - \frac{Zr}{a_0}\right)e^{-Zr/2a_0} \qquad (6.59)$$

and so forth for higher values of the quantum number n. The expressions for $|ns\rangle$ for $n = 1$, 2, and 3 are listed in Table 6.2.

All the s orbitals have the spherical harmonic $Y_{00}(\theta,\,\varphi)$ as a factor. This spherical harmonic is independent of the angles θ and φ, having a value $(2\sqrt{\pi})^{-1}$. Thus, the s orbitals depend only on the radial variable r and are spherically symmetric about the origin. Likewise, the electronic probability density $|\psi|^2$ is spherically symmetric for s orbitals.

p orbitals
The wave functions for $n = 2$, $l = 1$ obtained from equation (6.56) are as follows:

$$|2p_0\rangle = |210\rangle = \frac{(Z/a_0)^{5/2}}{4\sqrt{2\pi}}\,re^{-Zr/2a_0}\cos\theta \qquad (6.60a)$$

$$|2p_1\rangle = |211\rangle = \frac{1}{8\pi^{1/2}}\left(\frac{Z}{a_0}\right)^{5/2} re^{-Zr/2a_0}\sin\theta\,e^{i\varphi} \qquad (6.60b)$$

$$|2p_{-1}\rangle = |21-1\rangle = \frac{1}{8\pi^{1/2}}\left(\frac{Z}{a_0}\right)^{5/2} re^{-Zr/2a_0}\sin\theta\,e^{-i\varphi} \qquad (6.60c)$$

The 2s and $2p_0$ orbitals are real, but the $2p_1$ and $2p_{-1}$ orbitals are complex. Since the four orbitals have the same eigenvalue E_2, any linear combination of them also satisfies the Schrödinger equation (6.12) with eigenvalue E_2. Thus, we may replace the two complex orbitals by the following linear combinations to obtain two new real orbitals

Table 6.2. *Real wave functions for the hydrogen-like atom. The parameter a_μ has been replaced by a_0*

State	Wave function	
	Spherical coordinates	Cartesian coordinates
1s	$\dfrac{(Z/a_0)^{3/2}}{\sqrt{\pi}}\,\mathrm{e}^{-Zr/a_0}$	
2s	$\dfrac{(Z/a_0)^{3/2}}{4\sqrt{2\pi}}\left(2-\dfrac{Zr}{a_0}\right)\mathrm{e}^{-Zr/2a_0}$	
$2p_z$	$\dfrac{(Z/a_0)^{5/2}}{4\sqrt{2\pi}}\,r\mathrm{e}^{-Zr/2a_0}\cos\theta$	$\dfrac{(Z/a_0)^{5/2}}{4\sqrt{2\pi}}\,z\mathrm{e}^{-Zr/2a_0}$
$2p_x$	$\dfrac{(Z/a_0)^{5/2}}{4\sqrt{2\pi}}\,r\mathrm{e}^{-Zr/2a_0}\sin\theta\cos\varphi$	$\dfrac{(Z/a_0)^{5/2}}{4\sqrt{2\pi}}\,x\mathrm{e}^{-Zr/2a_0}$
$2p_y$	$\dfrac{(Z/a_0)^{5/2}}{4\sqrt{2\pi}}\,r\mathrm{e}^{-Zr/2a_0}\sin\theta\sin\varphi$	$\dfrac{(Z/a_0)^{5/2}}{4\sqrt{2\pi}}\,y\mathrm{e}^{-Zr/2a_0}$
3s	$\dfrac{(Z/a_0)^{3/2}}{81\sqrt{3\pi}}\left(27-18\dfrac{Zr}{a_0}+2\dfrac{Z^2r^2}{a_0^2}\right)\mathrm{e}^{-Zr/3a_0}$	
$3p_z$	$\dfrac{2(Z/a_0)^{5/2}}{81\sqrt{2\pi}}\left(6-\dfrac{Zr}{a_0}\right)r\mathrm{e}^{-Zr/3a_0}\cos\theta$	$\dfrac{2(Z/a_0)^{5/2}}{81\sqrt{2\pi}}\left(6-\dfrac{Zr}{a_0}\right)z\mathrm{e}^{-Zr/3a_0}$
$3p_x$	$\dfrac{2(Z/a_0)^{5/2}}{81\sqrt{2\pi}}\left(6-\dfrac{Zr}{a_0}\right)r\mathrm{e}^{-Zr/3a_0}\sin\theta\cos\varphi$	$\dfrac{2(Z/a_0)^{5/2}}{81\sqrt{2\pi}}\left(6-\dfrac{Zr}{a_0}\right)x\mathrm{e}^{-Zr/3a_0}$
$3p_y$	$\dfrac{2(Z/a_0)^{5/2}}{81\sqrt{2\pi}}\left(6-\dfrac{Zr}{a_0}\right)r\mathrm{e}^{-Zr/3a_0}\sin\theta\sin\varphi$	$\dfrac{2(Z/a_0)^{5/2}}{81\sqrt{2\pi}}\left(6-\dfrac{Zr}{a_0}\right)y\mathrm{e}^{-Zr/3a_0}$
$3d_{z^2}$	$\dfrac{(Z/a_0)^{7/2}}{81\sqrt{6\pi}}\,r^2\mathrm{e}^{-Zr/3a_0}(3\cos^2\theta-1)$	$\dfrac{(Z/a_0)^{7/2}}{81\sqrt{6\pi}}\,(3z^2-r^2)\mathrm{e}^{-Zr/3a_0}$
$3d_{xz}$	$\dfrac{2(Z/a_0)^{7/2}}{81\sqrt{2\pi}}\,r^2\mathrm{e}^{-Zr/3a_0}\sin\theta\cos\theta\cos\varphi$	$\dfrac{2(Z/a_0)^{7/2}}{81\sqrt{2\pi}}\,xz\mathrm{e}^{-Zr/3a_0}$
$3d_{yz}$	$\dfrac{2(Z/a_0)^{7/2}}{81\sqrt{2\pi}}\,r^2\mathrm{e}^{-Zr/3a_0}\sin\theta\cos\theta\sin\varphi$	$\dfrac{2(Z/a_0)^{7/2}}{81\sqrt{2\pi}}\,yz\mathrm{e}^{-Zr/3a_0}$
$3d_{x^2-y^2}$	$\dfrac{(Z/a_0)^{7/2}}{81\sqrt{2\pi}}\,r^2\mathrm{e}^{-Zr/3a_0}\sin^2\theta\cos2\varphi$	$\dfrac{(Z/a_0)^{7/2}}{81\sqrt{2\pi}}\,(x^2-y^2)\mathrm{e}^{-Zr/3a_0}$
$3d_{xy}$	$\dfrac{(Z/a_0)^{7/2}}{81\sqrt{2\pi}}\,r^2\mathrm{e}^{-Zr/3a_0}\sin^2\theta\sin2\varphi$	$\dfrac{2(Z/a_0)^{7/2}}{81\sqrt{2\pi}}\,xy\mathrm{e}^{-Zr/3a_0}$

$$|2p_x\rangle \equiv 2^{-1/2}(|2p_1\rangle + |2p_{-1}\rangle) = \frac{1}{4(2\pi)^{1/2}} \left(\frac{Z}{a_0}\right)^{5/2} re^{-Zr/2a_0} \sin\theta \cos\varphi \quad (6.61a)$$

$$|2p_y\rangle \equiv -i2^{-1/2}(|2p_1\rangle - |2p_{-1}\rangle) = \frac{1}{4(2\pi)^{1/2}} \left(\frac{Z}{a_0}\right)^{5/2} re^{-Zr/2a_0} \sin\theta \sin\varphi$$

$$(6.61b)$$

where equations (A.32) and (A.33) have been used. These new orbitals $|2p_x\rangle$ and $|2p_y\rangle$ are orthogonal to each other and to all the other eigenfunctions $|nlm\rangle$. The factor $2^{-1/2}$ ensures that they are normalized as well. Although these new orbitals are simultaneous eigenfunctions of the Hamiltonian operator $\hat{H}$ and of the operator $\hat{L}^2$, they are not eigenfunctions of the operator $\hat{L}_z$.

If we now substitute equations (5.29a), (5.29b), and (5.29c) into (6.61a), (6.61b), and (6.60a), respectively, we obtain for the set of three real 2p orbitals

$$|2p_x\rangle = \frac{1}{4(2\pi)^{1/2}} \left(\frac{Z}{a_0}\right)^{5/2} xe^{-Zr/2a_0} \qquad (6.62a)$$

$$|2p_y\rangle = \frac{1}{4(2\pi)^{1/2}} \left(\frac{Z}{a_0}\right)^{5/2} ye^{-Zr/2a_0} \qquad (6.62b)$$

$$|2p_z\rangle = \frac{1}{\pi^{1/2}} \left(\frac{Z}{2a_0}\right)^{5/2} ze^{-Zr/2a_0} \qquad (6.62c)$$

The subscript x, y, or z on a 2p orbital indicates that the angular part of the orbital has its maximum value along that axis. Graphs of the square of the angular part of these three functions are presented in Figure 6.2. The mathematical expressions for the real 2p and 3p atomic orbitals are given in Table 6.2.

d orbitals

The five wave functions for $n = 3$, $l = 2$ are

$$|3d_0\rangle = |320\rangle = \frac{1}{81\sqrt{6\pi}} \left(\frac{Z}{a_0}\right)^{7/2} r^2 e^{-(Zr/3a_0)}(3\cos^2\theta - 1) \qquad (6.63a)$$

$$|3d_{\pm 1}\rangle = |32 \pm 1\rangle = \frac{1}{81\sqrt{\pi}} \left(\frac{Z}{a_0}\right)^{7/2} r^2 e^{-(Zr/3a_0)} \sin\theta \cos\theta\, e^{\pm i\varphi} \qquad (6.63b)$$

$$|3d_{\pm 2}\rangle = |32 \pm 2\rangle = \frac{1}{162\sqrt{\pi}} \left(\frac{Z}{a_0}\right)^{7/2} r^2 e^{-(Zr/3a_0)} \sin^2\theta\, e^{\pm i2\varphi} \qquad (6.63c)$$

The orbital $|3d_0\rangle$ is real. Substitution of equation (5.29c) into (6.63a) and a change in notation for the subscript give

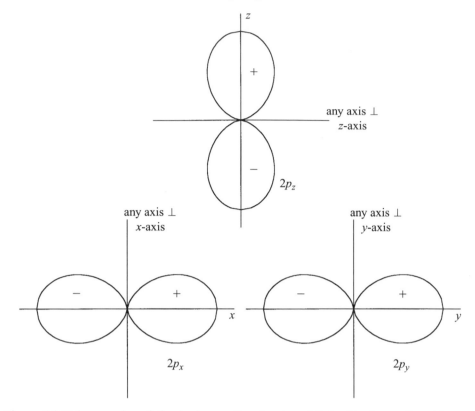

Figure 6.2 Polar graphs of the hydrogen 2p atomic orbitals. Regions of positive and negative values of the orbitals are indicated by $+$ and $-$ signs, respectively. The distance of the curve from the origin is proportional to the square of the angular part of the atomic orbital.

$$|3d_{z^2}\rangle = \frac{1}{81\sqrt{6\pi}}\left(\frac{Z}{a_0}\right)^{7/2}(3z^2 - r^2)e^{-(Zr/3a_0)} \qquad (6.64a)$$

From the four complex orbitals $|3d_1\rangle$, $|3d_{-1}\rangle$, $|3d_2\rangle$, and $|3d_{-2}\rangle$, we construct four equivalent real orbitals by the relations

$$|3d_{xz}\rangle \equiv 2^{-1/2}(|3d_1\rangle + |3d_{-1}\rangle) = \frac{2^{1/2}}{81\pi^{1/2}}\left(\frac{Z}{a_0}\right)^{7/2}xze^{-(Zr/3a_0)} \qquad (6.64b)$$

$$|3d_{yz}\rangle \equiv -i2^{-1/2}(|3d_1\rangle - |3d_{-1}\rangle) = \frac{2^{1/2}}{81\pi^{1/2}}\left(\frac{Z}{a_0}\right)^{7/2}yze^{-(Zr/3a_0)} \qquad (6.64c)$$

$$|3d_{x^2-y^2}\rangle \equiv 2^{-1/2}(|3d_2\rangle + |3d_{-2}\rangle) = \frac{1}{81(2\pi)^{1/2}} \left(\frac{Z}{a_0}\right)^{7/2} (x^2 - y^2)e^{-(Zr/3a_0)}$$

$$(6.64d)$$

$$|3d_{xy}\rangle \equiv -i2^{-1/2}(|3d_2\rangle - |3d_{-2}\rangle) = \frac{2^{1/2}}{81\pi^{1/2}} \left(\frac{Z}{a_0}\right)^{7/2} xye^{-(Zr/3a_0)} \qquad (6.64e)$$

In forming $|3d_{x^2-y^2}\rangle$ and $|3d_{xy}\rangle$, equations (A.37) and (A.38) were used. Graphs of the square of the angular part of these five real functions are shown in Figure 6.3 and the mathematical expressions are listed in Table 6.2.

Radial functions and expectation values

The radial functions $R_{nl}(r)$ for the 1s, 2s, 2p, 3s, 3p, and 3d atomic orbitals are shown in Figure 6.4. For states with $l \neq 0$, the radial functions vanish at the origin. For states with no angular momentum ($l = 0$), however, the radial function $R_{n0}(r)$ has a non-zero value at the origin. The function $R_{nl}(r)$ has $(n - l - 1)$ nodes between 0 and ∞, i.e., the function crosses the r-axis $(n - l - 1)$ times, not counting the origin.

The probability of finding the electron in the hydrogen-like atom, with the distance r from the nucleus between r and $r + dr$, with angle θ between θ and $\theta + d\theta$, and with the angle φ between φ and $\varphi + d\varphi$ is

$$|\psi_{nlm}|^2 \, d\tau - [R_{nl}(r)]^2 |Y_{lm}(\theta, \varphi)|^2 r^2 \sin\theta \, dr \, d\theta \, d\varphi$$

To find the probability $D_{nl}(r) \, dr$ that the electron is between r and $r + dr$ regardless of the direction, we integrate over the angles θ and φ to obtain

$$D_{nl}(r) \, dr = r^2 [R_{nl}(r)]^2 \, dr \int_0^\pi \int_0^{2\pi} |Y_{lm}(\theta, \varphi)|^2 \sin\theta \, d\theta \, d\varphi = r^2 [R_{nl}(r)]^2 \, dr$$

$$(6.65)$$

Since the spherical harmonics are normalized, the value of the double integral is unity.

The *radial distribution function* $D_{nl}(r)$ is the probability density for the electron being in a spherical shell with inner radius r and outer radius $r + dr$. For the 1s, 2s, and 2p states, these functions are

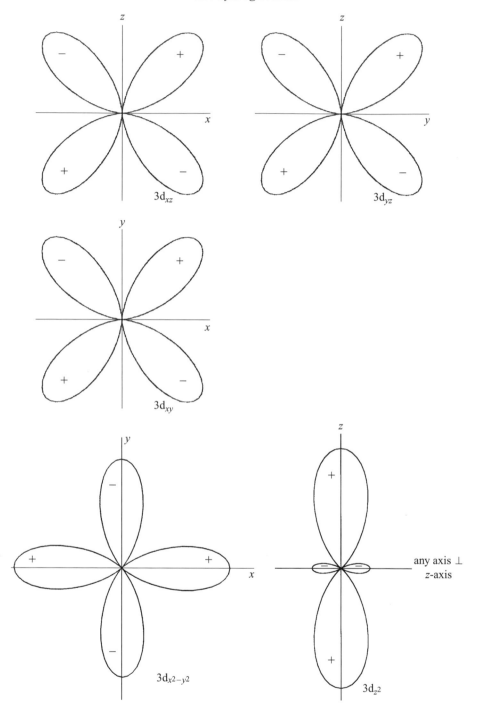

Figure 6.3 Polar graphs of the hydrogen 3d atomic orbitals. Regions of positive and negative values of the orbitals are indicated by $+$ and $-$ signs, respectively. The distance of the curve from the origin is proportional to the square of the angular part of the atomic orbital.

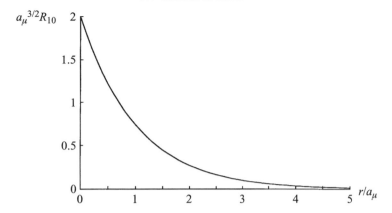

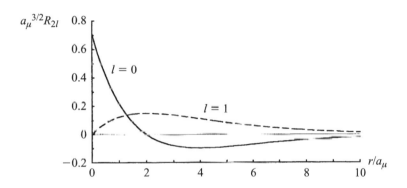

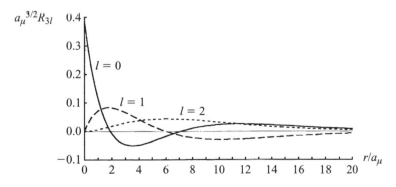

Figure 6.4 The radial functions $R_{nl}(r)$ for the hydrogen-like atom.

$$D_{10}(r) = 4 \left(\frac{Z}{a_0} \right)^3 r^2 e^{-2Zr/a_0}$$

$$D_{20}(r) = \frac{1}{8} \left(\frac{Z}{a_0} \right)^3 r^2 \left(2 - \frac{Zr}{a_0} \right)^2 e^{-Zr/a_0} \tag{6.66}$$

$$D_{21}(r) = \frac{1}{24} \left(\frac{Z}{a_0} \right)^5 r^4 e^{-Zr/a_0}$$

Higher-order functions are readily determined from Table 6.1. The radial distribution functions for the 1s, 2s, 2p, 3s, 3p, and 3d states are shown in Figure 6.5.

The most probable value r_{mp} of r for the 1s state is found by setting the derivative of $D_{10}(r)$ equal to zero

$$\frac{dD_{10}(r)}{dr} = 8 \left(\frac{Z}{a_0} \right)^3 r \left(1 - \frac{Zr}{a_0} \right) e^{-2Zr/a_0} = 0$$

which gives

$$r_{mp} = a_0/Z \tag{6.67}$$

Thus, for the hydrogen atom ($Z = 1$) the most probable distance of the electron from the nucleus is equal to the radius of the first Bohr orbit.

The radial distribution functions may be used to calculate expectation values of functions of the radial variable r. For example, the average distance of the electron from the nucleus for the 1s state is given by

$$\langle r \rangle_{1s} = \int_0^\infty r D_{10}(r) \, dr = 4 \left(\frac{Z}{a_0} \right)^3 \int_0^\infty r^3 e^{-2Zr/a_0} \, dr = \frac{3a_0}{2Z} \tag{6.68}$$

where equations (A.26) and (A.28) were used to evaluate the integral. By the same method, we find

$$\langle r \rangle_{2s} = \frac{6a_0}{Z}, \qquad \langle r \rangle_{2p} = \frac{5a_0}{Z}$$

The expectation values of powers and inverse powers of r for any arbitrary state of the hydrogen-like atom are defined by

$$\langle r^k \rangle_{nl} = \int_0^\infty r^k D_{nl}(r) \, dr = \int_0^\infty r^k [R_{nl}(r)]^2 r^2 \, dr \tag{6.69}$$

In Appendix H we show that these expectation values obey the recurrence relation

$$\frac{k+1}{n^2} \langle r^k \rangle_{nl} - (2k+1) \frac{a_0}{Z} \langle r^{k-1} \rangle_{nl} + k \left[l(l+1) + \frac{1-k^2}{4} \right] \frac{a_0^2}{Z^2} \langle r^{k-2} \rangle_{nl} = 0$$

$$\tag{6.70}$$

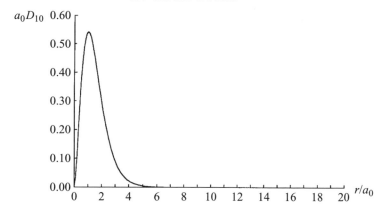

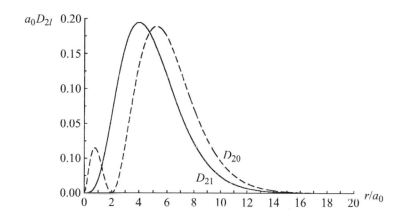

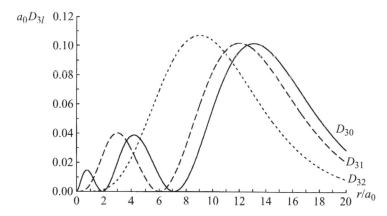

Figure 6.5 The radial distribution functions $D_{nl}(r)$ for the hydrogen-like atom.

For $k = 0$, equation (6.70) gives

$$\langle r^{-1} \rangle_{nl} = \frac{Z}{n^2 a_0} \qquad (6.71)$$

For $k = 1$, equation (6.70) gives

$$\frac{2}{n^2} \langle r \rangle_{nl} - \frac{3a_0}{Z} + l(l+1) \frac{a_0^2}{Z^2} \langle r^{-1} \rangle_{nl} = 0$$

or

$$\langle r \rangle_{nl} = \frac{a_0}{2Z} [3n^2 - l(l+1)] \qquad (6.72)$$

For $k = 2$, equation (6.70) gives

$$\frac{3}{n^2} \langle r^2 \rangle_{nl} - \frac{5a_0}{Z} \langle r \rangle_{nl} + 2[l(l+1) - \tfrac{3}{4}] \frac{a_0^2}{Z^2} = 0$$

or

$$\langle r^2 \rangle_{nl} = \frac{n^2 a_0^2}{2Z^2} [5n^2 - 3l(l+1) + 1] \qquad (6.73)$$

For higher values of k, equation (6.70) leads to $\langle r^3 \rangle_{nl}$, $\langle r^4 \rangle_{nl}$, ...

For $k = -1$, equation (6.70) relates $\langle r^{-3} \rangle_{nl}$ to $\langle r^{-2} \rangle_{nl}$

$$\langle r^{-3} \rangle_{nl} = \frac{Z}{l(l+1)a_0} \langle r^{-2} \rangle_{nl} \qquad (6.74)$$

For $k = -2, -3, \ldots$, equation (6.70) gives successively $\langle r^{-4} \rangle_{nl}$, $\langle r^{-5} \rangle_{nl}$, ... expressed in terms of $\langle r^{-2} \rangle_{nl}$.

Although the expectation value $\langle r^{-2} \rangle_{nl}$ cannot be obtained from equation (6.70), it can be evaluated by regarding the azimuthal quantum number l as the parameter in the Hellmann–Feynman theorem (equation (3.71)). Thus, we have

$$\frac{\partial E_n}{\partial l} = \left\langle \frac{\partial \hat{H}_l}{\partial l} \right\rangle \qquad (6.75)$$

where the Hamiltonian operator $\hat{H}_l$ is given by equation (6.18) and the energy levels E_n by equation (6.57). The derivative $\partial \hat{H}_l / \partial l$ is just

$$\frac{\partial \hat{H}_l}{\partial l} = \frac{\hbar^2}{2\mu r^2} (2l + 1) \qquad (6.76)$$

In the derivation of (6.57), the quantum number n is shown to be the value of l plus a positive integer. Accordingly, we have $\partial n / \partial l = 1$ and

$$\frac{\partial E_n}{\partial l} = -\frac{Z^2 e'^2}{2a_0} \frac{\partial}{\partial l} n^{-2} = -\frac{Z^2 e'^2}{2a_0} \frac{\partial n}{\partial l} \frac{\partial}{\partial n} n^{-2} = \frac{Z^2 \hbar^2}{\mu a_0^2} n^{-3} \qquad (6.77)$$

where $a_\mu = \hbar^2 / \mu e'^2$ has been replaced by $a_0 = \hbar^2 / m_e e'^2$. Substitution of equations (6.76) and (6.77) into (6.75) gives the desired result

$$\langle r^{-2}\rangle_{nl} = \frac{Z^2}{n^3(l+\frac{1}{2})a_0^2} \tag{6.78}$$

Expression (6.71) for the expectation value of r^{-1} may be used to calculate the average potential energy of the electron in the state $|nlm\rangle$. The potential energy $V(r)$ is given by equation (6.13). Its expectation value is

$$\langle V\rangle_{nl} = -Ze'^2\langle r^{-1}\rangle_{nl} = -\frac{Z^2 e'^2}{a_0 n^2} \tag{6.79}$$

The result depends only on the principal quantum number n, so we may drop the subscript l. A comparison with equation (6.57) shows that the total energy is equal to one-half of the average potential energy

$$E_n = \tfrac{1}{2}\langle V\rangle_n \tag{6.80}$$

Since the total energy is the sum of the kinetic energy T and the potential energy V, we also have the expression

$$T_n = -E_n = \frac{Z^2 e'^2}{2a_0 n^2} \tag{6.81}$$

The relationship $E_n = -T_n = (V_n/2)$ is an example of the quantum-mechanical *virial theorem*.

6.5 Spectra

The theoretical results for the hydrogen-like atom may be related to experimentally measured spectra. Observed spectral lines arise from transitions of the atom from one electronic energy level to another. The frequency ν of any given spectral line is given by the Planck relation

$$\nu = (E_2 - E_1)/h$$

where E_1 is the lower energy level and E_2 the higher one. In an *absorption spectrum*, the atom absorbs a photon of frequency ν and undergoes a transition from a lower to a higher energy level ($E_1 \rightarrow E_2$). In an *emission spectrum*, the process is reversed; the transition is from a higher to a lower energy level ($E_2 \rightarrow E_1$) and a photon is emitted. A spectral line is usually expressed as a wave number $\tilde{\nu}$, defined as the reciprocal of the wavelength λ

$$\tilde{\nu} \equiv \frac{1}{\lambda} = \frac{\nu}{c} = \frac{|E_2 - E_1|}{hc} \tag{6.82}$$

The hydrogen-like atomic energy levels are given in equation (6.48). If n_1 and n_2 are the principal quantum numbers of the energy levels E_1 and E_2, respectively, then the wave number of the spectral line is

Table 6.3. *Rydberg constant for hydrogen-like atoms*

Atom	R (cm^{-1})
^{1}H	109 677.58
^{2}H (D)	109 707.42
^{4}He$^+$	109 722.26
^{7}Li^{2+}	109 728.72
^{9}Be^{3+}	109 730.62
∞	109 737.31

$$\tilde{\nu} = RZ^2 \left(\frac{1}{n_1^2} - \frac{1}{n_2^2} \right), \qquad n_2 > n_1 \tag{6.83}$$

where the *Rydberg constant* R is given by

$$R = \frac{\mu e'^4}{4\pi\hbar^3 c} \tag{6.84}$$

The value of the Rydberg constant varies from one hydrogen-like atom to another because the reduced mass μ is a factor. It is not appropriate here to replace the reduced mass μ by the electronic mass m_e because the errors caused by this substitution are larger than the uncertainties in the experimental data. The measured values of the Rydberg constants for the atoms ^{1}H, ^{4}He$^+$, ^{7}Li^{2+}, and ^{9}Be^{3+} are listed in Table 6.3. Following the custom of the field of spectroscopy, we express the wave numbers in the unit cm^{-1} rather than the SI unit m^{-1}. Also listed in Table 6.3 is the extrapolated value of R for infinite nuclear mass. The calculated values from equation (6.84) are in agreement with the experimental values within the known number of significant figures for the fundamental constants m_e, e', and $\hbar$ and the nuclear masses m_N. The measured values of R have more significant figures than any of the quantities in equation (6.84) except the speed of light c.

The spectrum of hydrogen ($Z = 1$) is divided into a number of series of spectral lines, each series having a particular value for n_1. As many as six different series have been observed:

$n_1 = 1$, Lyman series   ultraviolet
$n_1 = 2$, Balmer series   visible
$n_1 = 3$, Paschen series   infrared
$n_1 = 4$, Brackett series   infrared
$n_1 = 5$, Pfund series   far infrared
$n_1 = 6$, Humphreys series   very far infrared

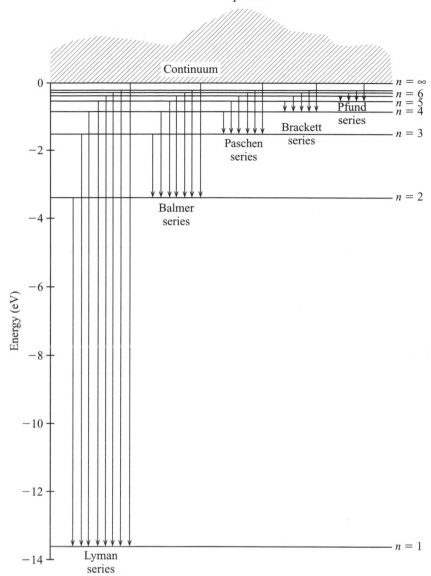

Figure 6.6 Energy levels for the hydrogen atom.

Thus, transitions from the lowest energy level $n_1 = 1$ to the higher energy levels $n_2 = 2, 3, 4, \ldots$ give the Lyman series, transitions from $n_1 = 2$ to $n_2 = 3, 4, 5, \ldots$ give the Balmer series, and so forth. An energy level diagram for the hydrogen atom is shown in Figure 6.6. The transitions corresponding to the spectral lines in the various series are shown as vertical lines between the energy levels.

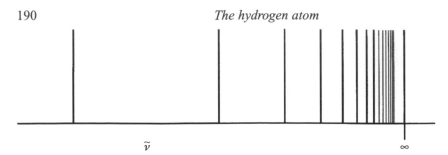

Figure 6.7 A typical series of spectral lines for a hydrogen-like atom shown in terms of the wave number $\tilde{\nu}$.

A typical series of spectral lines is shown schematically in Figure 6.7. The line at the lowest value of the wave number $\tilde{\nu}$ corresponds to the transition $n_1 \rightarrow (n_2 = n_1 + 1)$, the next line to $n_1 \rightarrow (n_2 = n_1 + 2)$, and so forth. These spectral lines are situated closer and closer together as n_2 increases and converge to the *series limit*, corresponding to $n_2 = \infty$. According to equation (6.83), the series limit is given by

$$\tilde{\nu} = \mathsf{R}/n_1^2 \tag{6.85}$$

Beyond the series limit is a continuous spectrum corresponding to transitions from the energy level n_1 to the continuous range of positive energies for the atom.

The reduced mass of the hydrogen isotope $^2\mathrm{H}$, known as deuterium, slightly differs from that of ordinary hydrogen $^1\mathrm{H}$. Accordingly, the Rydberg constants for hydrogen and for deuterium differ slightly as well. Since naturally occurring hydrogen contains about 0.02% deuterium, each observed spectral line in hydrogen is actually a doublet of closely spaced lines, the one for deuterium much weaker in intensity than the other. This effect of nuclear mass on spectral lines was used by Urey (1932) to prove the existence of deuterium.

Pseudo-Zeeman effect

The influence of an external magnetic field on the spectrum of an atom is known as the *Zeeman effect*. The magnetic field interacts with the magnetic moments within the atom and causes the atomic spectral lines to split into a number of closely spaced lines. In addition to a magnetic moment due to its orbital motion, an electron also possesses a magnetic moment due to an intrinsic angular momentum called *spin*. The concept of spin is discussed in Chapter 7. In the discussion here, we consider only the interaction of the external magnetic field with the magnetic moment due to the electronic orbital motion and neglect the effects of electron spin. Thus, the following analysis does not give results that correspond to actual observations. For this reason, we refer to this treatment as the *pseudo-Zeeman effect*.

When a magnetic field **B** is applied to a hydrogen-like atom with magnetic moment **M**, the resulting potential energy V is given by the classical expression

$$V = -\mathbf{M} \cdot \mathbf{B} = \frac{\mu_B}{\hbar} \mathbf{L} \cdot \mathbf{B} \tag{6.86}$$

where equation (5.81) has been introduced. If the z-axis is selected to be parallel to the vector **B**, then we have

$$V = \mu_B B L_z / \hbar \tag{6.87}$$

If we replace the z-component of the classical angular momentum in equation (6.87) by its quantum-mechanical operator, then the Hamiltonian operator $\hat{H}_B$ for the hydrogen-like atom in a magnetic field **B** becomes

$$\hat{H}_B = \hat{H} + \frac{\mu_B B}{\hbar} \hat{L}_z \tag{6.88}$$

where $\hat{H}$ is the Hamiltonian operator (6.14) for the atom in the absence of the magnetic field. Since the atomic orbitals ψ_{nlm} in equation (6.56) are simultaneous eigenfunctions of $\hat{H}$, $\hat{L}^2$, and $\hat{L}_z$, they are also eigenfunctions of the operator $\hat{H}_B$. Accordingly, we have

$$\hat{H}_B \psi_{nlm} = \left(\hat{H} + \frac{\mu_B B}{\hbar} \hat{L}_z \right) \psi_{nlm} = (E_n + m\mu_B B)\psi_{nlm} \tag{6.89}$$

where E_n is given by (6.48) and equation (6.15c) has been used. Thus, the energy levels of a hydrogen-like atom in an external magnetic field depend on the quantum numbers n and m and are given by

$$E_{nm} = -\frac{Z^2 e'^2}{2a_\mu n^2} + m\mu_B B, \qquad n = 1, 2, \ldots; \qquad m = 0, \pm 1, \ldots, \pm(n-1) \tag{6.90}$$

This dependence on m is the reason why m is called the *magnetic quantum number*.

The degenerate energy levels for the hydrogen atom in the absence of an external magnetic field are split by the magnetic field into a series of closely spaced levels, some of which are non-degenerate while others are still degenerate. For example, the energy level E_3 for $n = 3$ is nine-fold degenerate in the absence of a magnetic field. In the magnetic field, this energy level is split into five levels: E_3 (triply degenerate), $E_3 + \mu_B B$ (doubly degenerate), $E_3 - \mu_B B$ (doubly degenerate), $E_3 + 2\mu_B B$ (non-degenerate), and $E_3 - 2\mu_B B$ (non-degenerate). Energy levels for s orbitals ($l = 0$) are not affected by the application of the magnetic field. Energies for p orbitals ($l = 1$) are split by the magnetic field into three levels. For d orbitals ($l = 2$), the energies are split into five levels.

This splitting of the energy levels by the magnetic field leads to the splitting of the lines in the atomic spectrum. The wave number $\tilde{\nu}$ of the spectral line corresponding to a transition between the state $|n_1 l_1 m_1\rangle$ and the state $|n_2 l_2 m_2\rangle$ is

$$\tilde{\nu} = \frac{|\Delta E|}{hc} = \mathsf{R}Z^2 \left(\frac{1}{n_1^2} - \frac{1}{n_2^2} \right) + \frac{\mu_B B}{hc}(m_2 - m_1), \qquad n_2 > n_1 \qquad (6.91)$$

Transitions between states are subject to certain restrictions called *selection rules*. The conservation of angular momentum and the parity of the spherical harmonics limit transitions for hydrogen-like atoms to those for which $\Delta l = \pm 1$ and for which $\Delta m = 0, \pm 1$. Thus, an observed spectral line $\tilde{\nu}_0$ in the absence of the magnetic field, given by equation (6.83), is split into three lines with wave numbers $\tilde{\nu}_0 + (\mu_B B/hc)$, $\tilde{\nu}_0$, and $\tilde{\nu}_0 - (\mu_B B/hc)$.

Problems

6.1 Obtain equations (6.28) from equations (6.26).

6.2 Evaluate the commutator $[\hat{A}_\lambda, \hat{B}_\lambda]$ where the operators $\hat{A}_\lambda$ and $\hat{B}_\lambda$ are those in equations (6.26).

6.3 Show explicitly by means of integration by parts that the operator $\hat{H}_l$ in equation (6.18) is hermitian for a weighting function equal to r^2.

6.4 Demonstrate by means of integration by parts that the operator $\hat{H}'_l$ in equation (6.36) is hermitian for a weighting function $w(\rho) = \rho$.

6.5 Show that $(\hat{A}_\lambda + 1)S_{\lambda+1,l} = a_{\lambda+1,l}S_{\lambda l}$ and that $(\hat{B}_\lambda + 1)S_{\lambda l} = b_{\lambda+1,l}S_{\lambda+1,l}$.

6.6 Derive equation (6.45) from equation (6.34).

6.7 Derive the relationship

$$a_{nl} \int_0^\infty S_{nl}S_{n-1,l}\rho^2 \, d\rho - b_{n+1,l} \int_0^\infty S_{nl}S_{n+1,l}\rho^2 \, d\rho = 1$$

6.8 Evaluate $\langle r^{-1} \rangle_{nl}$ for the hydrogen-like atom using the properties of associated Laguerre polynomials. First substitute equations (6.22) and (6.55) into (6.69) for $k = -1$. Then apply equations (F.22) to obtain (6.71).

6.9 From equation (F.19) with $\nu = 2$, show that

$$\int_0^\infty \rho^{2l+3} e^{-\rho} [L_{n+l}^{2l+1}(\rho)]^2 \, d\rho = \frac{2[3n^2 - l(l+1)][(n+l)!]^3}{(n-l-1)!}$$

Then show that $\langle r \rangle_{nl}$ is given by equation (6.72).

6.10 Show that $\langle r \rangle_{2s} = 6a_0/Z$ using the appropriate radial distribution function in equations (6.66).

6.11 Set $\lambda = e'$ in the Hellmann–Feynman theorem (3.71) to obtain $\langle r^{-1} \rangle_{nl}$ for the hydrogen-like atom. Note that a_0 depends on e'.

6.12 Show explicitly for a hydrogen atom in the 1s state that the total energy E_1 is equal to one-half the expectation value of the potential energy of interaction between the electron and the nucleus. This result is an example of the quantum-mechanical virial theorem.

6.13 Calculate the frequency, wavelength, and wave number for the series limit of the Balmer series of the hydrogen-atom spectral lines.

6.14 The atomic spectrum of singly ionized helium He^+ with $n_1 = 4$, $n_2 = 5, 6, \ldots$ is known as the Pickering series. Calculate the energy differences, wave numbers, and wavelengths for the first three lines in this spectrum and for the series limit.

6.15 Calculate the frequency, wavelength, and wave number of the radiation emitted from an electronic transition from the third to the first electronic level of Li^{2+}. Calculate the ionization potential of Li^{2+} in electron volts.

6.16 Derive an expression in terms of R_∞ for the difference in wavelength, $\Delta\lambda = \lambda_H - \lambda_D$, between the first line of the Balmer series ($n_1 = 2$) for a hydrogen atom and the corresponding line for a deuterium atom? Assume that the masses of the proton and the neutron are the same.

7

Spin

7.1 Electron spin

In our development of quantum mechanics to this point, the behavior of a particle, usually an electron, is governed by a wave function that is dependent only on the cartesian coordinates x, y, z or, equivalently, on the spherical coordinates r, θ, φ. There are, however, experimental observations that cannot be explained by a wave function which depends on cartesian coordinates alone.

In a quantum-mechanical treatment of an alkali metal atom, the lone valence electron may be considered as moving in the combined field of the nucleus and the core electrons. In contrast to the hydrogen-like atom, the energy levels of this valence electron are found to depend on both the principal and the azimuthal quantum numbers. The experimental spectral line pattern corresponding to transitions between these energy levels, although more complex than the pattern for the hydrogen-like atom, is readily explained. However, in a highly resolved spectrum, an additional complexity is observed; most of the spectral lines are actually composed of two lines with nearly identical wave numbers. In an alkaline-earth metal atom, which has two valence electrons, many of the lines in a highly resolved spectrum are split into three closely spaced lines. The spectral lines for the hydrogen atom, as discussed in Section 6.5, are again observed to be composed of several very closely spaced lines, with equation (6.83) giving the average wave number of each grouping. The splitting of the spectral lines in the alkali and alkaline-earth metal atoms and in hydrogen cannot be explained in terms of the quantum-mechanical postulates that are presented in Section 3.7, i.e., they cannot be explained in terms of a wave function that is dependent only on cartesian coordinates.

G. E. Uhlenbeck and S. Goudsmit (1925) explained the splitting of atomic spectral lines by postulating that the electron possesses an intrinsic angular momentum, which is called *spin*. The component of the spin angular momentum in any direction has only the value $\hbar/2$ or $-\hbar/2$. This spin angular momentum is in addition to the orbital angular momentum of the electronic motion about the nucleus. They further assumed that the spin imparts to the electron a magnetic moment of magnitude $e\hbar/2m_e$, where $-e$ and m_e are the electronic charge and mass. The interaction of an electron's magnetic moment with its orbital motion accounts for the splitting of the spectral lines in the alkali and alkaline-earth metal atoms. A combination of spin and relativistic effects is needed to explain the fine structure of the hydrogen-atom spectrum.

The concept of spin as introduced by Uhlenbeck and Goudsmit may also be applied to the Stern–Gerlach experiment, which is described in detail in Section 1.7. The explanation for the splitting of the beam of silver atoms into two separate beams by the external inhomogeneous magnetic field requires the introduction of an additional parameter to describe the behavior of the odd electron. Thus, the magnetic moment of the silver atom is attributed to the odd electron possessing an intrinsic angular momentum which can have one of only two distinct values.

Following the hypothesis of electron spin by Uhlenbeck and Goudsmit, P. A. M. Dirac (1928) developed a quantum mechanics based on the theory of relativity rather than on Newtonian mechanics and applied it to the electron. He found that the spin angular momentum and the spin magnetic moment of the electron are obtained automatically from the solution of his relativistic wave equation without any further postulates. Thus, spin angular momentum is an intrinsic property of an electron (and of other elementary particles as well) just as are the charge and rest mass.

In classical mechanics, a sphere moving under the influence of a central force has two types of angular momentum, orbital and spin. Orbital angular momentum is associated with the motion of the center of mass of the sphere about the origin of the central force. Spin angular momentum refers to the motion of the sphere about an axis through its center of mass. It is tempting to apply the same interpretation to the motion of an electron and regard the spin as the angular momentum associated with the electron revolving on its axis. However, as Dirac's relativistic quantum theory shows, the spin angular momentum is an intrinsic property of the electron, not a property arising from any kind of motion. The electron is a structureless point particle, incapable of 'spinning' on an axis. In this regard, the term 'spin' in quantum mechanics can be misleading, but its use is well-established and universal.

Prior to Dirac's relativistic quantum theory, W. Pauli (1927) showed how spin could be incorporated into non-relativistic quantum mechanics. Since the subject of relativistic quantum mechanics is beyond the scope of this book, we present in this chapter Pauli's modification of the wave-function description so as to include spin. His treatment is equivalent to Dirac's relativistic theory in the limit of small electron velocities ($v/c \rightarrow 0$).

7.2 Spin angular momentum

The postulates of quantum mechanics discussed in Section 3.7 are incomplete. In order to explain certain experimental observations, Uhlenbeck and Goudsmit introduced the concept of spin angular momentum for the electron. This concept is not contained in our previous set of postulates; an additional postulate is needed. Further, there is no reason why the property of spin should be confined to the electron. As it turns out, other particles possess an intrinsic angular momentum as well. Accordingly, we now add a sixth postulate to the previous list of quantum principles.

6. A particle possesses an intrinsic angular momentum $\mathbf{S}$ and an associated magnetic moment $\mathbf{M_s}$. This spin angular momentum is represented by a hermitian operator $\hat{\mathbf{S}}$ which obeys the relation $\hat{\mathbf{S}} \times \hat{\mathbf{S}} = i\hbar\hat{\mathbf{S}}$. Each type of particle has a fixed spin quantum number or *spin s* from the set of values $s = 0, \frac{1}{2}, 1, \frac{3}{2}, 2, \ldots$ The spin s for the electron, the proton, or the neutron has a value $\frac{1}{2}$. The spin magnetic moment for the electron is given by $\mathbf{M_s} = -e\mathbf{S}/m_e$.

As noted in the previous section, spin is a purely quantum-mechanical concept; there is no classical-mechanical analog.

The spin magnetic moment $\mathbf{M_s}$ of an electron is proportional to the spin angular momentum $\mathbf{S}$,

$$\mathbf{M_s} = -\frac{g_s e}{2m_e}\mathbf{S} = -\frac{g_s \mu_B}{\hbar}\mathbf{S} \tag{7.1}$$

where g_s is the electron *spin gyromagnetic ratio* and the Bohr magneton μ_B is defined in equation (5.82). The experimental value of g_s is 2.002 319 304 and the value predicted by Dirac's relativistic quantum theory is exactly 2. The discrepancy is removed when the theory of quantum electrodynamics is applied. We adopt the value $g_s = 2$ here. A comparison of equations (5.81) and (7.1) shows that the proportionality constant between magnetic moment and angular momentum is twice as large in the case of spin. Thus, the spin gyromagnetic ratio for the electron is twice the orbital gyromagnetic ratio. The spin gyromagnetic ratios for the proton and the neutron differ from that of the electron.

The hermitian spin operator $\hat{\mathbf{S}}$ associated with the spin angular momentum $\mathbf{S}$ has components $\hat{S}_x, \hat{S}_y, \hat{S}_z$, so that

$$\hat{S} = i\hat{S}_x + j\hat{S}_y + k\hat{S}_z$$

$$\hat{S}^2 = \hat{S}_x^2 + \hat{S}_y^2 + \hat{S}_z^2$$

These components obey the commutation relations

$$[\hat{S}_x, \hat{S}_y] = i\hbar\hat{S}_z, \qquad [\hat{S}_y, \hat{S}_z] = i\hbar\hat{S}_x, \qquad [\hat{S}_z, \hat{S}_x] = i\hbar\hat{S}_y \qquad (7.2)$$

or, equivalently

$$\hat{\mathbf{S}} \times \hat{\mathbf{S}} = i\hbar\hat{\mathbf{S}} \qquad (7.3)$$

Thus, the quantum-mechanical treatment of generalized angular momentum presented in Section 5.2 may be applied to spin angular momentum. The spin operator $\hat{\mathbf{S}}$ is identified with the operator $\hat{\mathbf{J}}$ and its components $\hat{S}_x$, $\hat{S}_y$, $\hat{S}_z$ with $\hat{J}_x$, $\hat{J}_y$, $\hat{J}_z$. Equations (5.26) when applied to spin angular momentum are

$$\hat{S}^2|sm_s\rangle = s(s+1)\hbar^2|sm_s\rangle, \qquad s = 0, \tfrac{1}{2}, 1, \tfrac{3}{2}, 2, \ldots \qquad (7.4)$$

$$\hat{S}_z|sm_s\rangle = m_s\hbar|sm_s\rangle, \qquad m_s = -s - s + 1, \ldots, s - 1, s \qquad (7.5)$$

where the quantum numbers j and m are now denoted by s and m_s. The simultaneous eigenfunctions $|sm_s\rangle$ of the hermitian operators $\hat{S}^2$ and $\hat{S}_z$ are orthonormal

$$\langle s'm_s'|sm_s\rangle = \delta_{ss'}\delta_{m_s m_s'} \qquad (7.6)$$

The raising and lowering operators for spin angular momentum as defined by equations (5.18) are

$$\hat{S}_+ \equiv \hat{S}_x + i\hat{S}_y \qquad (7.7a)$$

$$\hat{S}_- = \hat{S}_x - i\hat{S}_y \qquad (7.7b)$$

and equations (5.27) take the form

$$\hat{S}_+|sm_s\rangle = \sqrt{(s - m_s)(s + m_s + 1)}\,\hbar|s, m_s + 1\rangle \qquad (7.8a)$$

$$\hat{S}_-|sm_s\rangle = \sqrt{(s + m_s)(s - m_s + 1)}\,\hbar|s, m_s - 1\rangle \qquad (7.8b)$$

In general, the spin quantum numbers s and m_s can have integer and half-integer values. Although the corresponding orbital angular-momentum quantum numbers l and m are restricted to integer values, there is no reason for such a restriction on s and m_s.

Every type of particle has a specific unique value of s, which is called the *spin* of that particle. The particle may be elementary, such as an electron, or composite but behaving as an elementary particle, such as an atomic nucleus. All ^{4}He nuclei, for example, have spin 0; all electrons, protons, and neutrons have spin $\tfrac{1}{2}$; all photons and deuterons (^{2}H nuclei) have spin 1; etc. Particles with spins 0, 1, 2, ... are called *bosons* and those with spins $\tfrac{1}{2}$, $\tfrac{3}{2}$, ... are *fermions*. A many particle system of bosons behaves differently from a many particle system of fermions. This quantum phenomenon is discussed in Chapter 8.

The state of a particle with zero spin ($s = 0$) may be represented by a state function $\Psi(\mathbf{r}, t)$ of the spatial coordinates $\mathbf{r}$ and the time t. However, the state of a particle having spin s ($s \neq 0$) must also depend on some spin variable. We select for this spin variable the component of the spin angular momentum along the z-axis and use the quantum number m_s to designate the state. Thus, for a particle in a specific spin state, the state function is denoted by $\Psi(\mathbf{r}, m_s, t)$, where m_s has only the $(2s + 1)$ possible values $-s\hbar$, $(-s + 1)\hbar$, ..., $(s - 1)\hbar$, $s\hbar$. While the variables $\mathbf{r}$ and t have a continuous range of values, the spin variable m_s has a finite number of discrete values.

For a particle that is not in a specific spin state, we denote the spin variable by σ. A general state function $\Psi(\mathbf{r}, \sigma, t)$ for a particle with spin s may be expanded in terms of the spin eigenfunctions $|s m_s\rangle$,

$$\Psi(\mathbf{r}, \sigma, t) = \sum_{m_s = -s}^{s} \Psi(\mathbf{r}, m_s, t)|s m_s\rangle \tag{7.9}$$

If $\Psi(\mathbf{r}, \sigma, t)$ is normalized, then we have

$$\langle \Psi | \Psi \rangle = \sum_{m_s = -s}^{s} \int |\Psi(\mathbf{r}, m_s, t)|^2 \, d\mathbf{r} = 1$$

where the orthonormal relations (7.6) have been used. The quantity $|\Psi(\mathbf{r}, m_s, t)|^2$ is the probability density for finding the particle at $\mathbf{r}$ at time t with the z-component of its spin equal to $m_s \hbar$. The integral $\int |\Psi(\mathbf{r}, m_s, t)|^2 \, d\mathbf{r}$ is the probability that at time t the particle has the value $m_s \hbar$ for the z-component of its spin angular momentum.

7.3 Spin one-half

Since electrons, protons, and neutrons are the fundamental constituents of atoms and molecules and all three elementary particles have spin one-half, the case $s = \frac{1}{2}$ is the most important for studying chemical systems. For $s = \frac{1}{2}$ there are only two eigenfunctions, $|\frac{1}{2}, \frac{1}{2}\rangle$ and $|\frac{1}{2}, -\frac{1}{2}\rangle$. For convenience, the state $s = \frac{1}{2}$, $m_s = \frac{1}{2}$ is often called *spin up* and the ket $|\frac{1}{2}, \frac{1}{2}\rangle$ is written as $|\uparrow\rangle$ or as $|\alpha\rangle$. Likewise, the state $s = \frac{1}{2}$, $m_s = -\frac{1}{2}$ is called *spin down* with the ket $|\frac{1}{2}, -\frac{1}{2}\rangle$ often expressed as $|\downarrow\rangle$ or $|\beta\rangle$. Equation (7.6) gives

$$\langle \alpha | \alpha \rangle = \langle \beta | \beta \rangle = 1, \qquad \langle \alpha | \beta \rangle = 0 \tag{7.10}$$

The most general spin state $|\chi\rangle$ for a particle with $s = \frac{1}{2}$ is a linear combination of $|\alpha\rangle$ and $|\beta\rangle$

$$|\chi\rangle = c_\alpha|\alpha\rangle + c_\beta|\beta\rangle \tag{7.11}$$

where c_α and c_β are complex constants. If the ket $|\chi\rangle$ is normalized, then equation (7.10) gives

$$|c_\alpha|^2 + |c_\beta|^2 = 1$$

The ket $|\chi\rangle$ may also be expressed as a column matrix, known as a *spinor*

$$|\chi\rangle = \begin{pmatrix} c_\alpha \\ c_\beta \end{pmatrix} = c_\alpha \begin{pmatrix} 1 \\ 0 \end{pmatrix} + c_\beta \begin{pmatrix} 0 \\ 1 \end{pmatrix} \tag{7.12}$$

where the eigenfunctions $|\alpha\rangle$ and $|\beta\rangle$ in spinor notation are

$$|\alpha\rangle = \begin{pmatrix} 1 \\ 0 \end{pmatrix}, \qquad |\beta\rangle = \begin{pmatrix} 0 \\ 1 \end{pmatrix} \tag{7.13}$$

Equations (7.4), (7.5), and (7.8) for the $s = \frac{1}{2}$ case are

$$\hat{S}^2|\alpha\rangle = \tfrac{3}{4}\hbar^2|\alpha\rangle, \qquad \hat{S}^2|\beta\rangle = \tfrac{3}{4}\hbar^2|\beta\rangle \tag{7.14}$$

$$\hat{S}_z|\alpha\rangle = \tfrac{1}{2}\hbar|\alpha\rangle, \qquad \hat{S}_z|\beta\rangle = -\tfrac{1}{2}\hbar|\beta\rangle \tag{7.15}$$

$$\hat{S}_+|\alpha\rangle = 0, \qquad \hat{S}_-|\beta\rangle = 0 \tag{7.16a}$$

$$\hat{S}_+|\beta\rangle = \hbar|\alpha\rangle, \qquad \hat{S}_-|\alpha\rangle = \hbar|\beta\rangle \tag{7.16b}$$

Equations (7.16) illustrate the behavior of $\hat{S}_+$ and $\hat{S}_-$ as ladder operators. The operator $\hat{S}_+$ 'raises' the state $|\beta\rangle$ to state $|\alpha\rangle$, but cannot raise $|\alpha\rangle$ any further, while $\hat{S}_-$ 'lowers' $|\alpha\rangle$ to $|\beta\rangle$, but cannot lower $|\beta\rangle$. From equations (7.7) and (7.16), we obtain the additional relations

$$\hat{S}_x|\alpha\rangle = \tfrac{1}{2}\hbar|\beta\rangle, \qquad \hat{S}_x|\beta\rangle = \tfrac{1}{2}\hbar|\alpha\rangle \tag{7.17a}$$

$$\hat{S}_y|\alpha\rangle = \tfrac{i}{2}\hbar|\beta\rangle, \qquad \hat{S}_y|\beta\rangle = \tfrac{i}{2}\hbar|\alpha\rangle \tag{7.17b}$$

We next introduce three operators $\sigma_x, \sigma_y, \sigma_z$ which satisfy the relations

$$\hat{S}_x = \tfrac{1}{2}\hbar\sigma_x, \qquad \hat{S}_y = \tfrac{1}{2}\hbar\sigma_y, \qquad \hat{S}_z = \tfrac{1}{2}\hbar\sigma_z \tag{7.18}$$

From equations (7.15) and (7.17), we find that the only eigenvalue for each of the operators $\sigma_x^2, \sigma_y^2, \sigma_z^2$ is 1. Thus, each squared operator is just the identity operator

$$\sigma_x^2 = \sigma_y^2 = \sigma_z^2 = 1 \tag{7.19}$$

According to equations (7.2) and (7.18), the commutation rules for $\sigma_x, \sigma_y, \sigma_z$ are

$$[\sigma_x, \sigma_y] = 2i\sigma_z, \qquad [\sigma_y, \sigma_z] = 2i\sigma_x, \qquad [\sigma_z, \sigma_x] = 2i\sigma_y \tag{7.20}$$

The set of operators $\sigma_x, \sigma_y, \sigma_z$ anticommute, a property which we demonstrate for the pair σ_x, σ_y as follows

$$2i(\sigma_x\sigma_y + \sigma_y\sigma_x) = (2i\sigma_x)\sigma_y + \sigma_y(2i\sigma_x)$$
$$= (\sigma_y\sigma_z - \sigma_z\sigma_y)\sigma_y + \sigma_y(\sigma_y\sigma_z - \sigma_z\sigma_y)$$
$$= -\sigma_z\sigma_y^2 + \sigma_y^2\sigma_z$$
$$= 0$$

where the second of equations (7.20) and equation (7.19) have been used. The same procedure may be applied to the pairs σ_y, σ_z and σ_x, σ_z, giving

$$(\sigma_x\sigma_y + \sigma_y\sigma_x) = (\sigma_y\sigma_z + \sigma_z\sigma_y) = (\sigma_z\sigma_x + \sigma_x\sigma_z) = 0 \tag{7.21}$$

Combining equations (7.20) and (7.21), we also have

$$\sigma_x\sigma_y = i\sigma_z, \qquad \sigma_y\sigma_z = i\sigma_x, \qquad \sigma_z\sigma_x = i\sigma_y \tag{7.22}$$

Pauli spin matrices

An explicit set of operators σ_x, σ_y, σ_z with the foregoing properties can be formed using 2×2 matrices. The properties of matrices are discussed in Appendix I. In matrix notation, equation (7.19) is

$$\sigma_x^2 = \sigma_y^2 = \sigma_z^2 = \begin{pmatrix} 1 & 0 \\ 0 & 1 \end{pmatrix} \tag{7.23}$$

We let σ_z be represented by the simplest 2×2 matrix with eigenvalues 1 and -1

$$\sigma_z = \begin{pmatrix} 1 & 0 \\ 0 & -1 \end{pmatrix} \tag{7.24}$$

To find σ_x and σ_y, we note that

$$\begin{pmatrix} a & b \\ c & d \end{pmatrix}\begin{pmatrix} 1 & 0 \\ 0 & -1 \end{pmatrix} = \begin{pmatrix} a & -b \\ c & -d \end{pmatrix}$$

and

$$\begin{pmatrix} 1 & 0 \\ 0 & -1 \end{pmatrix}\begin{pmatrix} a & b \\ c & d \end{pmatrix} = \begin{pmatrix} a & b \\ -c & -d \end{pmatrix}$$

Since σ_x and σ_y anticommute with σ_z as represented in (7.24), we must have

$$\begin{pmatrix} a & -b \\ c & -d \end{pmatrix} = \begin{pmatrix} -a & -b \\ c & d \end{pmatrix}$$

so that $a = d = 0$ and both σ_x and σ_y have the form

$$\begin{pmatrix} 0 & b \\ c & 0 \end{pmatrix}$$

Further, we have from (7.23)

$$\sigma_x^2 = \sigma_y^2 = \begin{pmatrix} 0 & b \\ c & 0 \end{pmatrix}\begin{pmatrix} 0 & b \\ c & 0 \end{pmatrix} = \begin{pmatrix} bc & 0 \\ 0 & bc \end{pmatrix} = \begin{pmatrix} 1 & 0 \\ 0 & 1 \end{pmatrix}$$

giving the relation $bc = 1$. If we select $b = c = 1$ for σ_x, then we have

$$\sigma_x = \begin{pmatrix} 0 & 1 \\ 1 & 0 \end{pmatrix}$$

The third of equations (7.22) determines that σ_y must be

$$\sigma_y = \begin{pmatrix} 0 & -i \\ i & 0 \end{pmatrix}$$

In summary, the three matrices are

$$\sigma_x = \begin{pmatrix} 0 & 1 \\ 1 & 0 \end{pmatrix}, \qquad \sigma_y = \begin{pmatrix} 0 & -i \\ i & 0 \end{pmatrix}, \qquad \sigma_z = \begin{pmatrix} 1 & 0 \\ 0 & -1 \end{pmatrix} \qquad (7.25)$$

and are known as the *Pauli spin matrices*.

The traces of the Pauli spin matrices vanish

$$\text{Tr}\,\sigma_x = \text{Tr}\,\sigma_y = \text{Tr}\,\sigma_z = 0$$

and their determinants equal -1

$$\det \sigma_x = \det \sigma_y = \det \sigma_z = -1$$

The unit matrix $\mathbf{I}$

$$\mathbf{I} = \begin{pmatrix} 1 & 0 \\ 0 & 1 \end{pmatrix}$$

and the three Pauli spin matrices in equation (7.25) form a complete set of 2×2 matrices. Any arbitrary 2×2 matrix $\mathbf{M}$ can always be expressed as the linear combination

$$\mathbf{M} = c_1\mathbf{I} + c_2\sigma_x + c_3\sigma_y + c_4\sigma_z$$

where c_1, c_2, c_3, c_4 are complex constants.

7.4 Spin–orbit interaction

The spin magnetic moment $\mathbf{M}_s$ of an electron interacts with its orbital magnetic moment to produce an additional term in the Hamiltonian operator and, therefore, in the energy. In this section, we derive the mathematical expression for this spin–orbit interaction and apply it to the hydrogen atom.

With respect to a coordinate system with the nucleus as the origin, the electron revolves about the fixed nucleus with angular momentum $\mathbf{L}$. However, with respect to a coordinate system with the electron as the origin, the nucleus revolves around the fixed electron. Since the revolving nucleus has an electric charge, it produces at the position of the electron a magnetic field $\mathbf{B}$ parallel to $\mathbf{L}$. The interaction of the spin magnetic moment $\mathbf{M}_s$ of the electron with this magnetic field $\mathbf{B}$ gives rise to the spin–orbit coupling with energy $-\mathbf{M}_s \cdot \mathbf{B}$.

According to the Biot and Savart law of electromagnetic theory,[1] the magnetic field **B** at the 'fixed' electron due to the revolving positively charged nucleus is given in SI units to first order in v/c by

$$\mathbf{B} = \frac{1}{c^2}(\mathbf{E} \times \mathbf{v}_n) \tag{7.26}$$

where **E** is the electric field due to the revolving nucleus, $\mathbf{v}_n$ is the velocity of the nucleus relative to the electron, and c is the speed of light. The electric force **F** is related to **E** and the potential energy $V(r)$ of interaction between the nucleus and the electron by

$$\mathbf{F} = -e\mathbf{E} = -\boldsymbol{\nabla}V$$

Thus, the electric field at the electron is

$$\mathbf{E} = \frac{\mathbf{r}_n}{er}\frac{dV(r)}{dr} \tag{7.27}$$

where $\mathbf{r}_n$ is the vector distance of the nucleus from the electron. The vector **r** from nucleus to electron is $-\mathbf{r}_n$ and the velocity **v** of the electron relative to the nucleus is $-\mathbf{v}_n$. Accordingly, the angular momentum **L** of the electron is

$$\mathbf{L} = \mathbf{r} \times \mathbf{p} = m_e(\mathbf{r} \times \mathbf{v}) = m_e(\mathbf{r}_n \times \mathbf{v}_n) \tag{7.28}$$

Combining equations (7.26), (7.27), and (7.28), we have

$$\mathbf{B} = \frac{1}{em_e c^2 r}\frac{dV(r)}{dr}\mathbf{L} \tag{7.29}$$

The spin–orbit energy $-\mathbf{M}_s \cdot \mathbf{B}$ may be related to the spin and orbital angular momenta through equations (7.1) and (7.29)

$$-\mathbf{M}_s \cdot \mathbf{B} = \frac{1}{m_e^2 c^2 r}\frac{dV(r)}{dr}\mathbf{L} \cdot \mathbf{S}$$

This expression is not quite correct, however, because of a relativistic effect in changing from the perspective of the electron to the perspective of the nucleus. The correction,[2] known as the *Thomas precession*, introduces the factor $\frac{1}{2}$ on the right-hand side to give

$$-\mathbf{M}_s \cdot \mathbf{B} = \frac{1}{2m_e^2 c^2 r}\frac{dV(r)}{dr}\mathbf{L} \cdot \mathbf{S}$$

The corresponding spin–orbit Hamiltonian operator $\hat{H}_{so}$ is, then,

$$\hat{H}_{so} = \frac{1}{2m_e^2 c^2 r}\frac{dV(r)}{dr}\hat{\mathbf{L}} \cdot \hat{\mathbf{S}} \tag{7.30}$$

[1] R. P. Feyman, R. B. Leighton, and M. Sands (1964) *The Feynman Lectures on Physics*, Vol. II (Addison-Wesley, Reading, MA) section 14-7.
[2] J. D. Jackson (1975) *Classical Electrodynamics*, 2nd edition (John Wiley & Sons, New York) pp. 541–2.

For a hydrogen atom, the potential energy $V(r)$ is given by equation (6.13) with $Z = 1$ and $\hat{H}_{so}$ becomes

$$\hat{H}_{so} = \xi(r)\hat{\mathbf{L}} \cdot \hat{\mathbf{S}} \tag{7.31}$$

where

$$\xi(r) = \frac{e^2}{8\pi\varepsilon_0 m_e^2 c^2 r^3} \tag{7.32}$$

Thus, the total Hamiltonian operator $\hat{H}$ for a hydrogen atom including spin–orbit coupling is

$$\hat{H} = \hat{H}_0 + \hat{H}_{so} = \hat{H}_0 + \xi(r)\hat{\mathbf{L}} \cdot \hat{\mathbf{S}} \tag{7.33}$$

where $\hat{H}_0$ is the Hamiltonian operator for the hydrogen atom without the inclusion of spin, as given in equation (6.14).

The effect of the spin–orbit interaction term on the total energy is easily shown to be small. The angular momenta $|\mathbf{L}|$ and $|\mathbf{S}|$ are each on the order of $\hbar$ and the distance r is of the order of the radius a_0 of the first Bohr orbit. If we also neglect the small difference between the electronic mass m_e and the reduced mass μ, the spin–orbit energy is of the order of

$$\frac{e^2\hbar^2}{8\pi\varepsilon_0 m_e^2 c^2 a_0^3} = \alpha^2|E_1|$$

where $|E_1|$ is the ground-state energy for the hydrogen atom with Hamiltonian operator $\hat{H}_0$ as given by equation (6.57) and α is the *fine structure constant*, defined by

$$\alpha = \frac{e^2}{4\pi\varepsilon_0 \hbar c} = \frac{\hbar}{m_e c a_0} = \frac{1}{137.036}$$

Thus, the spin–orbit interaction energy is about 5×10^{-5} times smaller than $|E_1|$.

While the Hamiltonian operator $\hat{H}_0$ for the hydrogen atom in the absence of the spin–orbit coupling term commutes with $\hat{\mathbf{L}}$ and with $\hat{\mathbf{S}}$, the total Hamiltonian operator $\hat{H}$ in equation (7.33) does not commute with either $\hat{\mathbf{L}}$ or $\hat{\mathbf{S}}$ because of the presence of the scalar product $\hat{\mathbf{L}} \cdot \hat{\mathbf{S}}$. To illustrate this feature, we consider the commutators $[\hat{L}_z, \hat{\mathbf{L}} \cdot \hat{\mathbf{S}}]$ and $[\hat{S}_z, \hat{\mathbf{L}} \cdot \hat{\mathbf{S}}]$,

$$[\hat{L}_z, \hat{\mathbf{L}} \cdot \hat{\mathbf{S}}] = [\hat{L}_z, (\hat{L}_x\hat{S}_x + \hat{L}_y\hat{S}_y + \hat{L}_z\hat{S}_z)] = [\hat{L}_z, \hat{L}_x]\hat{S}_x + [\hat{L}_z, \hat{L}_y]\hat{S}_y + 0$$

$$= i\hbar(\hat{L}_y\hat{S}_x - \hat{L}_x\hat{S}_y) \neq 0 \tag{7.34}$$

$$[\hat{S}_z, \hat{\mathbf{L}} \cdot \hat{\mathbf{S}}] = [\hat{S}_z, \hat{S}_x]\hat{L}_x + [\hat{S}_z, \hat{S}_y]\hat{L}_y = i\hbar(\hat{L}_x\hat{S}_y - \hat{L}_y\hat{S}_x) \neq 0 \tag{7.35}$$

where equations (5.10) and (7.2) have been used. Similar expressions apply to the other components of $\hat{\mathbf{L}}$ and $\hat{\mathbf{S}}$. Thus, the vectors $\mathbf{L}$ and $\mathbf{S}$ are no longer constants of motion. However, the operators $\hat{L}^2$ and $\hat{S}^2$ do commute with $\hat{\mathbf{L}} \cdot \hat{\mathbf{S}}$, which follows from equations (5.15), so that the quantities L^2 and S^2 are still constants of motion.

We now introduce the total angular momentum $\mathbf{J}$, which is the sum of $\mathbf{L}$ and $\mathbf{S}$

$$\mathbf{J} = \mathbf{L} + \mathbf{S} \qquad (7.36)$$

The operators $\hat{\mathbf{J}}$ and $\hat{J}^2$ commute with $\hat{H}_0$. The addition of equations (7.34) and (7.35) gives

$$[\hat{J}_z, \hat{\mathbf{L}} \cdot \hat{\mathbf{S}}] = [\hat{L}_z, \hat{\mathbf{L}} \cdot \hat{\mathbf{S}}] + [\hat{S}_z, \hat{\mathbf{L}} \cdot \hat{\mathbf{S}}] = 0$$

The addition of similar relations for the x- and y-components of these angular momentum vectors leads to the result that $[\hat{\mathbf{J}}, \hat{\mathbf{L}} \cdot \hat{\mathbf{S}}] = 0$, so that $\hat{\mathbf{J}}$ and $\hat{\mathbf{L}} \cdot \hat{\mathbf{S}}$ commute. Furthermore, we may easily show that $\hat{J}^2$ commutes with $\hat{\mathbf{L}} \cdot \hat{\mathbf{S}}$ because each term in $\hat{J}^2 = \hat{L}^2 + \hat{S}^2 + 2\hat{\mathbf{L}} \cdot \hat{\mathbf{S}}$ commutes with $\hat{\mathbf{L}} \cdot \hat{\mathbf{S}}$. Thus, $\hat{\mathbf{J}}$ and $\hat{J}^2$ commute with $\hat{H}$ in equation (7.33) and $\mathbf{J}$ and J^2 are constants of motion.

That the quantities L^2, S^2, J^2, and $\mathbf{J}$ are constants of motion, but $\mathbf{L}$ and $\mathbf{S}$ are not, is illustrated in Figure 7.1. The spin magnetic moment $\mathbf{M}_s$, which is antiparallel to $\mathbf{S}$, exerts a torque on the orbital magnetic moment $\mathbf{M}$, which is antiparallel to $\mathbf{L}$, and alters its direction, but not its magnitude. Thus, the orbital angular momentum vector $\mathbf{L}$ precesses about $\mathbf{J}$ and $\mathbf{L}$ is not a constant of motion. However, since the magnitude of $\mathbf{L}$ does not change, the quantity L^2 is a constant of motion. Likewise, the orbital magnetic moment $\mathbf{M}$ exerts a torque on $\mathbf{M}_s$, causing $\mathbf{S}$ to precess about $\mathbf{J}$. The vector $\mathbf{S}$ is, then, not a constant of

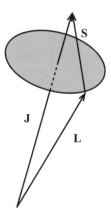

Figure 7.1 Precession of the orbital angular momentum vector $\mathbf{L}$ and the spin angular momentum vector $\mathbf{S}$ about their vector sum $\mathbf{J}$.

motion, but S^2 is. Since $\mathbf{J}$ is fixed in direction and magnitude, both $\mathbf{J}$ and J^2 are constants of motion.

If we form the cross product $\hat{\mathbf{J}} \times \hat{\mathbf{J}}$ and substitute equations (7.36), (5.11), and (7.3), we obtain

$$\hat{\mathbf{J}} \times \hat{\mathbf{J}} = (\hat{\mathbf{L}} + \hat{\mathbf{S}}) \times (\hat{\mathbf{L}} + \hat{\mathbf{S}}) = (\hat{\mathbf{L}} \times \hat{\mathbf{L}}) + (\hat{\mathbf{S}} \times \hat{\mathbf{S}}) = i\hbar\hat{\mathbf{L}} + i\hbar\hat{\mathbf{S}} = i\hbar\hat{\mathbf{J}}$$

where the cross terms $(\hat{\mathbf{L}} \times \hat{\mathbf{S}})$ and $(\hat{\mathbf{S}} \times \hat{\mathbf{L}})$ cancel each other. Thus, the operator $\hat{\mathbf{J}}$ obeys equation (5.12) and the quantum-mechanical treatment of Section 5.2 applies to the total angular momentum. Since $\hat{J}_x$, $\hat{J}_y$, and $\hat{J}_z$ each commute with $\hat{J}^2$ but do not commute with one another, we select $\hat{J}_z$ and seek the simultaneous eigenfunctions $|nlsjm_j\rangle$ of the set of mutually commuting operators $\hat{H}$, L^2, S^2, J^2, and $\hat{J}_z$

$$\hat{H}|nlsjm_j\rangle = E_n|nlsjm_j\rangle \tag{7.37a}$$

$$\hat{L}^2|nlsjm_j\rangle = l(l+1)\hbar^2|nlsjm_j\rangle \tag{7.37b}$$

$$\hat{S}^2|nlsjm_j\rangle = s(s+1)\hbar^2|nlsjm_j\rangle \tag{7.37c}$$

$$\hat{J}^2|nlsjm_j\rangle = j(j+1)\hbar^2|nlsjm_j\rangle \tag{7.37d}$$

$$\hat{J}_z|nlsjm_j\rangle = m_j\hbar|nlsjm_j\rangle, \qquad m_j = -j, -j+1, \ldots, j-1, j \tag{7.37e}$$

From the expression

$$\hat{J}_z|nlsjm_j\rangle = (\hat{L}_z = \hat{S}_z)|nlsjm_j\rangle = (m + m_s)\hbar|nlsjm_j\rangle$$

obtained from (7.36), (5.28b), and (7.5), we see that

$$m_j = m + m_s \tag{7.38}$$

The quantum number j takes on the values

$$l+s, l+s-1, l+s-2, \ldots, |l-s|$$

The argument leading to this conclusion is somewhat complicated and may be found elsewhere.[3] In the application being considered here, the spin s equals $\frac{1}{2}$ and the quantum number j can have only two values

$$j = l \pm \tfrac{1}{2} \tag{7.39}$$

The resulting vectors $\mathbf{J}$ are shown in Figure 7.2.

The scalar product $\hat{\mathbf{L}} \cdot \hat{\mathbf{S}}$ in equation (7.33) may be expressed in terms of operators that commute with $\hat{H}$ by

$$\hat{\mathbf{L}} \cdot \hat{\mathbf{S}} = \tfrac{1}{2}(\hat{\mathbf{L}} + \hat{\mathbf{S}}) \cdot (\hat{\mathbf{L}} + \hat{\mathbf{S}}) - \tfrac{1}{2}\hat{\mathbf{L}} \cdot \hat{\mathbf{L}} - \tfrac{1}{2}\hat{\mathbf{S}} \cdot \hat{\mathbf{S}} = \tfrac{1}{2}(\hat{J}^2 - \hat{L}^2 - \hat{S}^2) \tag{7.40}$$

[3] B. H. Brandsen and C. J. Joachain (1989) *Introduction to Quantum Mechanics* (Addison Wesley Longman, Harlow, Essex), pp. 299, 301; R. N. Zare (1988) *Angular Momentum* (John Wiley & Sons, New York), pp. 45–8.

Spin

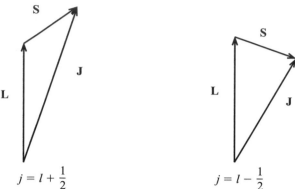

$$j = l + \frac{1}{2} \qquad\qquad j = l - \frac{1}{2}$$

Figure 7.2 The total angular momentum vectors **J** obtained from the sum of **L** and **S** for $s = \frac{1}{2}$ and $s = -\frac{1}{2}$.

so that $\hat{H}$ becomes

$$\hat{H} = \hat{H}_0 + \tfrac{1}{2}\xi(r)(\hat{J}^2 - \hat{L}^2 - \hat{S}^2) \tag{7.41}$$

Equation (7.37a) then takes the form

$$\{\hat{H}_0 + \tfrac{1}{2}\hbar^2\xi(r)[j(j+1) - l(l+1) - s(s+1)]\}|nlsjm_j\rangle = E_n|nlsjm_j\rangle \tag{7.42}$$

or

$$\left[\hat{H}_0 + \frac{l\hbar^2}{2}\xi(r)\right]|n,\ l,\ \tfrac{1}{2},\ l+\tfrac{1}{2},\ m_j\rangle = E_n|n,\ l,\ \tfrac{1}{2},\ l+\tfrac{1}{2},\ m_j\rangle \ \text{if}\ j = l + \tfrac{1}{2}$$

$$\tag{7.43a}$$

$$\left[\hat{H}_0 - \frac{(l+1)\hbar^2}{2}\xi(r)\right]|n,\ l,\ -\tfrac{1}{2},\ l-\tfrac{1}{2},\ m_j\rangle = E_n|n,\ l,\ -\tfrac{1}{2},\ l-\tfrac{1}{2},\ m_j\rangle$$

$$\text{if}\ j = l - \tfrac{1}{2} \tag{7.43b}$$

where equations (7.37b), (7.37c), (7.37d), and (7.39) have also been introduced.

Since the spin–orbit interaction energy is small, the solution of equations (7.43) to obtain E_n is most easily accomplished by means of perturbation theory, a technique which is presented in Chapter 9. The evaluation of E_n is left as a problem at the end of Chapter 9.

Problems

7.1 Determine the angle between the spin vector **S** and the z-axis for an electron in spin state $|\alpha\rangle$.

7.2 Prove equation (7.19) from equations (7.15) and (7.17).

7.3 Show that the pair of operators σ_y, σ_z anticommute.

7.4 Using the Pauli spin matrices in equation (7.25) and the spinors in (7.13),
(a) construct the operators σ_+ and σ_- corresponding to $\hat{S}_+$ and $\hat{S}_-$
(b) operate on $|\alpha\rangle$ and on $|\beta\rangle$ with σ^2, σ_z, σ_+, σ_-, σ_x, and σ_y and compare the results with equations (7.14), (7.15), (7.16), and (7.17).

7.5 Using the Pauli spin matrices in equation (7.25), verify the relationships in (7.19) and (7.22).

8

Systems of identical particles

The postulates 1 to 6 of quantum mechanics as stated in Sections 3.7 and 7.2 apply to multi-particle systems provided that each of the particles is distinguishable from the others. For example, the nucleus and the electron in a hydrogen-like atom are readily distinguishable by their differing masses and charges. When a system contains two or more identical particles, however, postulates 1 to 6 are not sufficient to predict the properties of the system. These postulates must be augmented by an additional postulate. This chapter introduces this new postulate and discusses its consequences.

8.1 Permutations of identical particles

Particles are *identical* if they cannot be distinguished one from another by any intrinsic property, such as mass, charge, or spin. There does not exist, in fact and in principle, any experimental procedure which can identify any one of the particles. In classical mechanics, even though all particles in the system may have the same intrinsic properties, each may be identified, at least in principle, by its precise trajectory as governed by Newton's laws of motion. This identification is not possible in quantum theory because each particle does not possess a trajectory; instead, the wave function gives the probability density for finding the particle at each point in space. When a particle is found to be in some small region, there is no way of determining either theoretically or experimentally which particle it is. Thus, all electrons are identical and therefore indistinguishable, as are all protons, all neutrons, all hydrogen atoms with ^{1}H nuclei, all hydrogen atoms with ^{2}H nuclei, all helium atoms with ^{4}He nuclei, all helium atoms with ^{3}He nuclei, etc.

Two-particle systems

For simplicity, we first consider a system composed of two identical particles of mass m. If we label one of the particles as particle 1 and the other as particle 2, then the Hamiltonian operator $\hat{H}(1, 2)$ for the system is

$$\hat{H}(1, 2) = \frac{\hat{\mathbf{p}}_1^2}{2m} + \frac{\hat{\mathbf{p}}_2^2}{2m} + V(\mathbf{q}_1, \mathbf{q}_2) \tag{8.1}$$

where $\mathbf{q}_i$ $(i = 1, 2)$ represents the three-dimensional (continuous) spatial coordinates $\mathbf{r}_i$ and the (discrete) spin coordinate σ_i of particle i. In order for these two identical particles to be indistinguishable from each other, the Hamiltonian operator must be *symmetric* with respect to particle interchange, i.e., if the coordinates (both spatial and spin) of the particles are interchanged, $\hat{H}(1, 2)$ must remain invariant

$$\hat{H}(1, 2) = \hat{H}(2, 1)$$

If $\hat{H}(1, 2)$ and $\hat{H}(2, 1)$ were to differ, then the corresponding Schrödinger equations and their solutions would also differ and this difference could be used to distinguish between the two particles.

The time-independent Schrödinger equation for the two-particle system is

$$\hat{H}(1, 2)\Psi_\nu(1, 2) = E_\nu\Psi_\nu(1, 2) \tag{8.2}$$

where ν delineates the various states. The notation $\Psi_\nu(1, 2)$ indicates that the first particle has coordinates $\mathbf{q}_1$ and the second particle has coordinates $\mathbf{q}_2$. If we exchange the two particles so that particles 1 and 2 now have coordinates $\mathbf{q}_2$ and $\mathbf{q}_1$, respectively, then the Schrödinger equation (8.2) becomes

$$\hat{H}(2, 1)\Psi_\nu(2, 1) = \hat{H}(1, 2)\Psi_\nu(2, 1) = E_\nu\Psi_\nu(2, 1) \tag{8.3}$$

where we have noted that $\hat{H}(1, 2)$ is symmetric. Equation (8.3) shows that $\Psi_\nu(2, 1)$ is also an eigenfunction of $\hat{H}(1, 2)$ belonging to the same eigenvalue E_ν. Thus, any linear combination of $\Psi_\nu(1, 2)$ and $\Psi_\nu(2, 1)$ is also an eigenfunction of $\hat{H}(1, 2)$ with eigenvalue E_ν. For simplicity of notation in the following presentation, we omit the index ν when it is clear that we are referring to a single quantum state.

The eigenfunction $\Psi(1, 2)$ has the form of a wave in six-dimensional space. The quantity $\Psi^*(1, 2)\Psi(1, 2)\,d\mathbf{r}_1\,d\mathbf{r}_2$ is the probability that particle 1 with spin function χ_1 is in the volume element $d\mathbf{r}_1$ centered at $\mathbf{r}_1$ and simultaneously particle 2 with spin function χ_2 is in the volume element $d\mathbf{r}_2$ at $\mathbf{r}_2$. The product $\Psi^*(1, 2)\Psi(1, 2)$ is, then, the probability density. The eigenfunction $\Psi(2, 1)$ also has the form of a six-dimensional wave. The quantity $\Psi^*(2, 1)\Psi(2, 1)$ is the probability density for particle 2 being at $\mathbf{r}_1$ with spin function χ_1 and simultaneously particle 1 being at $\mathbf{r}_2$ with spin function χ_2. In general, the two eigenfunctions $\Psi(1, 2)$ and $\Psi(2, 1)$ are not identical. As an example, if $\Psi(1, 2)$ is

$$\Psi(1, 2) = e^{-ar_1} e^{-br_2} (br_2 - 1)$$

where $r_1 = |\mathbf{r}_1|$ and $r_2 = |\mathbf{r}_2|$, then $\Psi(2, 1)$ would be

$$\Psi(2, 1) = e^{-ar_2} e^{-br_1} (br_1 - 1) \neq \Psi(1, 2)$$

Thus, the probability density of the pair of particles depends on how we label the two particles. Since the two particles are indistinguishable, we conclude that neither $\Psi(1, 2)$ nor $\Psi(2, 1)$ are desirable wave functions. We seek a wave function that does not make a distinction between the two particles and, therefore, does not designate which particle is at $\mathbf{r}_1$ and which is at $\mathbf{r}_2$.

To that end, we now introduce the linear hermitian *exchange operator* $\hat{P}$, which has the property

$$\hat{P} f(1, 2) = f(2, 1) \tag{8.4}$$

where $f(1, 2)$ is an arbitrary function of $\mathbf{q}_1$ and $\mathbf{q}_2$. If $\hat{P}$ operates on $\hat{H}(1, 2)\Psi(1, 2)$, we have

$$\hat{P}[\hat{H}(1, 2)\Psi(1, 2)] = \hat{H}(2, 1)\Psi(2, 1) = \hat{H}(1, 2)\Psi(2, 1) = \hat{H}(1, 2)\hat{P}\Psi(1, 2) \tag{8.5}$$

where we have used the fact that $\hat{H}(1, 2)$ is symmetric. From equation (8.5) we see that $\hat{P}$ and $\hat{H}(1, 2)$ commute

$$[\hat{P}, \hat{H}(1, 2)] = 0, \tag{8.6}$$

Consequently, the operators $\hat{P}$ and $\hat{H}(1, 2)$ have simultaneous eigenfunctions.

If $\Phi(1, 2)$ is an eigenfunction of $\hat{P}$, the corresponding eigenvalue λ is given by

$$\hat{P}\Phi(1, 2) = \lambda \Phi(1, 2) \tag{8.7}$$

We then have

$$\hat{P}^2\Phi(1, 2) = \hat{P}[\hat{P}\Phi(1, 2)] = \hat{P}[\lambda\Phi(1, 2)] = \lambda\hat{P}\Phi(1, 2) = \lambda^2\Phi(1, 2) \tag{8.8}$$

Moreover, operating on $\Phi(1, 2)$ twice in succession by $\hat{P}$ returns the two particles to their original order, so that

$$\hat{P}^2\Phi(1, 2) = \hat{P}\Phi(2, 1) = \Phi(1, 2) \tag{8.9}$$

From equations (8.8) and (8.9), we see that $\hat{P}^2 = 1$ and that $\lambda^2 = 1$. Since $\hat{P}$ is hermitian, the eigenvalue λ is real and we obtain $\lambda = \pm 1$.

There are only two functions which are simultaneous eigenfunctions of $\hat{H}(1, 2)$ and $\hat{P}$ with respective eigenvalues E and ± 1. These functions are the combinations

$$\Psi_S = 2^{-1/2}[\Psi(1, 2) + \Psi(2, 1)] \tag{8.10a}$$

$$\Psi_A = 2^{-1/2}[\Psi(1, 2) - \Psi(2, 1)] \tag{8.10b}$$

which satisfy the relations

$$\hat{P}\Psi_S = \Psi_S \tag{8.11a}$$

$$\hat{P}\Psi_A = -\Psi_A \tag{8.11b}$$

The factor $2^{-1/2}$ in equations (8.10) normalizes Ψ_S and Ψ_A if $\Psi(1, 2)$ is normalized. The combination Ψ_S is symmetric with respect to particle interchange because it remains unchanged when the two particles are exchanged. The function Ψ_A, on the other hand, is *antisymmetric* with respect to particle interchange because it changes sign, but is otherwise unchanged, when the particles are exchanged.

The functions Ψ_A and Ψ_S are orthogonal. To demonstrate this property, we note that the integral over all space of a function of two or more variables must be independent of the labeling of those variables

$$\int \cdots \int f(x_1, \ldots, x_N)\, dx_1 \ldots dx_N = \int \cdots \int f(y_1, \ldots, y_N)\, dy_1 \ldots dy_N \tag{8.12}$$

In particular, we have

$$\int\int f(1, 2)\, d\mathbf{q}_1\, d\mathbf{q}_2 = \int\int f(2, 1)\, d\mathbf{q}_1\, d\mathbf{q}_2$$

or

$$\langle \Psi(1, 2)|\Psi(2, 1)\rangle = \langle \Psi(2, 1)|\Psi(1, 2)\rangle \tag{8.13}$$

where $f(1, 2) - \Psi^*(1, 2)\Psi(2, 1)$. Application of equation (8.13) to $\langle \Psi_S|\Psi_A\rangle$ gives

$$\langle \Psi_S|\Psi_A\rangle = \langle \hat{P}\Psi_S|\hat{P}\Psi_A\rangle \tag{8.14}$$

Applying equations (8.11) to the right-hand side of (8.14), we obtain

$$\langle \Psi_S|\Psi_A\rangle = -\langle \Psi_S|\Psi_A\rangle$$

Thus, the scalar product $\langle \Psi_S|\Psi_A\rangle$ must vanish, showing that Ψ_A and Ψ_S are orthogonal.

If the wave function for the system is initially symmetric (antisymmetric), then it remains symmetric (antisymmetric) as time progresses. This property follows from the time-dependent Schrödinger equation

$$i\hbar \frac{\partial \Psi(1, 2)}{\partial t} = \hat{H}(1, 2)\Psi(1, 2) \tag{8.15}$$

Since $\hat{H}(1, 2)$ is symmetric, the time derivative $\partial\Psi/\partial t$ has the same symmetry as Ψ. During a small time interval Δt, therefore, the symmetry of Ψ does not change. By repetition of this argument, the symmetry remains the same over a succession of small time intervals, and by extension over all time.

Since Ψ_S does not change and only the sign of Ψ_A changes if particles 1 and 2 are interchanged, the respective probability densities $\Psi_S^*\Psi_S$ and $\Psi_A^*\Psi_A$ are independent of how the particles are labeled. Neither specifies which particle has coordinates $\mathbf{q}_1$ and which $\mathbf{q}_2$. Thus, only the linear combinations Ψ_S and Ψ_A are suitable wave functions for the two-identical-particle system. We note in passing that the two probability densities are not equal, even though Ψ_S and Ψ_A correspond to the same energy value E. We conclude that in order to incorporate into quantum theory the indistinguishability of the two identical particles, we must restrict the allowable wave functions to those that are symmetric and antisymmetric, i.e., to those that are simultaneous eigenfunctions of $\hat{H}(1, 2)$ and $\hat{P}$.

Three-particle systems

The treatment of a three-particle system introduces a new feature not present in a two-particle system. Whereas there are only two possible permutations and therefore only one exchange or permutation operator for two particles, the three-particle system requires several permutation operators.

We first label the particle with coordinates $\mathbf{q}_1$ as particle 1, the one with coordinates $\mathbf{q}_2$ as particle 2, and the one with coordinates $\mathbf{q}_3$ as particle 3. The Hamiltonian operator $\hat{H}(1, 2, 3)$ is dependent on the positions, momentum operators, and perhaps spin coordinates of each of the three particles. For identical particles, this operator must be symmetric with respect to particle interchange

$$\hat{H}(1, 2, 3) = \hat{H}(1, 3, 2) = \hat{H}(2, 3, 1) = \hat{H}(2, 1, 3) = \hat{H}(3, 1, 2) = \hat{H}(3, 2, 1)$$

If $\Psi(1, 2, 3)$ is a solution of the time-independent Schrödinger equation

$$\hat{H}(1, 2, 3)\Psi(1, 2, 3) = E\Psi(1, 2, 3) \qquad (8.16)$$

then $\Psi(1, 3, 2)$, $\Psi(2, 3, 1)$, etc., and any linear combinations of these wave functions are also solutions with the same eigenvalue E. The notation $\Psi(i, j, k)$ indicates that particle i has coordinates $\mathbf{q}_1$, particle j has coordinates $\mathbf{q}_2$, and particle k has coordinates $\mathbf{q}_3$. As in the two-particle case, we seek eigenfunctions of $\hat{H}(1, 2, 3)$ that do not specify which particle has coordinates $\mathbf{q}_i$, $i = 1, 2, 3$.

We define the six permutation operators $\hat{P}_{\alpha\beta\gamma}$ for $\alpha \neq \beta \neq \gamma = 1, 2, 3$ by the relations

$$\left. \begin{aligned} \hat{P}_{123}\Psi(i, j, k) &= \Psi(i, j, k) \\ \hat{P}_{132}\Psi(i, j, k) &= \Psi(i, k, j) \\ \hat{P}_{231}\Psi(i, j, k) &= \Psi(j, k, i) \\ \hat{P}_{213}\Psi(i, j, k) &= \Psi(j, i, k) \\ \hat{P}_{312}\Psi(i, j, k) &= \Psi(k, i, j) \\ \hat{P}_{321}\Psi(i, j, k) &= \Psi(k, j, i) \end{aligned} \right\} \quad i \neq j \neq k = 1, 2, 3 \qquad (8.17)$$

The operator $\hat{P}_{\alpha\beta\gamma}$ replaces the particle with coordinates $\mathbf{q}_1$ (the first position) by the particle with coordinates $\mathbf{q}_\alpha$, the particle with coordinates $\mathbf{q}_2$ (the second position) by that with $\mathbf{q}_\beta$, and the particle with coordinates $\mathbf{q}_3$ (the third position) by that with $\mathbf{q}_\gamma$. For example, we have

$$\hat{P}_{213}\Psi(1,\,2,\,3) = \Psi(2,\,1,\,3) \tag{8.18a}$$

$$\hat{P}_{213}\Psi(2,\,1,\,3) = \Psi(1,\,2,\,3) \tag{8.18b}$$

$$\hat{P}_{213}\Psi(3,\,2,\,1) = \Psi(2,\,3,\,1) \tag{8.18c}$$

$$\hat{P}_{231}\Psi(1,\,2,\,3) = \Psi(2,\,3,\,1) \tag{8.18d}$$

$$\hat{P}_{231}\Psi(2,\,3,\,1) = \Psi(3,\,1,\,2) \tag{8.18e}$$

The permutation operator $\hat{P}_{123}$ is an *identity operator* because it leaves the function $\Psi(i,\,j,\,k)$ unchanged. From (8.18a) and (8.18b), we obtain

$$\hat{P}_{213}^2\Psi(1,\,2,\,3) = \Psi(1,\,2,\,3)$$

so that $\hat{P}_{213}^2$ equals unity. The same relationship can be demonstrated to apply to the operators $\hat{P}_{132}$ and $\hat{P}_{321}$, as well as to the identity operator $\hat{P}_{123}$, giving

$$\hat{P}_{213}^2 = \hat{P}_{132}^2 = \hat{P}_{321}^2 = \hat{P}_{123}^2 = \hat{P}_{123} = 1 \tag{8.19}$$

Any permutation corresponding to one of the operators $\hat{P}_{\alpha\beta\gamma}$ other than $\hat{P}_{123}$ is equivalent to one or two pairwise exchanges. Accordingly, we introduce the linear hermitian exchange operators $\hat{P}_{12}$, $\hat{P}_{23}$, and $\hat{P}_{31}$ with the properties

$$\left.\begin{array}{l}\hat{P}_{12}\Psi(i,\,j,\,k) = \Psi(j,\,i,\,k) \\ \hat{P}_{23}\Psi(i,\,j,\,k) = \Psi(i,\,k,\,j) \\ \hat{P}_{31}\Psi(i,\,j,\,k) = \Psi(k,\,j,\,i)\end{array}\right\} \quad i \neq j \neq k = 1,\,2,\,3 \tag{8.20}$$

The exchange operator $\hat{P}_{\alpha\beta}$ interchanges the particles with coordinates $\mathbf{q}_\alpha$ and $\mathbf{q}_\beta$. It is obvious that the order of the subscripts in $\hat{P}_{\alpha\beta}$ is immaterial, so that $\hat{P}_{\alpha\beta} = \hat{P}_{\beta\alpha}$. The permutations from $\hat{P}_{213}$, $\hat{P}_{132}$, and $\hat{P}_{321}$ are the same as those from $\hat{P}_{12}$, $\hat{P}_{23}$, and $\hat{P}_{31}$, respectively, giving

$$\hat{P}_{213} = \hat{P}_{12}, \qquad \hat{P}_{132} = \hat{P}_{23}, \qquad \hat{P}_{321} = \hat{P}_{31}$$

The permutation from $\hat{P}_{231}$ may also be obtained by first applying the exchange operator $\hat{P}_{12}$ and then the operator $\hat{P}_{23}$. Alternatively, the same result may be obtained by first applying $\hat{P}_{23}$ followed by $\hat{P}_{31}$ or by first applying $\hat{P}_{31}$ followed by $\hat{P}_{12}$. This observation leads to the identities

$$\hat{P}_{231} = \hat{P}_{23}\hat{P}_{12} = \hat{P}_{31}\hat{P}_{23} = \hat{P}_{12}\hat{P}_{31} \tag{8.21}$$

A similar argument yields

$$\hat{P}_{312} = \hat{P}_{31}\hat{P}_{12} = \hat{P}_{23}\hat{P}_{31} = \hat{P}_{12}\hat{P}_{23} \tag{8.22}$$

These permutations of the three particles are expressed in terms of the minimum number of pairwise exchange operators. Less efficient routes can also be visualized. For example, the permutation operators $\hat{P}_{132}$ and $\hat{P}_{231}$ may also be expressed as

$$\hat{P}_{132} = \hat{P}_{31}\hat{P}_{23}\hat{P}_{12} = \hat{P}_{12}\hat{P}_{23}\hat{P}_{31}$$

$$\hat{P}_{231} = \hat{P}_{12}\hat{P}_{23}\hat{P}_{31}\hat{P}_{12} = \hat{P}_{31}\hat{P}_{12}\hat{P}_{31}\hat{P}_{12}$$

However, the number of pairwise exchanges for a given permutation is always either odd or even, so that $\hat{P}_{123}$, $\hat{P}_{231}$, $\hat{P}_{312}$ are *even permutations* and $\hat{P}_{132}$, $\hat{P}_{213}$, $\hat{P}_{321}$ are *odd permutations*.

Applying the same arguments regarding the exchange operator $\hat{P}$ for the two-particle system, we find that

$$\hat{P}_{12}^2 = \hat{P}_{23}^2 = \hat{P}_{31}^2 = 1$$

giving real eigenvalues ± 1 for each operator. We also find that each exchange operator commutes with the Hamiltonian operator $\hat{H}$

$$[\hat{P}_{12}, \hat{H}] = [\hat{P}_{23}, \hat{H}] = [\hat{P}_{31}, \hat{H}] = 0 \qquad (8.23)$$

so that $\hat{P}_{12}$ and $\hat{H}$ possess simultaneous eigenfunctions, $\hat{P}_{23}$ and $\hat{H}$ possess simultaneous eigenfunctions, and $\hat{P}_{31}$ and $\hat{H}$ possess simultaneous eigenfunctions. However, the operators $\hat{P}_{12}$, $\hat{P}_{23}$, $\hat{P}_{31}$ do not commute with each other. For example, if we operate on the wave function $\Psi(1, 2, 3)$ first with the product $\hat{P}_{31}\hat{P}_{12}$ and then with the product $\hat{P}_{12}\hat{P}_{31}$, we obtain

$$\hat{P}_{31}\hat{P}_{12}\Psi(1, 2, 3) = \hat{P}_{31}\Psi(2, 1, 3) = \Psi(3, 1, 2)$$
$$\hat{P}_{12}\hat{P}_{31}\Psi(1, 2, 3) = \hat{P}_{12}\Psi(3, 2, 1) = \Psi(2, 3, 1)$$

The wave function $\Psi(3, 1, 2)$ is not the same as $\Psi(2,3,1)$, leading to the conclusion that

$$\hat{P}_{31}\hat{P}_{12} \neq \hat{P}_{12}\hat{P}_{31}$$

Thus, a set of simultaneous eigenfunctions of $\hat{H}(1, 2, 3)$ and $\hat{P}_{12}$ and a set of simultaneous eigenfunctions of $\hat{H}(1, 2, 3)$ and $\hat{P}_{31}$ are not, in general, the same set. Likewise, neither set are simultaneous eigenfunctions of $\hat{H}(1, 2, 3)$ and $\hat{P}_{23}$.

There are, however, two eigenfunctions of $\hat{H}(1, 2, 3)$ which are also simultaneous eigenfunctions of all three pair exchange operators $\hat{P}_{12}$, $\hat{P}_{23}$, and $\hat{P}_{31}$. These eigenfunctions are Ψ_S and Ψ_A, which have the property

$$\hat{P}_{\alpha\beta}\Psi_S = \Psi_S, \qquad \alpha \neq \beta = 1, 2 \qquad (8.24a)$$

$$\hat{P}_{\alpha\beta}\Psi_A = -\Psi_A, \qquad \alpha \neq \beta = 1, 2 \qquad (8.24b)$$

To demonstrate this feature, we assume that $\Psi(1, 2, 3)$ is a simultaneous eigenfunction not only of $\hat{H}(1, 2, 3)$, but also of $\hat{P}_{12}$, $\hat{P}_{23}$, and $\hat{P}_{31}$. Therefore, we have

$$\hat{P}_{12}\Psi(1, 2, 3) = \lambda_1\Psi(1, 2, 3)$$
$$\hat{P}_{23}\Psi(1, 2, 3) = \lambda_2\Psi(1, 2, 3) \qquad (8.25)$$

$$\hat{P}_{31}\Psi(1, 2, 3) = \lambda_3\Psi(1, 2, 3)$$

where $\lambda_1 = \pm 1$, $\lambda_2 = \pm 1$, $\lambda_3 = \pm 1$ are the respective eigenvalues. From equations (8.21) and (8.25), we obtain

$$\hat{P}_{231}\Psi(1, 2, 3) = \hat{P}_{23}\hat{P}_{12}\Psi(1, 2, 3) = \hat{P}_{31}\hat{P}_{23}\Psi(1, 2, 3) = \hat{P}_{12}\hat{P}_{31}\Psi(1, 2, 3)$$

$$= \Psi(2, 3, 1)$$

or

$$\lambda_2\lambda_1\Psi(1, 2, 3) = \lambda_3\lambda_2\Psi(1, 2, 3) = \lambda_1\lambda_3\Psi(1, 2, 3)$$

from which it follows that

$$\lambda_1 = \lambda_2 = \lambda_3$$

Thus, the simultaneous eigenfunctions $\Psi(1, 2, 3)$ are either symmetric ($\lambda_1 = \lambda_2 = \lambda_3 = 1$) or antisymmetric ($\lambda_1 = \lambda_2 = \lambda_3 = -1$).

The symmetric Ψ_S or antisymmetric Ψ_A eigenfunctions may be constructed from $\Psi(1, 2, 3)$ by the relations

$$\Psi_S = 6^{-1/2}[\Psi(1, 2, 3) + \Psi(1, 3, 2) + \Psi(2, 3, 1) + \Psi(2, 1, 3) + \Psi(3, 1, 2)$$

$$+ \Psi(3, 2, 1)] \tag{8.26a}$$

$$\Psi_A = 6^{-1/2}[\Psi(1, 2, 3) - \Psi(1, 3, 2) + \Psi(2, 3, 1) - \Psi(2, 1, 3) + \Psi(3, 1, 2)$$

$$- \Psi(3, 2, 1)] \tag{8.26b}$$

where the factor $6^{-1/2}$ normalizes Ψ_S and Ψ_A if $\Psi(1, 2, 3)$ is normalized. As in the two-particle case, the functions Ψ_S and Ψ_A are orthogonal. Moreover, a wave function which is initially symmetric (antisymmetric) remains symmetric (antisymmetric) over time. The probability densities $\Psi_S^*\Psi_S$ and $\Psi_A^*\Psi_A$ are independent of how the three particles are labeled. The two functions Ψ_S and Ψ_A are, therefore, the eigenfunctions of $\hat{H}(1, 2, 3)$ that we are seeking.

Equations (8.26) may be expressed in another, equivalent way. If we let $\hat{P}$ be any one of the permutation operators $\hat{P}_{\alpha\beta\gamma}$ in equation (8.17), then we may write

$$\Psi_{S,A} = 6^{-1/2}\sum_P \delta_P\hat{P}\Psi(1, 2, 3) \tag{8.27}$$

where the summation is taken over the six different operators $\hat{P}_{\alpha\beta\gamma}$, and δ_P is either $+1$ or -1. For the symmetric wave function Ψ_S, δ_P is always $+1$, but for the antisymmetric wave function Ψ_A, δ_P is $+1$ (-1) if the permutation operator $\hat{P}$ involves the exchange of an even (odd) number of pairs of particles. Thus, δ_P is -1 for $\hat{P}_{132}$, $\hat{P}_{213}$ and $\hat{P}_{321}$.

N-particle systems

The treatment of a three-particle system may be generalized to an N-particle system. We begin by labeling the N particles, with each particle i having coordinates $\mathbf{q}_i$. For identical particles, the Hamiltonian operator must be symmetric with respect to particle permutations

$$\hat{H}(1, 2, \ldots, N) = \hat{H}(2, 1, \ldots, N) = \hat{H}(N, 2, \ldots, 1) = \cdots$$

There are $N!$ possible permutations of the N particles. If $\Psi(1, 2, \ldots, N)$ is a solution of the time-independent Schrödinger equation

$$\hat{H}(1, 2, \ldots, N)\Psi(1, 2, \ldots, N) = E\Psi(1, 2, \ldots, N) \qquad (8.28)$$

then $\Psi(2, 1, \ldots, N)$, $\Psi(N, 2, \ldots, 1)$, etc., and any linear combination of these wave functions are also solutions with eigenvalue E.

We next introduce the set of linear hermitian exchange operators $\hat{P}_{\alpha\beta}$ ($\alpha \neq \beta = 1, 2, \ldots, N$). The exchange operator $\hat{P}_{\alpha\beta}$ interchanges the pair of particles in positions α (with coordinates $\mathbf{q}_\alpha$) and β (with coordinates $\mathbf{q}_\beta$)

$$\hat{P}_{\alpha\beta}\Psi(i, \ldots, \underset{\alpha}{j}, \ldots, \underset{\beta}{k}, \ldots, l) = \Psi(i, \ldots, \underset{\alpha}{k}, \ldots, \underset{\beta}{j}, \ldots, l) \qquad (8.29)$$

As in the three-particle case, the order of the subscripts on $\hat{P}_{\alpha\beta}$ is immaterial. Since there are N choices for the first particle and $(N - 1)$ choices for the second particle ($\alpha \neq \beta$) and since each pair is to be counted only once ($\hat{P}_{\alpha\beta} = \hat{P}_{\beta\alpha}$), there are $N(N - 1)/2$ members of the set $\hat{P}_{\alpha\beta}$.

Applying the same arguments regarding the exchange operator $\hat{P}$ for the two-particle system, we find that $P_{\alpha\beta}^2 = 1$, giving real eigenvalues ± 1. We also find that $\hat{P}_{\alpha\beta}$ and $\hat{H}$ commute

$$[\hat{P}_{\alpha\beta}, \hat{H}] = 0, \qquad \alpha \neq \beta = 1, 2, \ldots, N \qquad (8.30)$$

so that they possess simultaneous eigenfunctions. However, the members of the set $\hat{P}_{\alpha\beta}$ do not commute with each other. There are only two functions, Ψ_S and Ψ_A, which are simultaneous eigenfunctions of $\hat{H}$ and all of the pairwise exchange operators $\hat{P}_{\alpha\beta}$. These two functions have the property

$$\hat{P}_{\alpha\beta}\Psi_S = \Psi_S, \qquad \alpha \neq \beta = 1, 2, \ldots, N \qquad (8.31a)$$

$$\hat{P}_{\alpha\beta}\Psi_A = -\Psi_A, \qquad \alpha \neq \beta = 1, 2, \ldots, N \qquad (8.31b)$$

and may be constructed from $\Psi(1, 2, \ldots, N)$ by the relation

$$\Psi_{S,A} = (N!)^{-1/2} \sum_P \delta_P \hat{P}\Psi(1, 2, \ldots, N) \qquad (8.32)$$

In equation (8.32) the operator $\hat{P}$ is any one of the $N!$ operators, including the identity operator, that permute a given order of particles to another order. The summation is taken over all $N!$ permutation operators. The quantity δ_P is always $+1$ for the symmetric wave function Ψ_S, but for the antisymmetric wave function Ψ_A, δ_P is $+1$ (-1) if the permutation operator $\hat{P}$ involves the exchange of an even (odd) number of particle pairs. The factor $(N!)^{-1/2}$ normalizes Ψ_S and Ψ_A if $\Psi(1, 2, \ldots, N)$ is normalized.

Using the same arguments as before, we can show that Ψ_S and Ψ_A in equation (8.32) are orthogonal and that, over time, Ψ_S remains symmetric and Ψ_A remains antisymmetric. Since the probability densities $\Psi_S^*\Psi_S$ and $\Psi_A^*\Psi_A$ are independent of how the N particles are labeled, the two functions Ψ_S and Ψ_A are the only suitable eigenfunctions of $\hat{H}(1, 2, \ldots, N)$ to represent a system of N indistinguishable particles.

8.2 Bosons and fermions

In quantum theory, identical particles must be indistinguishable in order for the theory to predict results that agree with experimental observations. Consequently, as shown in Section 8.1, the wave functions for a multi-particle system must be symmetric or antisymmetric with respect to the interchange of any pair of particles. If the wave functions are not either symmetric or antisymmetric, then the probability densities for the distribution of the particles over space are dependent on how the particles are labeled, a property that is inconsistent with indistinguishability. It turns out that these wave functions must be further restricted to be *either* symmetric *or* antisymmetric, but not both, depending on the identity of the particles.

In order to accommodate this feature into quantum mechanics, we must add a seventh postulate to the six postulates stated in Sections 3.7 and 7.2.

7. The wave function for a system of N identical particles is either symmetric or antisymmetric with respect to the interchange of any pair of the N particles. Elementary or composite particles with integral spins ($s = 0, 1, 2, \ldots$) possess symmetric wave functions, while those with half-integral spins ($s = \frac{1}{2}, \frac{3}{2}, \ldots$) possess antisymmetric wave functions.

The relationship between spin and the symmetry character of the wave function can be established in relativistic quantum theory. In non-relativistic quantum mechanics, however, this relationship must be regarded as a postulate.

As pointed out in Section 7.2, electrons, protons, and neutrons have spin $\frac{1}{2}$. Therefore, a system of N electrons, or N protons, or N neutrons possesses an antisymmetric wave function. A symmetric wave function is not allowed. Nuclei of ^{4}He and atoms of ^{4}He have spin 0, while photons and ^{2}H nuclei have spin 1. Accordingly, these particles possess symmetric wave functions, never antisymmetric wave functions. If a system is composed of several kinds of particles, then its wave function must be separately symmetric or antisymmetric with respect to each type of particle. For example, the wave function for the hydrogen molecule must be antisymmetric with respect to the interchange of the two nuclei (protons) and also antisymmetric with respect to the interchange of the two electrons. As another example, the wave function for the oxygen molecule with ^{16}O nuclei (each with spin 0) must be symmetric with respect to the interchange of the two nuclei and antisymmetric with respect to the interchange of any pair of the eight electrons.

The behavior of a multi-particle system with a symmetric wave function differs markedly from the behavior of a system with an antisymmetric wave function. Particles with integral spin and therefore symmetric wave functions satisfy *Bose–Einstein statistics* and are called *bosons*, while particles with antisymmetric wave functions satisfy *Fermi–Dirac statistics* and are called *fermions*. Systems of ^4He atoms (helium-4) and of ^3He atoms (helium-3) provide an excellent illustration. The ^4He atom is a boson with spin 0 because the spins of the two protons and the two neutrons in the nucleus and of the two electrons are paired. The ^3He atom is a fermion with spin $\frac{1}{2}$ because the single neutron in the nucleus is unpaired. Because these two atoms obey different statistics, the thermodynamic and other macroscopic properties of liquid helium-4 and liquid helium-3 are dramatically different.

8.3 Completeness relation

The completeness relation for a multi-dimensional wave function is given by equation (3.32). However, this expression does not apply to the wave functions $\Psi_{\nu S, A}$ for a system of identical particles because $\Psi_{\nu S, A}$ are either symmetric or antisymmetric, whereas the right-hand side of equation (3.32) is neither. Accordingly, we derive here[1] the appropriate expression for the completeness relation or, as it is often called, the closure property for $\Psi_{\nu S, A}$.

For compactness of notation, we introduce the $4N$-dimensional vector $\mathbf{Q}$ with components $\mathbf{q}_i$ for $i = 1, 2, \ldots, N$. The permutation operators $\hat{P}$ are allowed to operate on $\mathbf{Q}$ directly rather than on the wave functions. Thus, the expression $\hat{P}\Psi(1, 2, \ldots, N)$ is identical to $\Psi(\hat{P}\mathbf{Q})$. In this notation, equation (8.32) takes the form

$$\Psi_{\nu S, A} = (N!)^{-1/2} \sum_P \delta_P \Psi_\nu(\hat{P}\mathbf{Q}) \tag{8.33}$$

We begin by considering an arbitrary function $f(\mathbf{Q})$ of the $4N$-dimensional vector $\mathbf{Q}$. Following equation (8.33), we can construct from $f(\mathbf{Q})$ a function $F(\mathbf{Q})$ which is either symmetric or antisymmetric by the relation

---

[1] We follow the derivation of D. D. Fitts (1968) *Nuovo Cimento* **55B**, 557.

$$F(\mathbf{Q}) = (N!)^{-1/2} \sum_P \delta_P f(\hat{P}\mathbf{Q}) \tag{8.34}$$

Since $F(\mathbf{Q})$ is symmetric (antisymmetric), it may be expanded in terms of a complete set of symmetric (antisymmetric) wave functions $\Psi_\nu(\mathbf{Q})$ (we omit the subscript S, A)

$$F(\mathbf{Q}) = \sum_\nu c_\nu \Psi_\nu(\mathbf{Q}) \tag{8.35}$$

The coefficients c_ν are given by

$$c_\nu = \int \Psi_\nu^*(\mathbf{Q}') F(\mathbf{Q}') \, d\mathbf{Q}' \tag{8.36}$$

because the wave functions $\Psi_\nu(\mathbf{Q})$ are orthonormal. We use the integral notation to include summation over the spin coordinates as well as integration over the spatial coordinates. Substitution of equation (8.36) into (8.35) yields

$$F(\mathbf{Q}) = \int F(\mathbf{Q}') \left[\sum_\nu \Psi_\nu^*(\mathbf{Q}') \Psi_\nu(\mathbf{Q}) \right] d\mathbf{Q}' \tag{8.37}$$

where the order of summation and the integration over $\mathbf{Q}'$ have been interchanged. We next substitute equation (8.34) for $F(\mathbf{Q}')$ into (8.37) to obtain

$$F(\mathbf{Q}) = (N!)^{-1/2} \sum_P \delta_P \int f(\hat{P}\mathbf{Q}') \left[\sum_\nu \Psi_\nu^*(\mathbf{Q}') \Psi_\nu(\mathbf{Q}) \right] d\mathbf{Q}' \tag{8.38}$$

We now introduce the reciprocal or inverse operator $\hat{P}^{-1}$ to the permutation operator $\hat{P}$ (see Section 3.1) such that

$$\hat{P}^{-1}\hat{P} = \hat{P}\hat{P}^{-1} = 1$$

We observe that

$$\Psi_\nu(\hat{P}^{-1}\mathbf{Q}) = \delta_{P^{-1}} \Psi_\nu(\mathbf{Q}) = \delta_P \Psi_\nu(\mathbf{Q}) \tag{8.39}$$

The quantity $\delta_{P^{-1}}$ equals δ_P because both $\hat{P}^{-1}$ and $\hat{P}$ involve the interchange of the same number of particle pairs. We also note that

$$\sum_P \delta_P^2 = N! \tag{8.40}$$

because there are $N!$ terms in the summation and each term equals unity.

We next operate on each term on the right-hand side of equation (8.38) by $\hat{P}^{-1}$. Since $\hat{P}$ in equation (8.38) operates only on the variable $\mathbf{Q}'$ and since the order of integration over $\mathbf{Q}'$ is immaterial, we obtain

$$F(\mathbf{Q}) = (N!)^{-1/2} \sum_P \delta_P \int f(\mathbf{Q}') \left[\sum_\nu \Psi_\nu^*(\hat{P}^{-1}\mathbf{Q}') \Psi_\nu(\mathbf{Q}) \right] d\mathbf{Q}' \tag{8.41}$$

Application of equations (8.39) and (8.40) to (8.41) gives

$$F(\mathbf{Q}) = (N!)^{1/2} \int f(\mathbf{Q}') \left[\sum_{\nu} \Psi_{\nu}^*(\mathbf{Q}')\Psi_{\nu}(\mathbf{Q}) \right] d\mathbf{Q}' \qquad (8.42)$$

Since $f(\mathbf{Q}')$ is a completely arbitrary function of $\mathbf{Q}'$, we may compare equations (8.34) and (8.42) and obtain

$$\sum_{\nu} \Psi_{\nu}^*(\mathbf{Q}')\Psi_{\nu}(\mathbf{Q}) = (N!)^{-1} \sum_{P} \delta_P \delta(\hat{P}\mathbf{Q} - \mathbf{Q}') \qquad (8.43)$$

where $\delta(\mathbf{Q} - \mathbf{Q}')$ is the Dirac delta function

$$\delta(\mathbf{Q} - \mathbf{Q}') = \prod_{i=1}^{N} \delta(\mathbf{r}_i - \mathbf{r}_i')\delta_{\sigma_i\sigma_i'} \qquad (8.44)$$

Equation (8.43) is the completeness relation for a complete set of symmetric (antisymmetric) multi-particle wave functions.

8.4 Non-interacting particles

In this section we consider a many-particle system in which the particles act independently of each other. For such a system of N identical particles, the Hamiltonian operator $\hat{H}(1, 2, \ldots, N)$ may be written as the sum of one-particle Hamiltonian operators $\hat{H}(i)$ for $i = 1, 2, \ldots, N$

$$\hat{H}(1, 2, \ldots, N) = \hat{H}(1) + \hat{H}(2) + \cdots + \hat{H}(N) \qquad (8.45)$$

In this case, the operator $\hat{H}(1, 2, \ldots, N)$ is obviously symmetric with respect to particle interchanges. For the N particles to be identical, the operators $\hat{H}(i)$ must all have the same form, the same set of orthonormal eigenfunctions $\psi_n(i)$, and the same set of eigenvalues E_n, where

$$\hat{H}(i)\psi_n(i) = E_n\psi_n(i); \qquad i = 1, 2, \ldots, N \qquad (8.46)$$

As a consequence of equation (8.45), the eigenfunctions $\Psi_\nu(1, 2, \ldots, N)$ of $\hat{H}(1, 2, \ldots, N)$ are products of the one-particle eigenfunctions

$$\Psi_\nu(1, 2, \ldots, N) = \psi_a(1)\psi_b(2) \ldots \psi_p(N) \qquad (8.47)$$

and the eigenvalues E_ν of $\hat{H}(1, 2, \ldots, N)$ are sums of one-particle energies

$$E_\nu = E_a + E_b + \cdots + E_p \qquad (8.48)$$

In equations (8.47) and (8.48), the index ν represents the set of one-particle states $a, b, \ldots, p$ and indicates the state of the N-particle system.

The N-particle eigenfunctions $\Psi_\nu(1, 2, \ldots, N)$ in equation (8.47) are not properly symmetrized. For bosons, the wave function $\Psi_\nu(1, 2, \ldots, N)$ must be symmetric with respect to particle interchange and for fermions it must be antisymmetric. Properly symmetrized wave functions may be readily constructed by applying equation (8.32). For example, for a system of two identical particles, one particle in state ψ_a, the other in state ψ_b, the symmetrized two-particle wave functions are

$$\Psi_{ab,S}(1, 2) = 2^{-1/2}[\psi_a(1)\psi_b(2) + \psi_a(2)\psi_b(1)] \tag{8.49a}$$

$$\Psi_{ab,A}(1, 2) = 2^{-1/2}[\psi_a(1)\psi_b(2) - \psi_a(2)\psi_b(1)] \tag{8.49b}$$

The expression (8.49a) for two bosons is not quite right, however, if states ψ_a and ψ_b are the same state ($a = b$), for then the normalization constant is $\frac{1}{2}$ rather than $2^{-1/2}$, so that

$$\Psi_{aa,S}(1, 2) = \psi_a(1)\psi_a(2)$$

From equation (8.49b), we see that the wavefunction vanishes for two identical fermions in the same single-particle state

$$\Psi_{aa,A}(1, 2) = 0$$

In other words, *two identical fermions cannot simultaneously be in the same quantum state*. This statement is known as the *Pauli exclusion principle* because it was first postulated by W. Pauli (1925) in order to explain the periodic table of the elements.

For N identical non-interacting bosons, equation (8.32) needs to be modified in order for Ψ_S to be normalized when some particles are in identical single-particle states. The modified expression is

$$\Psi_S = \left(\frac{N_a!N_b!\cdots}{N!}\right)^{1/2} \sum_P \hat{P}\psi_a(1)\psi_b(2)\ldots\psi_p(N) \tag{8.50}$$

where N_n indicates the number of times the state n occurs in the product of the single-particle wave functions. Permutations which give the same product are included only once in the summation on the right-hand side of equation (8.50). For example, for three particles, with two in state a and one in state b, the products $\psi_a(1)\psi_a(2)\psi_b(3)$ and $\psi_a(2)\psi_a(1)\psi_b(3)$ are identical and only one is included in the summation.

For N identical non-interacting fermions, equation (8.32) may also be expressed as a *Slater determinant*

$$\Psi_A = (N!)^{-1/2} \begin{vmatrix} \psi_a(1) & \psi_a(2) & \cdots & \psi_a(N) \\ \psi_b(1) & \psi_b(2) & \cdots & \psi_b(N) \\ \cdots & \cdots & \cdots & \cdots \\ \psi_p(1) & \psi_p(2) & \cdots & \psi_p(N) \end{vmatrix} \tag{8.51}$$

The expansion of this determinant is identical to equation (8.32) with $\Psi(1, 2, \ldots, N)$ given by (8.47). The properties of determinants are discussed in Appendix I. The wave function Ψ_A in equation (8.51) is clearly antisymmetric because interchanging any pair of particles is equivalent to interchanging two columns and hence changes the sign of the determinant. Moreover, if any pair of particles are in the same single-particle state, then two rows of the Slater determinant are identical and the determinant vanishes, in agreement with the Pauli exclusion principle.

Although the concept of non-interacting particles is an idealization, the model may be applied to real systems as an approximation when the interactions between particles are small. Such an approximation is often useful as a starting point for more extensive calculations, such as those discussed in Chapter 9.

Probability densities

The difference in behavior between bosons and fermions is clearly demonstrated by their probability densities $|\Psi_S|^2$ and $|\Psi_A|^2$. For a pair of non-interacting bosons, we have from equation (8.49a)

$$|\Psi_S|^2 = \tfrac{1}{2}|\psi_a(1)|^2|\psi_b(2)|^2 + \tfrac{1}{2}|\psi_a(2)|^2|\psi_b(1)|^2 + \mathrm{Re}[\psi_a^*(1)\psi_b^*(2)\psi_a(2)\psi_b(1)]$$

$$(8.52)$$

For a pair of non-interacting fermions, equation (8.49b) gives

$$|\Psi_A|^2 = \tfrac{1}{2}|\psi_a(1)|^2|\psi_b(2)|^2 + \tfrac{1}{2}|\psi_a(2)|^2|\psi_b(1)|^2 - \mathrm{Re}[\psi_a^*(1)\psi_b^*(2)\psi_a(2)\psi_b(1)]$$

$$(8.53)$$

The probability density for a pair of distinguishable particles with particle 1 in state a and particle 2 in state b is $|\psi_a(1)|^2|\psi_b(2)|^2$. If the distinguishable particles are interchanged, the probability density is $|\psi_a(2)|^2|\psi_b(1)|^2$. The probability density for one distinguishable particle (either one) being in state a and the other in state b is, then

$$\tfrac{1}{2}|\psi_a(1)|^2|\psi_b(2)|^2 + \tfrac{1}{2}|\psi_a(2)|^2|\psi_b(1)|^2$$

which appears in both $|\Psi_S|^2$ and $|\Psi_A|^2$. The last term on the right-hand sides of equations (8.52) and (8.53) arises because the particles are indistinguishable and this term is known as the *exchange density* or *overlap density*. Since the exchange density is added in $|\Psi_S|^2$ and subtracted in $|\Psi_A|^2$, it is responsible for the different behavior of bosons and fermions.

The values of $|\Psi_S|^2$ and $|\Psi_A|^2$ when the two particles have the same coordinate value, say $\mathbf{q}_0$, so that $\mathbf{q}_1 = \mathbf{q}_2 = \mathbf{q}_0$, are

$$|\Psi_S|_0^2 = \tfrac{1}{2}|\psi_a(\mathbf{q}_0)|^2|\psi_b(\mathbf{q}_0)|^2 + \tfrac{1}{2}|\psi_a(\mathbf{q}_0)|^2|\psi_b(\mathbf{q}_0)|^2$$
$$+ \mathrm{Re}[\psi_a^*(\mathbf{q}_0)\psi_b^*(\mathbf{q}_0)\psi_a(\mathbf{q}_0 0\psi_b(\mathbf{q}_0)]$$
$$= 2|\psi_a(\mathbf{q}_0)|^2|\psi_b(\mathbf{q}_0)|^2$$
$$|\Psi_A|_0^2 = \tfrac{1}{2}|\psi_a(\mathbf{q}_0)|^2|\psi_b(\mathbf{q}_0)|^2 + \tfrac{1}{2}|\psi_a(\mathbf{q}_0)|^2|\psi_b(\mathbf{q}_0)|^2$$
$$- \mathrm{Re}[\psi_a^*(\mathbf{q}_0)\psi_b^*(\mathbf{q}_0)\psi_a(\mathbf{q}_0)\psi_b(\mathbf{q}_0)]$$
$$= 0$$

Thus, the two bosons have an increased probability density of being at the same point in space, while the two fermions have a vanishing probability density of being at the same point. This conclusion also applies to systems with N identical particles. Identical bosons (fermions) behave as though they are under the influence of mutually attractive (repulsive) forces. These apparent forces are called *exchange forces*, although they are not forces in the mechanical sense, but rather statistical results.

The exchange density in equations (8.52) and (8.53) is important only when the single-particle wave functions $\psi_a(\mathbf{q})$ and $\psi_b(\mathbf{q})$ overlap substantially. Suppose that the probability density $|\psi_a(\mathbf{q})|^2$ is negligibly small except in a region A and that $|\psi_b(\mathbf{q})|^2$ is negligibly small except in a region B, which does not overlap with region A. The quantities $\psi_a^*(1)\psi_b(1)$ and $\psi_b^*(2)\psi_a(2)$ are then negligibly small and the exchange density essentially vanishes. For $\mathbf{q}_1$ in region A and $\mathbf{q}_2$ in region B, only the first term $|\psi_a(1)|^2|\psi_b(2)|^2$ on the right-hand sides of equations (8.52) and (8.53) is important. This expression is just the probability density for particle 1 confined to region A and particle 2 confined to region B. The two particles become distinguishable by means of their locations and their joint wave function does not need to be made symmetric or antisymmetric. Thus, only particles whose probability densities overlap to a non-negligible extent need to be included in the symmetrization process. For example, electrons in a non-bonded atom and electrons within a molecule possess antisymmetric wave functions; electrons in neighboring atoms and molecules are too remote to be included.

Electron spin and the helium atom

We may express the single-particle wave function $\psi_n(\mathbf{q}_i)$ as the product of a spatial wave function $\phi_n(\mathbf{r}_i)$ and a spin function $\chi(i)$. For a fermion with spin $\tfrac{1}{2}$, such as an electron, there are just two spin states, which we designate by $\alpha(i)$ for $m_s = \tfrac{1}{2}$ and $\beta(i)$ for $m_s = -\tfrac{1}{2}$. Therefore, for two particles there are three symmetric spin wave functions

$$\alpha(1)\alpha(2)$$
$$\beta(1)\beta(2)$$
$$2^{-1/2}[\alpha(1)\beta(2) + \alpha(2)\beta(1)]$$

and one antisymmetric spin wave function

$$2^{-1/2}[\alpha(1)\beta(2) - \alpha(2)\beta(1)]$$

where the factors $2^{-1/2}$ are normalization constants. When the spatial and spin wave functions are combined, there are four antisymmetric combinations: a singlet state ($S = 0$)

$$\tfrac{1}{2}[\phi_a(1)\phi_b(2) + \phi_a(2)\phi_b(1)][\alpha(1)\beta(2) - \alpha(2)\beta(1)]$$

and three triplet states ($S = 1$)

$$2^{-1/2}[\phi_a(1)\phi_b(2) - \phi_a(2)\phi_b(1)]\begin{cases} \alpha(1)\alpha(2) \\ \beta(1)\beta(2) \\ 2^{-1/2}[\alpha(1)\beta(2) + \alpha(2)\beta(1)] \end{cases}$$

These four antisymmetric wave functions are normalized if the single-particle spatial wave functions $\phi_n(\mathbf{r}_i)$ are normalized. If the two fermions are in the same state $\phi_a(\mathbf{r}_i)$, then only the singlet state occurs

$$2^{-1/2}\phi_a(1)\phi_a(2)[\alpha(1)\beta(2) - \alpha(2)\beta(1)]$$

The helium atom serves as a simple example for the application of this construction. If the nucleus (for which $Z = 2$) is considered to be fixed in space, the Hamiltonian operator $\hat{H}$ for the two electrons is

$$\hat{H} = -\frac{\hbar^2}{2m_e}(\nabla_1^2 + \nabla_2^2) - \frac{Ze'^2}{r_1} - \frac{Ze'^2}{r_2} + \frac{e'^2}{r_{12}} \tag{8.54}$$

where r_1 and r_2 are the distances of electrons 1 and 2 from the nucleus, r_{12} is the distance between the two electrons, and $e' = e$ for CGS units or $e' = e/(4\pi\varepsilon_0)^{1/2}$ for SI units. Spin–orbit and spin–spin interactions of the electrons are small and have been neglected. The electron–electron interaction is relatively small in comparison with the interaction between an electron and a nucleus, so that as a crude first-order approximation the last term on the right-hand side of equation (8.54) may be neglected. The operator $\hat{H}$ then becomes the sum of two hydrogen-atom Hamiltonian operators with $Z = 2$. The corresponding single-particle states are the hydrogen-like atomic orbitals ψ_{nlm} discussed in Section 6.4. The energy of the helium atom depends on the principal quantum numbers n_1 and n_2 of the two electrons and is the sum of two hydrogen-like atomic energies with $Z = 2$

$$E_{n_1,n_2} = -\frac{m_e Z^2 e'^4}{2\hbar^2}\left(\frac{1}{n_1^2} + \frac{1}{n_2^2}\right) = -54.4\,\text{eV}\left(\frac{1}{n_1^2} + \frac{1}{n_2^2}\right)$$

In the ground state of helium, according to this model, the two electrons are in the 1s orbital with opposing spins. The ground-state wave function is

$$\Psi_0(1, 2) = 2^{-1/2} 1s(1)1s(2)[\alpha(1)\beta(2) - \alpha(2)\beta(1)]$$

and the ground-state energy is -108.8 eV. The energy of the ground state of the helium ion He^+, for which $n_1 = 1$ and $n_2 = \infty$, is -54.4 eV. In Section 9.6, we consider the contribution of the electron–electron repulsion term to the ground-state energy of helium and obtain more realistic values.

Although the orbital energies for a hydrogen-like atom depend only on the principal quantum number n, for a multi-electron atom these orbital energies increase as the azimuthal quantum number l increases. The reason is that the electron probability density near the nucleus decreases as l increases, as shown in Figure 6.5. Therefore, on average, an electron with a larger l value is screened from the attractive force of the nucleus by the inner electrons more than an electron with a smaller l value, thereby increasing its energy. Thus, the 2s orbital has a lower energy than the 2p orbitals.

Following this argument, in the first- and second-excited states, the electrons are placed in the 1s and 2s orbitals. The antisymmetric spatial wave function has the lower energy, so that the first-excited state $\Psi_1(1, 2)$ is a triplet state,

$$\Psi_1(1, 2) = 2^{-1/2}[1s(1)2s(2) - 1s(2)2s(1)] \begin{cases} \alpha(1)\alpha(2) \\ \beta(1)\beta(2) \\ 2^{-1/2}[\alpha(1)\beta(2) + \alpha(2)\beta(1)] \end{cases}$$

and the second-excited state $\Psi_2(1, 2)$ is a singlet state

$$\Psi_2(1, 2) = \tfrac{1}{2}[1s(1)2s(2) + 1s(2)2s(1)][\alpha(1)\beta(2) - \alpha(2)\beta(1)]$$

Similar constructions apply to higher excited states. The triplet states are called *orthohelium*, while the singlet states are called *parahelium*. For a given pair of atomic orbitals, the orthohelium has the lower energy. In constructing these excited states, we place one of the electrons in the 1s atomic orbital and the other in an excited atomic orbital. If both electrons were placed in excited orbitals ($n_1 \geq 2$, $n_2 \geq 2$), the resulting energy would be equal to or greater than -27.2 eV, which is greater than the energy of He^+, and the atom would ionize.

This same procedure may be used to explain, in a qualitative way, the chemical behavior of the elements in the periodic table. The application of the Pauli exclusion principle to the ground states of multi-electron atoms is discussed in great detail in most elementary textbooks on the principles of chemistry and, therefore, is not repeated here.

8.5 The free-electron gas

The concept of non-interacting fermions may be applied to electrons in a metal. A metal consists of an ordered three-dimensional array of atoms in which some of the valence electrons are so weakly bound to their parent atoms that they form an 'electron gas'. These mobile electrons then move in the Coulombic field produced by the array of ionized atoms. In addition, the mobile electrons repel each other according to Coulomb's law. For a given mobile electron, its Coulombic interactions with the ions and the other mobile electrons are long-ranged and are relatively constant over the range of the electron's position. Consequently, as a first-order approximation, the mobile electrons may be treated as a gas of identical non-interacting fermions in a constant potential energy field.

The free-electron gas was first applied to a metal by A. Sommerfeld (1928) and this application is also known as the Sommerfeld model. Although the model does not give results that are in quantitative agreement with experiments, it does predict the qualitative behavior of the electronic contribution to the heat capacity, electrical and thermal conductivity, and thermionic emission. The reason for the success of this model is that the quantum effects due to the antisymmetric character of the electronic wave function are very large and dominate the effects of the Coulombic interactions.

Each of the electrons in the free-electron gas may be regarded as a particle in a three-dimensional box, as discussed in Section 2.8. Energies may be defined relative to the constant potential energy field due to the electron–ion and electron–electron interactions in the metallic crystal, so that we may arbitrarily set this potential energy equal to zero without loss of generality. Since the mobile electrons are not allowed to leave the metal, the potential energy outside the metal is infinite. For simplicity, we assume that the metallic crystal is a cube of volume v with sides of length a, so that $v = a^3$. As given by equations (2.82) and (2.83), the single-particle wave functions and energy levels are

$$\psi_{n_x,n_y,n_z} = \sqrt{\frac{8}{v}} \sin \frac{n_x \pi x}{a} \sin \frac{n_y \pi y}{a} \sin \frac{n_z \pi z}{a} \qquad (8.55)$$

$$E_{n_x,n_y,n_z} = \frac{h^2}{8 m_e a^2} (n_x^2 + n_y^2 + n_z^2) \qquad (8.56)$$

where m_e is the electronic mass and the quantum numbers n_x, n_y, n_z have values n_x, n_y, $n_z = 1, 2, 3, \ldots$.

We next consider a three-dimensional cartesian space with axes n_x, n_y, n_z. Each point in this n-space with positive (but non-zero) integer values of n_x, n_y, and n_z corresponds to a single-particle state ψ_{n_x,n_y,n_z}. These points all lie in the positive octant of this space. If we divide the octant into unit cubic cells, every point representing a single-particle state lies at the corner of one of these unit cells. Accordingly, we may associate a volume of unit size with each single-particle state. Equation (8.56) may be rewritten in the form

$$n_x^2 + n_y^2 + n_z^2 = \frac{8m_e a^2 E}{h^2}$$

which we recognize as the equation in n-space of a sphere with radius R equal to $\sqrt{8m_e a^2 E/h^2}$. The number $\mathcal{N}(E)$ of single-particle states with energy less than or equal to E is then the volume of the octant of a sphere of radius R

$$\mathcal{N}(E) = \frac{1}{8}\frac{4\pi}{3}R^3 = \frac{\pi}{6}\left(\frac{8m_e a^2 E}{h^2}\right)^{3/2} = \frac{4\pi v}{3h^3}(2m_e E)^{3/2} \qquad (8.57)$$

The number of single-particle states with energies between E and $E + dE$ is $\omega(E)\,dE$, where $\omega(E)$ is the density of single-particle states and is related to $\mathcal{N}(E)$ by

$$\omega(E) = \frac{d\mathcal{N}(E)}{dE} = \frac{2\pi v}{h^3}(2m_e)^{3/2}E^{1/2} \qquad (8.58)$$

According to the Pauli exclusion principle, no more than two electrons, one spin up, the other spin down, can have the same set of quantum numbers n_x, n_y, n_z. At a temperature of absolute zero, two electrons can be in the ground state with energy $3h^2/8m_e a^2$, two in each of the three states with energy $6h^2/8m_e a^2$, two in each of the three states with energy $9h^2/8m_e a^2$, etc. The states with the lowest energies are filled, each with two electrons, until the spherical octant in n-space is filled up to a value E_F, which is called the *Fermi energy*. If there are N electrons in the free-electron gas, then we have

$$N = 2\mathcal{N}(E_F) = \frac{8\pi v}{3h^3}(2m_e E_F)^{3/2} \qquad (8.59)$$

or

$$E_F = \frac{h^2}{8m_e}\left(\frac{3N}{\pi v}\right)^{2/3}. \qquad (8.60)$$

where equation (8.57) has been used. The Fermi energy is dependent on the density N/v of the free-electron gas, but not on the size of the metallic crystal.

The total energy E_{tot} of the N particles is given by

$$E_{tot} = 2\int_0^{E_F} E\omega(E)\,dE \qquad (8.61)$$

where the factor 2 in front of the integral arises because each single-particle state is doubly occupied. Substitution of equation (8.58) into (8.61) gives

$$E_{tot} = \frac{8\pi \upsilon}{5h^3}(2m_e)^{3/2}E_F^{5/2}$$

which may be simplified to

$$E_{tot} = \frac{3}{5}NE_F \qquad (8.62)$$

The average energy $\overline{E}$ per electron is, then

$$\overline{E} = \frac{E_{tot}}{N} = \frac{3}{5}E_F \qquad (8.63)$$

Equations (8.57) and (8.58) are valid only for values of E sufficiently large and for energy levels sufficiently close together that E can be treated as a continuous variable. For a metallic crystal of volume 1 cm^3, the lowest energy level is about 10^{-14} eV and the spacing between levels is likewise of the order of 10^{-14} eV. Since metals typically possess about 10^{22} to 10^{23} free electrons per cm^3, the Fermi energy E_F is about 1.5 to 8 eV and the average energy $\overline{E}$ per electron is about 1 to 5 eV. Thus, for all practical purposes, the energy of the lowest level may be taken as zero and the energy values may be treated as continuous.

The smooth surface of the spherical octant in n-space which defines the Fermi energy cuts through some of the unit cubic cells that represent single-particle states. The replacement of what should be a ragged surface by a smooth surface results in a negligible difference because the density of single-particle states near the Fermi energy E_F is so large that E is essentially continuous. At the Fermi energy E_F, the density of single-particle states is

$$\omega(E_F) = \frac{2\pi\upsilon m_e}{h^2}\left(\frac{3N}{\pi\upsilon}\right)^{1/3} \qquad (8.64)$$

which typically is about 10^{22} to 10^{23} states per eV. Thus, near the Fermi energy E_F, a differential energy range dE of 10^{-10} eV contains about 10^{11} to 10^{12} doubly occupied single-particle states.

Since the potential energy of the electrons in the free-electron gas is assumed to be zero, all the energy of the mobile electrons is kinetic. The electron velocity u_F at the Fermi level E_F is given by

$$\tfrac{1}{2}m_e u_F^2 = E_F \qquad (8.65)$$

and the average electron velocity $\overline{u}$ is given by

$$\tfrac{1}{2}m_e\overline{u}^2 = \overline{E} = \tfrac{3}{5}E_F \qquad (8.66)$$

For electrons in a metal, these velocities are on the order of 10^8 cm s^{-1}.

The Fermi temperature T_F is defined by the relation

$$E_F = k_B T_F \qquad (8.67)$$

where k_B is Boltzmann's constant, and typically ranges from 18 000 K to 90 000 K for metals. At temperatures up to the melting temperature, we have the relationship

$$k_B T \ll E_F$$

Thus, even at temperatures well above absolute zero, the electrons are essentially all in the lowest possible energy states. As a result, the electronic heat capacity at constant volume, which equals dE_{tot}/dT, is small at ordinary temperatures and approaches zero at low temperatures.

The free-electron gas exerts a pressure on the walls of the infinite potential well in which it is contained. If the volume v of the gas is increased slightly by an amount dv, then the energy levels E_{n_x,n_y,n_z} in equation (8.56) decrease slightly and consequently the Fermi energy E_F in equation (8.60) and the total energy E_{tot} in (8.62) also decrease. The change in total energy of the gas is equal to the work $-P\,dv$ done on the gas by the surroundings, where P is the pressure of the gas. Thus, we have

$$P = -\frac{dE_{tot}}{dv} = -\frac{3N}{5}\frac{dE_F}{dv} = \frac{2NE_F}{5v} = \frac{2E_{tot}}{3v} \tag{8.68}$$

where equations (8.60) and (8.62) have been used. For a typical metal, the pressure P is of the order of 10^6 atm.

8.6 Bose–Einstein condensation

The behavior of a system of identical bosons is in sharp contrast to that for fermions. At low temperatures, non-interacting fermions of spin s fill the single-particle states with the lowest energies, $2s + 1$ particles in each state. Non-interacting bosons, on the other hand, have no restrictions on the number of particles that can occupy any given single-particle state. Therefore, at extremely low temperatures, all of the bosons drop into the ground single-particle state. This phenomenon is known as *Bose–Einstein condensation*.

Although A. Einstein predicted this type of behavior in 1924, only recently has Bose–Einstein condensation for weakly interacting bosons been observed experimentally. In one study,[2] a cloud of rubidium-87 atoms was cooled to a temperature of 170×10^{-9} K (170 nK), at which some of the atoms began to condense into the single-particle ground state. The condensation continued as the temperature was lowered to 20 nK, finally giving about 2000 atoms in the ground state. In other studies, small gaseous samples of sodium atoms[3] and of

[2] M. H. Anderson, J. R. Ensher, M. R. Matthews, C. E. Wieman, and E. A. Cornell (1995) *Science* **269**, 198.
[3] K. B. Davis, M.-O. Mewes, M. R. Andrews, N. J. van Druten, D. S. Durfee, D. M. Kurn, and W. Ketterle (1995) *Phys. Rev. Lett.* **75**, 3969.

lithium-7 atoms[4,5] have also been cooled sufficiently to undergo Bose–Einstein condensation.

Although we have explained Bose–Einstein condensation as a characteristic of an ideal or nearly ideal gas, i.e., a system of non-interacting or weakly interacting particles, systems of strongly interacting bosons also undergo similar transitions. Liquid helium-4, as an example, has a phase transition at 2.18 K and below that temperature exhibits very unusual behavior. The properties of helium-4 at and near this phase transition correlate with those of an ideal Bose–Einstein gas at and near its condensation temperature. Although the actual behavior of helium-4 is due to a combination of the effects of quantum statistics and interparticle forces, its qualitative behavior is related to Bose–Einstein condensation.

Problems

8.1 Show that the exchange operators $\hat{P}$ in equation (8.4) and $\hat{P}_{\alpha\beta}$ in (8.20) are hermitian.

8.2 Noting from equation (8.10) that

$$\Psi(1, 2) = 2^{-1/2}(\Psi_S + \Psi_A)$$
$$\Psi(2, 1) = 2^{-1/2}(\Psi_S - \Psi_A)$$

show that $\Psi(1, 2)$ and $\Psi(2, 1)$ are orthogonal if Ψ_S and Ψ_A are normalized.

8.3 Verify the validity of the relationships in equation (8.19).

8.4 Verify the validity of the relationships in equation (8.22).

8.5 Apply equation (8.12) to show that Ψ_S and Ψ_A in (8.26) are normalized.

8.6 Consider two identical non-interacting particles, each of mass m, in a one-dimensional box of length a. Suppose that they are in the same spin state so that spin may be ignored.

   (a) What are the four lowest energy levels, their degeneracies, and their corresponding wave functions if the two particles are distinguishable?

   (b) What are the four lowest energy levels, their degeneracies, and their corresponding wave functions if the two particles are identical fermions?

   (c) What are the four lowest energy levels, their degeneracies, and their corresponding wave functions if the two particles are identical bosons?

8.7 Consider a crude approximation to the ground state of the lithium atom in which the electron–electron repulsions are neglected. Construct the ground-state wave function in terms of the hydrogen-like atomic orbitals.

[4] C. C. Bradley, C. A. Sackett, J. J. Tollett, and R. G. Hulet (1995) *Phys. Rev. Lett.* **75**, 1687.
[5] C. C. Bradley, C. A. Sackett, and R. G. Hulet (1997) *Phys. Rev. Lett.* **78**, 985.

8.8 The atomic weight of silver is $107.9 \mathrm{\,g\,mol^{-1}}$ and its density is $10.49 \mathrm{\,g\,cm^{-3}}$. Assuming that each silver atom has one conduction electron, calculate
  (a) the Fermi energy and the average electronic energy (in joules and in eV),
  (b) the average electronic velocity,
  (c) the Fermi temperature,
  (d) the pressure of the electron gas.

8.9 The bulk modulus or modulus of compression B is defined by

$$B = -v\left(\frac{\partial P}{\partial v}\right)_T$$

Show that B for a free-electron gas is given by $B = 5P/3$.

9

Approximation methods

In the preceding chapters we solved the time-independent Schrödinger equation for a few one-particle and pseudo-one-particle systems: the particle in a box, the harmonic oscillator, the particle with orbital angular momentum, and the hydrogen-like atom. There are other one-particle systems, however, for which the Schrödinger equation cannot be solved exactly. Moreover, exact solutions of the Schrödinger equation cannot be obtained for any system consisting of two or more particles if there is a potential energy of interaction between the particles. Such systems include all atoms except hydrogen, all molecules, non-ideal gases, liquids, and solids. For this reason we need to develop approximation methods to solve the Schrödinger equation with sufficient accuracy to explain and predict the properties of these more complicated systems. Two of these approximation methods are the *variation method* and *perturbation theory*. These two methods are developed and illustrated in this chapter.

9.1 Variation method

Variation theorem

The variation method gives an approximation to the ground-state energy E_0 (the lowest eigenvalue of the Hamiltonian operator $\hat{H}$) for a system whose time-independent Schrödinger equation is

$$\hat{H}\psi_n = E_n\psi_n, \qquad n = 0, 1, 2, \ldots \tag{9.1}$$

In many applications of quantum mechanics to chemical systems, a knowledge of the ground-state energy is sufficient. The method is based on the *variation theorem*: if ϕ is *any* normalized, well-behaved function of the same variables as ψ_n and satisfies the same boundary conditions as ψ_n, then the quantity $\mathscr{E} = \langle\phi|\hat{H}|\phi\rangle$ is always greater than or equal to the ground-state energy E_0

$$\mathscr{E} \equiv \langle\phi|\hat{H}|\phi\rangle \geqslant E_0 \tag{9.2}$$

Except for the restrictions stated above, the function ϕ, called the *trial function*, is completely arbitrary. If ϕ is identical with the ground-state eigenfunction ψ_0, then of course the quantity $\mathscr{E}$ equals E_0. If ϕ is one of the excited-state eigenfunctions, then $\mathscr{E}$ is equal to the corresponding excited-state energy and is obviously greater than E_0. However, no matter what trial function ϕ is selected, the quantity $\mathscr{E}$ is never less than E_0.

To prove the variation theorem, we assume that the eigenfunctions ψ_n form a complete, orthonormal set and expand the trial function ϕ in terms of that set

$$\phi = \sum_n a_n \psi_n \tag{9.3}$$

where, according to equation (3.28)

$$a_n = \langle \psi_n | \phi \rangle \tag{9.4}$$

Since the trial function ϕ is normalized, we have

$$\langle \phi | \phi \rangle = \left\langle \sum_k a_k \psi_k \middle| \sum_n a_n \psi_n \right\rangle = \sum_k \sum_n a_k^* a_n \langle \psi_k | \psi_n \rangle$$

$$= \sum_k \sum_n a_k^* a_n \delta_{kn} = \sum_n |a_n|^2 = 1$$

We next substitute equation (9.3) into the integral for $\mathscr{E}$ in (9.2) and subtract the ground-state energy E_0, giving

$$\mathscr{E} - E_0 = \langle \phi | \hat{H} - E_0 | \phi \rangle = \sum_k \sum_n a_k^* a_n \langle \psi_k | \hat{H} - E_0 | \psi_n \rangle$$

$$= \sum_k \sum_n a_k^* a_n (E_n - E_0) \langle \psi_k | \psi_n \rangle = \sum_n |a_n|^2 (E_n - E_0) \tag{9.5}$$

where equation (9.1) has been used. Since E_n is greater than or equal to E_0 and $|a_n|^2$ is always positive or zero, we have $\mathscr{E} - E_0 \geqslant 0$ and the theorem is proved.

In the event that ϕ is not normalized, then ϕ in equation (9.2) is replaced by $A\phi$, where A is the normalization constant, and this equation becomes

$$\mathscr{E} \equiv |A|^2 \langle \phi | \hat{H} | \phi \rangle \geqslant E_0$$

The normalization relation is

$$\langle A\phi | A\phi \rangle = |A|^2 \langle \phi | \phi \rangle = 1$$

giving

$$\mathscr{E} \equiv \frac{\langle \phi | \hat{H} | \phi \rangle}{\langle \phi | \phi \rangle} \geqslant E_0 \tag{9.6}$$

In practice, the trial function ϕ is chosen with a number of parameters λ_1,

$\lambda_2, \ldots$, which can be varied. The quantity $\mathscr{E}$ is then a function of these parameters: $\mathscr{E}(\lambda_1, \lambda_2, \ldots)$. For each set of parameter values, the corresponding value of $\mathscr{E}(\lambda_1, \lambda_2, \ldots)$ is always greater than or equal to the true ground-state energy E_0. The value of $\mathscr{E}(\lambda_1, \lambda_2, \ldots)$ closest to E_0 is obtained, therefore, by minimizing $\mathscr{E}$ with respect to each of these parameters. Selecting a sufficiently large number of parameters in a well-chosen analytical form for the trial function ϕ yields an approximation very close to E_0.

Ground-state eigenfunction

If the quantity $\mathscr{E}$ is identical to the ground-state energy E_0, which is usually non-degenerate, then the trial function ϕ is identical to the ground-state eigenfunction ψ_0. This identity follows from equation (9.5), which becomes

$$\sum_{n(\neq 0)} |a_n|^2 (E_n - E_0) = 0$$

where the term for $n = 0$ vanishes because $E_n - E_0$ vanishes. This relationship is valid only if each coefficient a_n equals zero for $n \neq 0$. From equation (9.3), the normalized trial function ϕ is then equal to ψ_0. Should the ground-state energy be degenerate, then the function ϕ is identical to one of the ground-state eigenfunctions.

When the quantity $\mathscr{E}$ is not identical to E_0, we assume that the trial function ϕ which minimizes $\mathscr{E}$ is an approximation to the ground-state eigenfunction ψ_0. However, in general, $\mathscr{E}$ is a closer approximation to E_0 than ϕ is to ψ_0.

Example: particle in a box

As a simple application of the variation method to determine the ground-state energy, we consider a particle in a one-dimensional box. The Schrödinger equation for this system and its exact solution are presented in Section 2.5. The ground-state eigenfunction is shown in Figure 2.2 and is observed to have no nodes and to vanish at $x = 0$ and $x = a$. As a trial function ϕ we select

$$\phi = x(a - x), \qquad 0 \leqslant x \leqslant a$$
$$= 0, \qquad\qquad x < 0, \, x > a$$

which has these same properties. Since we have

$$\langle \phi | \phi \rangle = \int_0^a x^2 (a - x)^2 \, dx = \frac{a^5}{30}$$

the normalized trial function is

$$\phi = \frac{\sqrt{30}}{a^{5/2}} x(a - x), \qquad 0 \leqslant x \leqslant a$$

The quantity $\mathscr{E}$ is, then

$$\mathscr{E} = \langle \phi | \hat{H} | \phi \rangle = \frac{30}{a^5} \int_0^a (ax - x^2) \left(\frac{-\hbar^2}{2m} \frac{d^2}{dx^2} \right) (ax - x^2) \, dx$$

$$= \frac{30\hbar^2}{ma^5} \int_0^a (ax - x^2) \, dx = \frac{5\hbar^2}{ma^2}$$

The exact ground-state energy E_1 is shown in equation (2.39) to be $\pi^2\hbar^2/2ma^2$. Thus, we have

$$\mathscr{E} = \frac{10}{\pi^2} E_1 = 1.013 E_1 > E_1$$

giving a 1.3% error.

Example: harmonic oscillator

We next consider an example with a variable parameter. For the harmonic oscillator, discussed in Chapter 4, we select

$$\phi = e^{-cx^2}$$

as the trial function, where c is a parameter to be varied so as to minimize $\mathscr{E}(c)$. This function has no nodes and approaches zero in the limits $x \to \pm\infty$. Since the integral $\langle \phi | \phi \rangle$ is

$$\langle \phi | \phi \rangle = \int_{-\infty}^{\infty} e^{-2cx^2} \, dx = \left(\frac{\pi}{2c} \right)^{1/2}$$

where equation (A.5) is used, the normalized trial function is

$$\phi = \left(\frac{2c}{\pi} \right)^{1/4} e^{-cx^2}$$

The Hamiltonian operator $\hat{H}$ for the harmonic oscillator is given in equation (4.12). The quantity $\mathscr{E}(c)$ is then determined as follows

$$\mathscr{E}(c) = -\left(\frac{2c}{\pi} \right)^{1/2} \frac{\hbar^2}{2m} \int_{-\infty}^{\infty} e^{-cx^2} \frac{d^2}{dx^2} e^{-cx^2} \, dx + \left(\frac{2c}{\pi} \right)^{1/2} \frac{m\omega^2}{2} \int_{-\infty}^{\infty} x^2 e^{-2cx^2} \, dx$$

$$= \left(\frac{2c}{\pi} \right)^{1/2} \frac{\hbar^2 c}{m} \int_{-\infty}^{\infty} (1 - 2cx^2) e^{-2cx^2} \, dx + \left(\frac{c}{2\pi} \right)^{1/2} m\omega^2 \int_{-\infty}^{\infty} x^2 e^{-2cx^2} \, dx$$

$$= \left(\frac{2c}{\pi} \right)^{1/2} \frac{\hbar^2 c}{m} \left[\left(\frac{\pi}{2c} \right)^{1/2} - c \left(\frac{\pi}{8c^3} \right)^{1/2} \right] + \left(\frac{c}{2\pi} \right)^{1/2} \frac{m\omega^2}{2} \left(\frac{\pi}{8c^3} \right)^{1/2}$$

$$= \frac{\hbar^2 c}{2m} + \frac{m\omega^2}{8c}$$

where equations (A.5) and (A.7) have been used.

To find the minimum value of $\mathscr{E}(c)$, we set the derivative $d\mathscr{E}/dc$ equal to zero and obtain

$$\frac{d\mathscr{E}}{dc} = \frac{\hbar^2}{2m} - \frac{m\omega^2}{8c^2} = 0$$

so that

$$c = \frac{m\omega}{2\hbar}$$

We have taken the positive square root because the parameter c must be positive for ϕ to be well-behaved. The best estimate of the ground-state energy is then

$$\mathscr{E} = \frac{\hbar^2}{2m}\left(\frac{m\omega}{2\hbar}\right) + \frac{m\omega^2}{8}\left(\frac{2\hbar}{m\omega}\right) = \frac{\hbar\omega}{2}$$

which is the exact result.

The reason why we obtain the exact ground-state energy in this simple example is that the trial function ϕ has the same mathematical form as the exact ground-state eigenfunction, given by equation (4.39). When the parameter c is evaluated to give a minimum value for $\mathscr{E}$, the function ϕ becomes identical to the exact eigenfunction.

Excited-state energies

The variation theorem may be extended in some cases to estimate the energies of excited states. Under special circumstances it may be possible to select a trial function ϕ for which the first few coefficients in the expansion (9.3) vanish: $a_0 = a_1 = \cdots = a_{k-1} = 0$, in which case we have

$$\phi = \sum_{n(\geq k)} a_n \psi_n$$

and

$$\sum_{n(\geq k)} |a_n|^2 = 1$$

We assume here that the eigenfunctions ψ_n in equation (9.1) are labeled in order of increasing energy, so that

$$E_0 \leq E_1 \leq E_2 \leq \cdots$$

Following the same procedure used to prove the variation theorem, we obtain

$$\mathscr{E} - E_k = \sum_{n(\geq k)} |a_n|^2 (E_n - E_k)$$

from which it follows that

$$\mathscr{E} \geq E_k \tag{9.7}$$

Thus, the quantity $\mathscr{E}$ is an upper bound to the energy E_k corresponding to the state ψ_k. For situations in which ϕ can be made orthogonal to each exact eigenfunction ψ_0, ψ_1, ..., ψ_{k-1}, the coefficients a_0, a_1, ..., a_{k-1} vanish according to equation (9.4) and the inequality (9.7) applies.

An example is a one-dimensional system for which the potential energy $V(x)$ is an even function of the position variable x. The eigenfunction ψ_0 with the lowest eigenvalue E_0 has no nodes and therefore must be an even function of x. The eigenfunction ψ_1 has one node, located at the origin, and therefore must be an odd function of x. If we select for ϕ any odd function of x, then ϕ is orthogonal to any even function of x, including ψ_0, and the coefficient a_0 vanishes. Thus, the integral $\mathscr{E} = \langle \phi | \hat{H} | \phi \rangle$ gives an upper bound to E_1 even though the ground-state eigenfunction ψ_0 may not be known.

When the exact eigenfunctions ψ_0, ψ_1, ..., ψ_{k-1} are not known, they may be approximated by trial functions ϕ_0, ϕ_1, ..., ϕ_{k-1} which successively give upper bounds for E_0, E_1, ..., E_{k-1}, respectively. In this case, the function ϕ_1 is constructed to be orthogonal to ϕ_0, ϕ_2 constructed orthogonal to both ϕ_0 and ϕ_1, and so forth. In general, this method is difficult to apply and gives increasingly less accurate results with increasing n.

9.2 Linear variation functions

A convenient and widely used form for the trial function ϕ is the linear variation function

$$\phi = \sum_{i=1}^{N} c_i \chi_i \tag{9.8}$$

where χ_1, χ_2, ..., χ_N are an incomplete set of linearly independent functions which have the same variables and which satisfy the same boundary conditions as the exact eigenfunctions ψ_n of equation (9.1). The functions χ_i are selected to be real and are not necessarily orthogonal to one another. Thus, the *overlap integral* S_{ij}, defined as

$$S_{ij} \equiv \langle \chi_i | \chi_j \rangle \tag{9.9}$$

is not generally equal to δ_{ij}. The coefficients c_i are also restricted to real values and are variation parameters to be determined by the minimization of the variation integral $\mathscr{E}$.

If we substitute equation (9.8) into (9.6) and define H_{ij} by

$$H_{ij} \equiv \langle \chi_i | \hat{H} | \chi_j \rangle \tag{9.10}$$

we obtain

$$\mathscr{E} = \frac{\displaystyle\sum_{i=1}^{N}\sum_{j=1}^{N} c_i c_j H_{ij}}{\displaystyle\sum_{i=1}^{N}\sum_{j=1}^{N} c_i c_j S_{ij}}$$

or

$$\mathscr{E} \sum_{i=1}^{N}\sum_{j=1}^{N} c_i c_j S_{ij} = \sum_{i=1}^{N}\sum_{j=1}^{N} c_i c_j H_{ij} \tag{9.11}$$

To find the values of the parameters c_i in equation (9.8) which minimize $\mathscr{E}$, we differentiate equation (9.11) with respect to each coefficient c_k ($k = 1, 2, \ldots, N$)

$$\frac{\partial \mathscr{E}}{\partial c_k} \sum_{i=1}^{N}\sum_{j=1}^{N} c_i c_j S_{ij} + \mathscr{E} \frac{\partial}{\partial c_k}\left(\sum_{i=1}^{N}\sum_{j=1}^{N} c_i c_j S_{ij}\right) = \frac{\partial}{\partial c_k}\left(\sum_{i=1}^{N}\sum_{j=1}^{N} c_i c_j H_{ij}\right)$$

and set $(\partial \mathscr{E}/\partial c_k) = 0$ for each value of k. The first term on the left-hand side vanishes. The remaining two terms may be combined to give

$$\frac{\partial}{\partial c_k}\left(\sum_{i=1}^{N}\sum_{j=1}^{N} c_i c_j (H_{ij} - \mathscr{E}S_{ij})\right) = \sum_{i=1}^{N}\sum_{j=1}^{N}\left(\frac{\partial c_i}{\partial c_k} c_j + c_i \frac{\partial c_j}{\partial c_k}\right)(H_{ij} - \mathscr{E}S_{ij})$$

$$= \sum_{i=1}^{N}\sum_{j=1}^{N}(\delta_{ik} c_j + c_i \delta_{jk})(H_{ij} - \mathscr{E}S_{ij})$$

$$= \sum_{j=1}^{N} c_j(H_{kj} - \mathscr{E}S_{kj}) + \sum_{i=1}^{N} c_i(H_{ik} - \mathscr{E}S_{ik})$$

$$= 0$$

where we have noted that $(\partial c_i/\partial c_k) = \delta_{ik}$ because the coefficients c_i in equation (9.8) are independent of each other. If we replace the dummy index j by i and note that $H_{ik} = H_{ki}$ and $S_{ik} = S_{ki}$ because the functions χ_i are real, we obtain a set of N linear homogeneous simultaneous equations

$$\sum_{i=1}^{N} c_i(H_{ki} - \mathscr{E}S_{ki}) = 0, \qquad k = 1, 2, \ldots, N \tag{9.12}$$

Equation (9.12) has the form

$$\sum_{i=1}^{N} a_{ki} x_i = 0, \qquad k = 1, 2, \ldots, N \qquad (9.13)$$

for which a trivial solution is $x_i = 0$ for all i. A non-trivial solution exists if, and only if, the determinant of the coefficients a_{ki} vanishes

$$\begin{vmatrix} a_{11} & a_{12} & \cdots & a_{1N} \\ a_{21} & a_{22} & \cdots & a_{2N} \\ \cdots & \cdots & \cdots & \cdots \\ a_{N1} & a_{N2} & \cdots & a_{NN} \end{vmatrix} = 0$$

This determinant or its equivalent algebraic expansion is known as the *secular equation*. In equation (9.12) the parameters c_i correspond to the unknown quantities x_i in equation (9.13) and the terms $(H_{ki} - \mathscr{E} S_{ki})$ correspond to the coefficients a_{ki}. Thus, a non-trivial solution for the N parameters c_i exists only if the determinant with elements $(H_{ki} - \mathscr{E} S_{ki})$ vanishes

$$\begin{vmatrix} H_{11} - \mathscr{E} S_{11} & H_{12} - \mathscr{E} S_{12} & \cdots & H_{1N} - \mathscr{E} S_{1N} \\ H_{21} - \mathscr{E} S_{21} & H_{22} - \mathscr{E} S_{22} & \cdots & H_{2N} - \mathscr{E} S_{2N} \\ \cdots & \cdots & \cdots & \cdots \\ H_{N1} - \mathscr{E} S_{N1} & H_{N2} - \mathscr{E} S_{N2} & \cdots & H_{NN} - \mathscr{E} S_{NN} \end{vmatrix} = 0 \qquad (9.14)$$

The secular equation (9.14) is satisfied only for certain values of $\mathscr{E}$. Since this equation is of degree N in $\mathscr{E}$, there are N real roots

$$\mathscr{E}_0 \le \mathscr{E}_1 \le \mathscr{E}_2 \le \cdots \le \mathscr{E}_{N-1}$$

According to the variation theorem, the lowest root $\mathscr{E}_0$ is an upper bound to the ground-state energy E_0: $E_0 \le \mathscr{E}_0$. The other roots may be shown[1] to be upper bounds for the excited-state energy levels

$$E_1 \le \mathscr{E}_1, E_2 \le \mathscr{E}_2, \ldots, E_{N-1} \le \mathscr{E}_{N-1}$$

9.3 Non-degenerate perturbation theory

Perturbation theory provides a procedure for finding approximate solutions to the Schrödinger equation for a system which differs only slightly from a system for which the solutions are known. The Hamiltonian operator $\hat{H}$ for the system of interest is given by

$$\hat{H} = -\frac{\hbar^2}{2} \sum_{i=1}^{N} \frac{1}{m_i} \nabla_i^2 + V(\mathbf{r}_1, \mathbf{r}_2, \ldots, \mathbf{r}_N)$$

[1] J. K. L. MacDonald (1933) *Phys. Rev.* **43**, 830.

where N is the number of particles in the system. We suppose that the Schrödinger equation for this Hamiltonian operator

$$\hat{H}\psi_n = E_n\psi_n \tag{9.15}$$

cannot be solved exactly by known mathematical techniques.

In perturbation theory we assume that $\hat{H}$ may be expanded in terms of a small parameter λ

$$\hat{H} = \hat{H}^{(0)} + \lambda\hat{H}^{(1)} + \lambda^2\hat{H}^{(2)} + \cdots = \hat{H}^{(0)} + \hat{H}' \tag{9.16}$$

where

$$\hat{H}' = \lambda\hat{H}^{(1)} + \lambda^2\hat{H}^{(2)} + \cdots \tag{9.17}$$

The quantity $\hat{H}^{(0)}$ is the *unperturbed* Hamiltonian operator whose orthonormal eigenfunctions $\psi_n^{(0)}$ and eigenvalues $E_n^{(0)}$ are known exactly, so that

$$\hat{H}^{(0)}\psi_n^{(0)} = E_n^{(0)}\psi_n^{(0)} \tag{9.18}$$

The operator $\hat{H}'$ is called the *perturbation* and is small. Thus, the operator $\hat{H}$ differs only slightly from $\hat{H}^{(0)}$ and the eigenfunctions and eigenvalues of $\hat{H}$ do not differ greatly from those of the unperturbed Hamiltonian operator $\hat{H}^{(0)}$. The parameter λ is introduced to facilitate the comparison of the orders of magnitude of various terms. In the limit $\lambda \to 0$, the *perturbed system* reduces to the *unperturbed system*. For many systems there are no terms in the perturbed Hamiltonian operator higher than $\hat{H}^{(1)}$ and for convenience the parameter λ in equations (9.16) and (9.17) may then be set equal to unity.

The mathematical procedure that we present here for solving equation (9.15) is known as *Rayleigh–Schrödinger perturbation theory*. There are other procedures, but they are seldom used. In the Rayleigh–Schrödinger method, the eigenfunctions ψ_n and the eigenvalues E_n are expanded as power series in λ

$$\psi_n = \psi_n^{(0)} + \lambda\psi_n^{(1)} + \lambda^2\psi_n^{(2)} + \cdots \tag{9.19}$$

$$E_n = E_n^{(0)} + \lambda E_n^{(1)} + \lambda^2 E_n^{(2)} + \cdots \tag{9.20}$$

The quantities $\psi_n^{(1)}$ and $E_n^{(1)}$ are the *first-order corrections* to ψ_n and E_n, the quantities $\psi_n^{(2)}$ and $E_n^{(2)}$ are the *second-order corrections*, and so forth. If the perturbation $\hat{H}'$ is small, then equations (9.19) and (9.20) converge rapidly for all values of λ where $0 \leq \lambda \leq 1$.

We next substitute the expansions (9.16), (9.19), and (9.20) into equation (9.15) and collect coefficients of like powers of λ to obtain

$$\hat{H}^{(0)}\psi_n^{(0)} + \lambda(\hat{H}^{(1)}\psi_n^{(0)} + \hat{H}^{(0)}\psi_n^{(1)}) + \lambda^2(\hat{H}^{(2)}\psi_n^{(0)} + \hat{H}^{(1)}\psi_n^{(1)} + \hat{H}^{(0)}\psi_n^{(2)}) + \cdots$$
$$= E_n^{(0)}\psi_n^{(0)} + \lambda(E_n^{(1)}\psi_n^{(0)} + E_n^{(0)}\psi_n^{(1)}) + \lambda^2(E_n^{(2)}\psi_n^{(0)} + E_n^{(1)}\psi_n^{(1)} + E_n^{(0)}\psi_n^{(2)}) + \cdots$$

$$\tag{9.21}$$

This equation has the form

$$f(\varepsilon) = \sum_k b_k \varepsilon^k = 0$$

where

$$b_k = \frac{1}{k!}\left(\frac{\partial^k f}{\partial \varepsilon^k}\right)_{\varepsilon=0}$$

Since $f(\varepsilon)$ is identically zero, the coefficients b_k all vanish. Thus, the coefficients of λ^k on the left-hand side of equation (9.21) are equal to the coefficients of λ^k on the right-hand side. The coefficients of λ^0 give equation (9.18) for the unperturbed system. The coefficients of λ yield

$$(\hat{H}^{(0)} - E_n^{(0)})\psi_n^{(1)} = -(\hat{H}^{(1)} - E_n^{(1)})\psi_n^{(0)} \tag{9.22}$$

while the coefficients of λ^2 give

$$(\hat{H}^{(0)} - E_n^{(0)})\psi_n^{(2)} + (\hat{H}^{(1)} - E_n^{(1)})\psi_n^{(1)} = -(\hat{H}^{(2)} - E_n^{(2)})\psi_n^{(0)} \tag{9.23}$$

and so forth.

First-order corrections

To find the first-order correction $E_n^{(1)}$ to the eigenvalue E_n, we multiply equation (9.22) by the complex conjugate of $\psi_n^{(0)}$ and integrate over all space to obtain

$$\langle\psi_n^{(0)}|\hat{H}^{(0)}|\psi_n^{(1)}\rangle - E_n^{(0)}\langle\psi_n^{(0)}|\psi_n^{(1)}\rangle = -\langle\psi_n^{(0)}|\hat{H}^{(1)}|\psi_n^{(0)}\rangle + E_n^{(1)}$$

where we have noted that $\psi_n^{(0)}$ is normalized. Since $\hat{H}^{(0)}$ is hermitian, the first integral on the left-hand side takes the form

$$\langle\psi_n^{(0)}|\hat{H}^{(0)}|\psi_n^{(1)}\rangle = \langle\hat{H}^{(0)}\psi_n^{(0)}|\psi_n^{(1)}\rangle = E_n^{(0)}\langle\psi_n^{(0)}|\psi_n^{(1)}\rangle$$

and therefore cancels the second integral on the left-hand side. The first-order correction $E_n^{(1)}$ is, then, the expectation value of the perturbation $\hat{H}^{(1)}$ in the unperturbed state

$$E_n^{(1)} = \langle\psi_n^{(0)}|\hat{H}^{(1)}|\psi_n^{(0)}\rangle \equiv \hat{H}_{nn}^{(1)} \tag{9.24}$$

The first-order correction $\psi_n^{(1)}$ to the eigenfunction is obtained by multiplying equation (9.22) by the complex conjugate of $\psi_k^{(0)}$ for $k \neq n$ and integrating over all space to give

$$\langle\psi_k^{(0)}|\hat{H}^{(0)}|\psi_n^{(1)}\rangle - E_n^{(0)}\langle\psi_k^{(0)}|\psi_n^{(1)}\rangle = -\langle\psi_k^{(0)}|\hat{H}^{(1)}|\psi_n^{(0)}\rangle + E_n^{(1)}\langle\psi_k^{(0)}|\psi_n^{(0)}\rangle \tag{9.25}$$

Noting that the unperturbed eigenfunctions are orthogonal

$$\langle\psi_k^{(0)}|\psi_n^{(0)}\rangle = \delta_{kn} \tag{9.26}$$

applying the hermitian property of $\hat{H}^{(0)}$ to the first term on the left-hand side, and writing $H_{kn}^{(1)}$ for $\langle \psi_k^{(0)} | \hat{H}^{(1)} | \psi_n^{(0)} \rangle$, we may express equation (9.25) as

$$(E_k^{(0)} - E_n^{(0)})\langle \psi_k^{(0)} | \psi_n^{(1)} \rangle = -\hat{H}_{kn}^{(1)} \tag{9.27}$$

The orthonormal eigenfunctions $\psi_j^{(0)}$ for the unperturbed system are assumed to form a complete set. Thus, the perturbation corrections $\psi_n^{(1)}$ may be expanded in terms of the set $\psi_j^{(0)}$

$$\psi_n^{(1)} = \sum_j a_{nj} \psi_j^{(0)} = a_{nn} \psi_j^{(0)} + \sum_{j(\neq n)} a_{nj} \psi_j^{(0)}$$

where a_{nj} are complex constants given by

$$a_{nj} = \langle \psi_j^{(0)} | \psi_n^{(1)} \rangle \tag{9.28}$$

If the complete set of eigenfunctions for the unperturbed system includes a continuous range of functions, then the expansion of $\psi_n^{(1)}$ must include these functions. The inclusion of this continuous range is implied in the summation notation. The total eigenfunction ψ_n for the perturbed system to first order in λ is, then

$$\psi_n = (1 + \lambda a_{nn})\psi_n^{(0)} + \lambda \sum_{j(\neq n)} a_{nj} \psi_j^{(0)} \tag{9.29}$$

Since the function $\psi_n^{(0)}$ is already included in zero order in the expansion of ψ_n, we may, without loss of generality, set a_{nn} equal to zero, so that

$$\psi_n^{(1)} = \sum_{j(\neq n)} a_{nj} \psi_j^{(0)} \tag{9.30}$$

This choice affects the normalization constant of ψ_n, but has no other consequence. Furthermore, equation (9.28) for $j = n$ becomes

$$\langle \psi_n^{(0)} | \psi_n^{(1)} \rangle = 0 \tag{9.31}$$

showing that with $a_{nn} = 0$, the first-order correction $\psi_n^{(1)}$ is orthogonal to the unperturbed eigenfunction $\psi_n^{(0)}$.

With the choice $a_{nn} = 0$, the total eigenfunction ψ_n to first order is normalized. To show this, we form the scalar product $\langle \psi_n | \psi_n \rangle$ using equation (9.29) and retain only zero-order and first-order terms to obtain

$$\langle \psi_n | \psi_n \rangle = \langle \psi_n^{(0)} | \psi_n^{(0)} \rangle + \lambda \sum_{j(\neq n)} (a_{nj}\langle \psi_n^{(0)} | \psi_j^{(0)} \rangle + a_{nj}^*\langle \psi_j^{(0)} | \psi_n^{(0)} \rangle)$$

$$= 1 + \lambda \sum_{j(\neq n)} (a_{nj} + a_{nj}^*)\delta_{nj} = 1$$

where equation (9.26) has been used.

Substitution of equation (9.30) into (9.27) gives

$$(E_k^{(0)} - E_n^{(0)}) \sum_{j(\neq n)} a_{nj} \langle \psi_k^{(0)} | \psi_j^{(0)} \rangle = (E_k^{(0)} - E_n^{(0)}) a_{nk} = -\hat{H}_{kn}^{(1)}$$

where again equation (9.26) is utilized. If the eigenvalue $E_n^{(0)}$ is non-degenerate, then $E_k^{(0)}$ cannot equal $E_n^{(0)}$ for all k and n and we can divide by $(E_k^{(0)} - E_n^{(0)})$ to solve for a_{nk}

$$a_{nk} = \frac{-\hat{H}_{kn}^{(1)}}{E_k^{(0)} - E_n^{(0)}} \tag{9.32}$$

The situation where $E_n^{(0)}$ is degenerate requires a more complex treatment, which is presented in Section 9.5. The first-order correction $\psi_n^{(1)}$ is obtained by combining equations (9.30) and (9.32)

$$\psi_n^{(1)} = -\sum_{k(\neq n)} \frac{\hat{H}_{kn}^{(1)}}{E_k^{(0)} - E_n^{(0)}} \psi_k^{(0)} \tag{9.33}$$

Second-order corrections

The second-order correction $E_n^{(2)}$ to the eigenvalue E_n is obtained by multiplying equation (9.23) by $\psi_n^{(0)*}$ and integrating over all space

$$\langle \psi_n^{(0)} | \hat{H}^{(0)} - E_n^{(0)} | \psi_n^{(2)} \rangle + \langle \psi_n^{(0)} | \hat{H}^{(1)} | \psi_n^{(1)} \rangle - E_n^{(1)} \langle \psi_n^{(0)} | \psi_n^{(1)} \rangle$$
$$= -\langle \psi_n^{(0)} | \hat{H}^{(2)} | \psi_n^{(0)} \rangle + E_n^{(2)}$$

where the normalization of $\psi_n^{(0)}$ has been noted. Application of the hermitian property of $\hat{H}^{(0)}$ cancels the first term on the left-hand side. The third term on the left-hand side vanishes according to equation (9.31). Writing $\hat{H}_{nn}^{(2)}$ for $\langle \psi_n^{(0)} | \hat{H}^{(2)} | \psi_n^{(0)} \rangle$ and substituting equation (9.33) then give

$$E_n^{(2)} = \hat{H}_{nn}^{(2)} + \langle \psi_n^{(0)} | \hat{H}^{(1)} | \psi_n^{(1)} \rangle$$

$$= \hat{H}_{nn}^{(2)} - \sum_{k(\neq n)} \frac{\hat{H}_{nk}^{(1)} \hat{H}_{kn}^{(1)}}{E_k^{(0)} - E_n^{(0)}} = \hat{H}_{nn}^{(2)} - \sum_{k(\neq n)} \frac{|\hat{H}_{kn}^{(1)}|^2}{E_k^{(0)} - E_n^{(0)}} \tag{9.34}$$

where we have also noted that $\hat{H}_{nk}^{(1)}$ equals $\hat{H}_{kn}^{(1)*}$ because $\hat{H}^{(1)}$ is hermitian.

In many applications there is no second-order term in the perturbed Hamiltonian operator so that $\hat{H}_{nn}^{(2)}$ is zero. In such cases each unperturbed eigenvalue $E_n^{(0)}$ is raised by the terms in the summation corresponding to eigenvalues $E_k^{(0)}$ less than $E_n^{(0)}$ and lowered by the terms with eigenvalues $E_k^{(0)}$ greater than $E_n^{(0)}$. The eigenvalue $E_n^{(0)}$ is perturbed to the greatest extent by the terms with eigenvalues $E_k^{(0)}$ close to $E_n^{(0)}$. The contribution to the second-order correction $E_n^{(2)}$ of terms with eigenvalues far removed from $E_n^{(0)}$ is small. For the lowest eigenvalue $E_0^{(0)}$, all of the terms are negative so that $E_0^{(2)}$ is negative.

We also see that, in these cases, the first-order correction $\psi_n^{(1)}$ to the eigenfunction determines the second-order correction $E_n^{(2)}$ to the eigenvalue.

To obtain the second-order perturbation correction $\psi_n^{(2)}$ to the eigenfunction, we multiply equation (9.23) by $\psi_{kn}^{(0)*}$ for $k \neq n$ and integrate over all space

$$\langle \psi_k^{(0)} | \hat{H}^{(0)} - E_n^{(0)} | \psi_n^{(2)} \rangle + \langle \psi_k^{(0)} | \hat{H}^{(1)} | \psi_n^{(1)} \rangle - E_n^{(1)} \langle \psi_k^{(0)} | \psi_n^{(1)} \rangle$$

$$= -\langle \psi_k^{(0)} | \hat{H}^{(2)} | \psi_n^{(0)} \rangle + E_n^{(2)} \langle \psi_k^{(0)} | \psi_n^{(0)} \rangle$$

As before, we apply the hermitian property of $\hat{H}^{(0)}$, introduce the abbreviation $\hat{H}_{kn}^{(2)}$, and use the orthogonality relation (9.26) to obtain

$$(E_k^{(0)} - E_n^{(0)})\langle \psi_k^{(0)} | \psi_n^{(2)} \rangle + \langle \psi_k^{(0)} | \hat{H}^{(1)} | \psi_n^{(1)} \rangle - E_n^{(1)} \langle \psi_k^{(0)} | \psi_n^{(1)} \rangle = -\hat{H}_{kn}^{(2)} \quad (9.35)$$

We next expand the function $\psi_n^{(2)}$ in terms of the complete set of unperturbed eigenfunctions $\psi_j^{(0)}$

$$\psi_n^{(2)} = \sum_{j(\neq n)} b_{nj} \psi_j^{(0)} \qquad (9.36)$$

where, without loss of generality, the term $j = n$ may be omitted for the same reason that $\psi_n^{(0)}$ is omitted in equation (9.30). The coefficients b_{nj} are complex constants given by

$$b_{nj} = \langle \psi_j^{(0)} | \psi_n^{(2)} \rangle \qquad (9.37)$$

Substitution of equations (9.24), (9.28), (9.30), and (9.37) into (9.35) gives

$$(E_k^{(0)} - E_n^{(0)})b_{nk} + \sum_{j(\neq n)} a_{nj} \hat{H}_{kj}^{(1)} - a_{nk} \hat{H}_{nn}^{(1)} = -\hat{H}_{kn}^{(2)}$$

or

$$b_{nk} = -\frac{\hat{H}_{kn}^{(2)} + \displaystyle\sum_{j(\neq n)} a_{nj} \hat{H}_{kj}^{(1)} - a_{nk} \hat{H}_{nn}^{(1)}}{(E_k^{(0)} - E_n^{(0)})} \qquad (9.38)$$

Combining equations (9.32), (9.36), and (9.38), we obtain the final result

$$\psi_n^{(2)} = \sum_{k(\neq n)} \left[\frac{-\hat{H}_{kn}^{(2)}}{E_k^{(0)} - E_n^{(0)}} + \sum_{j(\neq n)} \frac{\hat{H}_{kj}^{(1)} \hat{H}_{jn}^{(1)}}{(E_k^{(0)} - E_n^{(0)})(E_j^{(0)} - E_n^{(0)})} - \frac{\hat{H}_{kn}^{(1)} \hat{H}_{nn}^{(1)}}{(E_k^{(0)} - E_n^{(0)})^2} \right] \psi_k^{(0)}$$

$$(9.39)$$

Summary

The non-degenerate eigenvalue E_n for the perturbed system to second order is obtained by substituting equations (9.24) and (9.34) into (9.20) to give

$$E_n \approx E_n^{(0)} + \lambda \hat{H}_{nn}^{(1)} + \lambda^2 \left[\hat{H}_{nn}^{(2)} - \sum_{k(\neq n)} \frac{|\hat{H}_{kn}^{(1)}|^2}{E_k^{(0)} - E_n^{(0)}} \right] \tag{9.40}$$

The corresponding eigenfunction ψ_n to second order is obtained by combining equations (9.19), (9.33), and (9.39)

$$\psi_n \approx \psi_n^{(0)} - \lambda \sum_{k(\neq n)} \frac{\hat{H}_{kn}^{(1)}}{E_k^{(0)} - E_n^{(0)}} \psi_k^{(0)}$$

$$+ \lambda^2 \sum_{k(\neq n)} \left[\frac{-\hat{H}_{kn}^{(2)}}{E_k^{(0)} - E_n^{(0)}} + \sum_{j(\neq n)} \frac{\hat{H}_{kj}^{(1)} \hat{H}_{jn}^{(1)}}{(E_k^{(0)} - E_n^{(0)})(E_j^{(0)} - E_n^{(0)})} - \frac{\hat{H}_{kn}^{(1)} \hat{H}_{nn}^{(1)}}{(E_k^{(0)} - E_n^{(0)})^2} \right] \psi_k^{(0)} \tag{9.41}$$

While the eigenvalue $E_n^{(0)}$ for the unperturbed system must be non-degenerate for these expansions to be valid, some or all of the other eigenvalues $E_k^{(0)}$ for $k \neq n$ may be degenerate. The summations in equations (9.40) and (9.41) are to be taken over all *states* of the unperturbed system other than the state $\psi_n^{(0)}$. If an eigenvalue $E_i^{(0)}$ is g_i-fold degenerate, then it is included g_i times in the summations. If the unperturbed eigenfunctions have a continuous range, then the summations in equations (9.40) and (9.41) must include an integration over those states as well.

Relation to variation method
If we use the wave function $\psi_0^{(0)}$ for the unperturbed ground state as a trial function ϕ in the variation method of Section 9.1 and set $\hat{H}$ equal to $\hat{H}^{(0)} + \lambda \hat{H}^{(1)}$, then we have from equations (9.2), (9.18), and (9.24)

$$\mathscr{E} = \langle \phi | \hat{H} | \phi \rangle = \langle \psi_0^{(0)} | \hat{H}^{(0)} + \lambda \hat{H}^{(1)} | \psi_0^{(0)} \rangle = E_0^{(0)} + \lambda E_0^{(1)}$$

and $\mathscr{E}$ is equal to the first-order energy as determined by perturbation theory. If we instead use a trial function ϕ which contains some parameters and which equals $\psi_0^{(0)}$ for some set of parameter values, then the corresponding energy $\mathscr{E}$ from equation (9.2) is at least as good an approximation as $E_0^{(0)} + \lambda E_0^{(1)}$ to the true ground-state energy.

Moreover, if the wave function $\psi_0^{(0)} + \lambda \psi_0^{(1)}$ is used as a trial function ϕ, then the quantity $\mathscr{E}$ from equation (9.2) is equal to the second-order energy determined by perturbation theory. Any trial function ϕ with parameters which reduces to $\psi_0^{(0)} + \lambda \psi_0^{(1)}$ for some set of parameter values yields an approximate energy $\mathscr{E}$ from equation (9.2) which is no less accurate than the second-order perturbation value.

9.4 Perturbed harmonic oscillator

As illustrations of the application of perturbation theory we consider two examples of a perturbed harmonic oscillator. In the first example, we suppose that the potential energy V of the oscillator is

$$V = \tfrac{1}{2}kx^2 + cx^4 = \tfrac{1}{2}m\omega^2 x^2 + cx^4$$

where c is a small quantity. The units of V are those of $\hbar\omega$ (energy), while the units of x are shown in equation (4.14) to be those of $(\hbar/m\omega)^{1/2}$. Accordingly, the units of c are those of $m^2\omega^3/\hbar$ and we may express c as

$$c = \lambda\frac{m^2\omega^3}{\hbar}$$

where λ is dimensionless. The potential energy then takes the form

$$V = \tfrac{1}{2}m\omega^2 x^2 + \lambda\frac{m^2\omega^3 x^4}{\hbar} \qquad (9.42)$$

The Hamiltonian operator $\hat{H}^{(0)}$ for the unperturbed harmonic oscillator is given by equation (4.12) and its eigenvalues $E_n^{(0)}$ and eigenfunctions $\psi_n^{(0)}$ are shown in equations (4.30) and (4.41). The perturbation $\hat{H}'$ is

$$\hat{H}' = \hat{H}^{(1)} = \lambda\frac{m^2\omega^3 x^4}{\hbar} \qquad (9.43)$$

Higher-order terms $\hat{H}^{(2)}$, $\hat{H}^{(3)}$, ... in the perturbed Hamiltonian operator do not appear in this example.

To find the perturbation corrections to the eigenvalues and eigenfunctions, we require the matrix elements $\langle n'|x^4|n\rangle$ for the unperturbed harmonic oscillator. These matrix elements are given by equations (4.51). The first-order correction $E_n^{(1)}$ to the eigenvalue E_n is evaluated using equations (9.24), (9.43), and (4.51c)

$$E_n^{(1)} = \hat{H}_{nn}^{(1)} = \frac{\lambda m^2\omega^3}{\hbar}\langle n|x^4|n\rangle = \tfrac{3}{2}(n^2 + n + \tfrac{1}{2})\lambda\hbar\omega \qquad (9.44)$$

The second-order correction $E_n^{(2)}$ is obtained from equations (9.34), (9.43), and (4.51) as follows

$E_n^{(2)}$

$$= -\frac{|\hat{H}_{n-4,n}^{(1)}|^2}{E_{n-4}^{(0)} - E_n^{(0)}} - \frac{|\hat{H}_{n-2,n}^{(1)}|^2}{E_{n-2}^{(0)} - E_n^{(0)}} - \frac{|\hat{H}_{n+2,n}^{(1)}|^2}{E_{n+2}^{(0)} - E_n^{(0)}} - \frac{|\hat{H}_{n+4,n}^{(1)}|^2}{E_{n+4}^{(0)} - E_n^{(0)}}$$

$$= -\frac{\lambda^2 m^4 \omega^6}{\hbar^2} \left[\frac{\langle n-4|x^4|n\rangle^2}{(-4\hbar\omega)} + \frac{\langle n-2|x^4|n\rangle^2}{(-2\hbar\omega)} + \frac{\langle n+2|x^4|n\rangle^2}{2\hbar\omega} + \frac{\langle n+4|x^4|n\rangle^2}{4\hbar\omega} \right]$$

$$= -\tfrac{1}{8}(34n^3 + 51n^2 + 59n + 21)\lambda^2 \hbar\omega \tag{9.45}$$

The perturbed energy E_n to second order is, then

$$E_n \approx E_n^{(0)} + E_n^{(1)} + E_n^{(2)}$$

$$= (n + \tfrac{1}{2})\hbar\omega + \tfrac{3}{2}(n^2 + n + \tfrac{1}{2})\lambda\hbar\omega - \tfrac{1}{8}(34n^3 + 51n^2 + 59n + 21)\lambda^2\hbar\omega \tag{9.46}$$

In the expression (9.45) for the second-order correction $E_n^{(2)}$, the summation on the right-hand side includes all states k other than the state n, but only for the states $(n-4)$, $(n-2)$, $(n+2)$, and $(n+4)$ are the contributions to the summation non-vanishing. For the two lowest values of n, giving $E_0^{(2)}$ and $E_1^{(2)}$, only the two terms $k = (n+2)$ and $k = (n+4)$ should be included in the summation. However, the terms for the meaningless values $k = (n-2)$ and $k = (n-4)$ vanish identically, so that their inclusion in equation (9.45) is valid. A similar argument applies to $E_2^{(2)}$ and $E_3^{(2)}$, wherein the term for the meaningless value $k = (n-4)$ is identically zero. Thus, equation (9.46) applies to all values of n and the perturbed ground-state energy E_0, for example, is

$$E_0 \approx (\tfrac{1}{2} + \tfrac{3}{4}\lambda - \tfrac{21}{8}\lambda^2)\hbar\omega$$

The evaluation of the first- and second-order corrections to the eigenfunctions is straightforward, but tedious. Consequently, we evaluate here only the first-order correction $\psi_0^{(1)}$ for the ground state. According to equations (9.33), (9.43), and (4.51), this correction term is given by

$$\psi_0^{(1)} = -\frac{\hat{H}_{20}^{(1)}}{E_2^{(0)} - E_0^{(0)}} \psi_2^{(0)} - \frac{\hat{H}_{40}^{(1)}}{E_4^{(0)} - E_0^{(0)}} \psi_4^{(0)}$$

$$= -\frac{\lambda m^2 \omega^3}{\hbar} \left[\frac{\langle 2|x^4|0\rangle}{2\hbar\omega} \psi_2^{(0)} + \frac{\langle 4|x^4|0\rangle}{4\hbar\omega} \psi_4^{(0)} \right]$$

$$= -\frac{\lambda}{4\sqrt{2}} [6\psi_2^{(0)} + \sqrt{3}\psi_4^{(0)}] \tag{9.47}$$

If the unperturbed eigenfunctions $\psi_2^{(0)}$ and $\psi_4^{(0)}$ as given by equation (4.41) are explicitly introduced, then the perturbed ground-state eigenfunction ψ_0 to first order is

$$\psi_0 \approx \psi_0^{(0)} + \psi_0^{(1)} = \left(\frac{m\omega}{\pi\hbar}\right)^{1/4} \left[1 - \frac{\lambda}{16}(4\xi^4 + 12\xi^2 - 9)\right] e^{-\xi^2/2} \qquad (9.48)$$

As a second example, we suppose that the potential energy V for the perturbed harmonic oscillator is

$$V = \tfrac{1}{2}kx^2 + cx^3 = \tfrac{1}{2}m\omega^2 x^2 + \lambda\left(\frac{m^3\omega^5}{\hbar}\right)^{1/2} x^3 \qquad (9.49)$$

where $c = \lambda(m^3\omega^5/\hbar)^{1/2}$ is again a small quantity and λ is dimensionless. The perturbation $\hat{H}'$ for this example is

$$\hat{H}' = \hat{H}^{(1)} = \lambda\left(\frac{m^3\omega^5}{\hbar}\right)^{1/2} x^3 \qquad (9.50)$$

The matrix elements $\langle n'|x^3|n\rangle$ for the unperturbed harmonic oscillator are given by equations (4.50). The first-order correction term $E_n^{(1)}$ is obtained by substituting equations (9.50) and (4.50e) into (9.24), giving the result

$$E_n^{(1)} = \lambda\left(\frac{m^3\omega^5}{\hbar}\right)^{1/2} \langle n|x^3|n\rangle = 0 \qquad (9.51)$$

Thus, the first-order perturbation to the eigenvalue is zero. The second-order term $E_n^{(2)}$ is evaluated using equations (9.34), (9.50), and (4.50), giving the result

$$E_n \approx E_n^{(2)}$$

$$= -\frac{|\hat{H}_{n-3,n}^{(1)}|^2}{E_{n-3}^{(0)} - E_n^{(0)}} - \frac{|\hat{H}_{n-1,n}^{(1)}|^2}{E_{n-1}^{(0)} - E_n^{(0)}} - \frac{|\hat{H}_{n+1,n}^{(1)}|^2}{E_{n+1}^{(0)} - E_n^{(0)}} - \frac{|\hat{H}_{n+3,n}^{(1)}|^2}{E_{n+3}^{(0)} - E_n^{(0)}}$$

$$= -\frac{\lambda^2 m^3 \omega^5}{\hbar}\left[\frac{\langle n-3|x^3|n\rangle^2}{(-3\hbar\omega)} + \frac{\langle n-1|x^3|n\rangle^2}{(-\hbar\omega)} + \frac{\langle n+1|x^3|n\rangle^2}{\hbar\omega} + \frac{\langle n+3|x^3|n\rangle^2}{3\hbar\omega}\right]$$

$$= -\tfrac{1}{8}(30n^2 + 30n + 11)\lambda^2\hbar\omega \qquad (9.52)$$

9.5 Degenerate perturbation theory

The perturbation method presented in Section 9.3 applies only to non-degenerate eigenvalues $E_n^{(0)}$ of the unperturbed system. When $E_n^{(0)}$ is degenerate, the denominators vanish for those terms in equations (9.40) and (9.41) in which $E_k^{(0)}$ is equal to $E_n^{(0)}$, making the perturbations to E_n and ψ_n indeterminate. In this section we modify the perturbation method to allow for degenerate eigenvalues. In view of the complexity of this new procedure, we consider only the first-order perturbation corrections to the eigenvalues and eigenfunctions.

The eigenvalues and eigenfunctions for the unperturbed system are given by equation (9.18), but now the eigenvalue $E_n^{(0)}$ is g_n-fold degenerate. Accordingly, there are g_n eigenfunctions with the same eigenvalue $E_n^{(0)}$. For greater clarity, we change the notation here and denote the eigenfunctions corresponding to $E_n^{(0)}$ as $\psi_{n\alpha}^{(0)}$, $\alpha = 1, 2, \ldots, g_n$. Equation (9.18) is then replaced by the equivalent expression

$$H^{(0)}\psi_{n\alpha}^{(0)} = E_n^{(0)}\psi_{n\alpha}^{(0)}, \qquad \alpha = 1, 2, \ldots, g_n \qquad (9.53)$$

Each of the eigenfunctions $\psi_{n\alpha}^{(0)}$ is orthogonal to all the other unperturbed eigenfunctions $\psi_{ka}^{(0)}$ for $k \neq n$, but is not necessarily orthogonal to the other eigenfunctions for $E_n^{(0)}$. Any linear combination $\phi_{n\alpha}$ of the members of the set $\psi_{n\alpha}^{(0)}$

$$\phi_{n\alpha} = \sum_{\beta=1}^{g_n} c_{\alpha\beta}\psi_{n\beta}^{(0)}, \qquad \alpha = 1, 2, \ldots, g_n \qquad (9.54)$$

is also a solution of equation (9.53) with the same eigenvalue $E_n^{(0)}$. As discussed in Section 3.3, the members of the set $\psi_{n\alpha}^{(0)}$ may be constructed to be orthonormal and we assume that this construction has been carried out, so that

$$\langle \psi_{n\beta}^{(0)} | \psi_{n\alpha}^{(0)} \rangle = \delta_{\alpha\beta}, \qquad \alpha, \beta = 1, 2, \ldots, g_n \qquad (9.55)$$

By suitable choices for the coefficients $c_{\alpha\beta}$ in equation (9.54), the functions $\phi_{n\alpha}$ may also be constructed as an orthonormal set

$$\langle \phi_{n\beta} | \phi_{n\alpha} \rangle = \delta_{\alpha\beta}, \qquad \alpha, \beta = 1, 2, \ldots, g_n \qquad (9.56)$$

Substitution of equation (9.54) into (9.56) and application of (9.55) give

$$\sum_{\gamma=1}^{g_n} c_{\beta\gamma}^* c_{\alpha\gamma} = \delta_{\alpha\beta}, \qquad \alpha, \beta = 1, 2, \ldots, g_n \qquad (9.57)$$

The Schrödinger equation for the perturbed system is

$$\hat{H}\psi_{n\alpha} = E_{n\alpha}\psi_{n\alpha}, \qquad \alpha = 1, 2, \ldots, g_n \qquad (9.58)$$

where the Hamiltonian operator $\hat{H}$ is given by equation (9.16), $E_{n\alpha}$ are the eigenvalues for the perturbed system, and $\psi_{n\alpha}$ are the corresponding eigenfunctions. While the unperturbed eigenvalue $E_n^{(0)}$ is g_n-fold degenerate, the perturbation $\hat{H}'$ in the Hamiltonian operator often splits the eigenvalue $E_n^{(0)}$ into g_n different values. For this reason, the perturbed eigenvalues $E_{n\alpha}$ require the additional index α. The perturbation expansions of $E_{n\alpha}$ and $\psi_{n\alpha}$ in powers of λ are

$$E_{na} = E_n^{(0)} + \lambda E_{na}^{(1)} + \lambda^2 E_{na}^{(2)} + \cdots, \qquad \alpha = 1, 2, \ldots, g_n \qquad (9.59)$$

$$\psi_{na} = \psi_{na}^{(0)} + \lambda \psi_{na}^{(1)} + \lambda^2 \psi_{na}^{(2)} + \cdots, \qquad \alpha = 1, 2, \ldots, g_n \qquad (9.60)$$

Note that in equation (9.59) the zero-order term is the same for all values of α.

In the limit $\lambda \to 0$, the Hamiltonian operator $\hat{H}$ approaches the unperturbed operator $\hat{H}^{(0)}$ and the perturbed eigenvalue E_{na} approaches the degenerate unperturbed eigenvalue $E_n^{(0)}$. The perturbed eigenfunction ψ_{na} approaches a function which satisfies equation (9.53), but this limiting eigenfunction may not be any one of the initial functions $\psi_{na}^{(0)}$. In general, this limiting function is some linear combination of the initial unperturbed eigenfunctions $\psi_{na}^{(0)}$, as expressed in equation (9.54). Thus, along with the determination of the first-order correction terms $E_{na}^{(1)}$ and $\psi_{na}^{(1)}$, we must find the set of unperturbed eigenfunctions $\phi_{na}^{(0)}$ to which the perturbed eigenfunctions reduce in the limit $\lambda \to 0$. In other words, we need to evaluate the coefficients $c_{\alpha\beta}$ in the linear combinations (9.54) which transform the initial set of unperturbed eigenfunctions $\psi_{na}^{(0)}$ into the 'correct' set $\phi_{na}^{(0)}$. Equation (9.60) is then replaced by

$$\psi_{na} = \phi_{na}^{(0)} + \lambda \psi_{na}^{(1)} + \lambda^2 \psi_{na}^{(2)} + \cdots, \qquad \alpha = 1, 2, \ldots, g_n \qquad (9.61)$$

The first-order equations (9.22) and (9.24) apply here provided the additional index α and the 'correct' unperturbed eigenfunctions are used

$$(\hat{H}^{(0)} - E_n^{(0)})\psi_{na}^{(1)} = -(\hat{H}^{(1)} - E_{na}^{(1)})\phi_{na}^{(0)} \qquad (9.62)$$

$$E_{na}^{(1)} = \langle \phi_{na}^{(0)} | \hat{H}^{(1)} | \phi_{na}^{(0)} \rangle \qquad (9.63)$$

However, equation (9.63) for the first-order corrections to the eigenvalues cannot be used directly at this point because the functions $\phi_{na}^{(0)}$ are not known.

To find $E_{na}^{(1)}$ we multiply equation (9.62) by $\psi_{n\beta}^{(0)*}$, the complex conjugate of one of the initial unperturbed eigenfunctions belonging to the degenerate eigenvalue $E_n^{(0)}$, and integrate over all space to obtain

$$\langle \psi_{n\beta}^{(0)} | \hat{H}^{(0)} - E_n^{(0)} | \psi_{na}^{(1)} \rangle = -\langle \psi_{n\beta}^{(0)} | \hat{H}^{(1)} - E_{na}^{(1)} | \phi_{na}^{(0)} \rangle$$

Applying the hermitian property of $\hat{H}^{(0)}$, we see that the left-hand side vanishes. Substitution of the expansion (9.54) for $\phi_{na}^{(0)}$ using γ as the dummy expansion index gives

$$\sum_{\gamma=1}^{g_n} c_{\alpha\gamma} \langle \psi_{n\beta}^{(0)} | \hat{H}^{(1)} - E_{na}^{(1)} | \psi_{n\gamma}^{(0)} \rangle = 0, \qquad \alpha, \beta = 1, 2, \ldots, g_n$$

If we introduce the abbreviation

$$\hat{H}_{n\beta,n\gamma}^{(1)} \equiv \langle \psi_{n\beta}^{(0)} | \hat{H}^{(1)} | \psi_{n\gamma}^{(0)} \rangle, \qquad \beta, \gamma = 1, 2, \ldots, g_n$$

and apply the orthonormality condition (9.55), this equation takes the form

$$\sum_{\gamma=1}^{g_n} c_{\alpha\gamma}(\hat{H}^{(1)}_{n\beta,n\gamma} - E^{(1)}_{n\alpha}\delta_{\beta\gamma}) = 0, \qquad \alpha, \beta = 1, 2, \ldots, g_n \qquad (9.64)$$

Note that the integrals $\hat{H}^{(1)}_{n\alpha,n\gamma}$ are evaluated with the known initial set of unperturbed eigenfunctions, in contrast to the integrals in equation (9.63), which require the unknown functions $\phi^{(0)}_{n\alpha}$. For a given eigenvalue $E^{(1)}_{n\alpha}$, the expression (9.64) is a set of g_n linear homogeneous simultaneous equations, one for each value of β ($\beta = 1, 2, \ldots, g_n$)

$$\beta = 1: c_{\alpha1}(\hat{H}^{(1)}_{n1,n1} - E^{(1)}_{n\alpha}) + c_{\alpha2}\hat{H}^{(1)}_{n1,n2} + c_{\alpha3}\hat{H}^{(1)}_{n1,n3} + \cdots + c_{\alpha g_n}\hat{H}^{(1)}_{n1,ng_n} = 0$$

$$\beta = 2: c_{\alpha1}\hat{H}^{(1)}_{n2,n1} + c_{\alpha2}(\hat{H}^{(1)}_{n2,n2} - E^{(1)}_{n\alpha}) + c_{\alpha3}\hat{H}^{(1)}_{n2,n3} + \cdots + c_{\alpha g_n}\hat{H}^{(1)}_{n2,ng_n} = 0$$

$$\vdots$$

$$\beta = g_n: c_{\alpha1}\hat{H}^{(1)}_{ng_n,n1} + c_{\alpha2}\hat{H}^{(1)}_{ng_n,n2} + c_{\alpha3}\hat{H}^{(1)}_{ng_n,n3} + \cdots + c_{\alpha g_n}(\hat{H}^{(1)}_{ng_n,ng_n} - E^{(1)}_{n\alpha})$$
$$= 0$$

Equation (9.64) has the form of (9.13) with the coefficients $c_{\alpha\gamma}$ corresponding to the unknown quantities x_i and the terms $(\hat{H}^{(1)}_{n\beta,n\gamma} - E^{(1)}_{n\alpha}\delta_{\beta\gamma})$ corresponding to the coefficients a_{ki}. Thus, a non-trivial solution for the g_n coefficients $c_{\alpha\gamma}$ ($\gamma = 1, 2, \ldots, g_n$) exists only if the determinant with elements $(\hat{H}^{(1)}_{n\beta,n\gamma} - E^{(1)}_{n\alpha}\delta_{\beta\gamma})$ vanishes

$$\begin{vmatrix} H^{(1)}_{n1,n1} - E^{(1)}_{n\alpha} & H^{(1)}_{n1,n2} & \cdots & H^{(1)}_{n1,ng_n} \\ H^{(1)}_{n2,n1} & H^{(1)}_{n2,n2} - E^{(1)}_{n\alpha} & \cdots & H^{(1)}_{n2,ng_n} \\ \cdots & \cdots & \cdots & \cdots \\ H^{(1)}_{ng_n,n1} & H^{(1)}_{ng_n,n2} & \cdots & H^{(1)}_{ng_n,ng_n} - E^{(1)}_{n\alpha} \end{vmatrix} = 0 \qquad (9.65)$$

Only for some values of the first-order correction term $E^{(1)}_{n\alpha}$ is the secular equation (9.65) satisfied. This secular equation is of degree g_n in $E^{(1)}_{n\alpha}$, giving g_n roots

$$E^{(1)}_{n1}, E^{(1)}_{n2}, \ldots, E^{(1)}_{ng_n}$$

all of which are real because $\hat{H}^{(1)}$ is hermitian. The perturbed eigenvalues to first order are, then

$$E_{n1} = E^{(0)}_n + \lambda E^{(1)}_{n1}$$

$$\vdots$$

$$E_{ng_n} = E^{(0)}_n + \lambda E^{(1)}_{ng_n}$$

If the g_n roots are all different, then in first order the g_n-fold degenerate unperturbed eigenvalue $E^{(0)}_n$ is split into g_n different perturbed eigenvalues. In this case, the degeneracy is removed in first order by the perturbation. We assume in the continuing presentation that all the roots are indeed different from each other.

'Correct' zero-order eigenfunctions

The determination of the coefficients $c_{\alpha\gamma}$ is not necessary for finding the first-order perturbation corrections to the eigenvalues, but is required to obtain the 'correct' zero-order eigenfunctions and their first-order corrections. The coefficients $c_{\alpha\gamma}$ for each value of α ($\alpha = 1, 2, \ldots, g_n$) are obtained by substituting the value found for $E_{n\alpha}^{(1)}$ from the secular equation (9.65) into the set of simultaneous equations (9.64) and solving for the coefficients $c_{\alpha 2}, \ldots, c_{\alpha g_n}$ in terms of $c_{\alpha 1}$. The normalization condition (9.57) is then used to determine $c_{\alpha 1}$. This procedure uniquely determines the complete set of coefficients $c_{\alpha\gamma}$ (α, $\gamma = 1, 2, \ldots, g_n$) because we have assumed that all the roots $E_{n\alpha}^{(1)}$ are different.

If by accident or by clever choice, the initial set of unperturbed eigenfunctions $\psi_{n\alpha}^{(0)}$ is actually the 'correct' set, i.e., if in the limit $\lambda \to 0$ the perturbed eigenfunction $\psi_{n\alpha}$ reduces to $\psi_{n\alpha}^{(0)}$ for all values of α, then the coefficients $c_{\alpha\gamma}$ are given by $c_{\alpha\gamma} = \delta_{\alpha\gamma}$ and the secular determinant is diagonal

$$\begin{vmatrix} H_{n1,n1}^{(1)} - E_{n\alpha}^{(1)} & 0 & \cdots & 0 \\ 0 & H_{n2,n2}^{(1)} - E_{n\alpha}^{(1)} & \cdots & 0 \\ \cdots & \cdots & \cdots & \cdots \\ 0 & 0 & \cdots & H_{ng_n,ng_n}^{(1)} - E_{n\alpha}^{(1)} \end{vmatrix} = 0$$

The first-order corrections to the eigenvalues are then given by

$$E_{n\alpha}^{(1)} = \hat{H}_{n\alpha,n\alpha}^{(1)}, \qquad \alpha = 1, 2, \ldots, g_n \tag{9.66}$$

It is obviously a great advantage to select the 'correct' set of unperturbed eigenfunctions as the initial set, so that the simpler equation (9.66) may be used. A general procedure for achieving this goal is to find a hermitian operator $\hat{A}$ that commutes with both $\hat{H}^{(0)}$ and $\hat{H}^{(1)}$ and has eigenfunctions χ_α with non-degenerate eigenvalues μ_α, so that

$$[\hat{A}, \hat{H}^{(0)}] = [\hat{A}, \hat{H}^{(1)}] = 0 \tag{9.67}$$

and

$$\hat{A}\chi_\alpha = \mu_\alpha \chi_\alpha$$

Since $\hat{A}$ and $\hat{H}^{(0)}$ commute, they have simultaneous eigenfunctions. Therefore, we may select $\chi_1, \chi_2, \ldots, \chi_{g_n}$ as the initial set of unperturbed eigenfunctions

$$\psi_{n\alpha}^{(0)} = \chi_\alpha, \qquad \alpha = 1, 2, \ldots, g_n$$

We next form the integral $\langle \chi_\beta | [\hat{A}, \hat{H}^{(1)}] | \chi_\alpha \rangle$ ($\beta \neq \alpha$), which of course vanishes according to equation (9.67),

$$\langle \chi_\beta | [\hat{A}, \hat{H}^{(1)}] | \chi_\alpha \rangle = \langle \chi_\beta | \hat{A} \hat{H}^{(1)} | \chi_\alpha \rangle - \langle \chi_\beta | \hat{H}^{(1)} \hat{A} | \chi_\alpha \rangle$$

$$= \langle \hat{A} \chi_\beta | \hat{H}^{(1)} | \chi_\alpha \rangle - \mu_\alpha \langle \chi_\beta | \hat{H}^{(1)} | \chi_\alpha \rangle$$

$$= (\mu_\beta - \mu_\alpha) \langle \psi_{n\beta}^{(0)} | \hat{H}^{(1)} | \psi_{n\alpha}^{(0)} \rangle$$

$$= (\mu_\beta - \mu_\alpha) \hat{H}_{n\beta, n\alpha}^{(1)}$$

$$= 0$$

Since $\mu_\beta \neq \mu_\alpha$, the off-diagonal elements $\hat{H}_{n\beta, n\alpha}^{(1)}$ equal zero and the set of functions $\psi_{n\alpha}^{(0)} = \chi_\alpha$ is the 'correct' set. The parity operator $\hat{\Pi}$ discussed in Section 3.8 can often be used in this context for selecting 'correct' unperturbed eigenfunctions.

First-order corrections to the eigenfunctions

To obtain the first-order corrections $\psi_{n\alpha}^{(1)}$ to the eigenfunctions $\psi_{n\alpha}$, we multiply equation (9.62) by $\psi_{k\beta}^{(0)*}$ for $k \neq n$ and integrate over all space

$$\langle \psi_{k\beta}^{(0)} | \hat{H}^{(0)} - E_n^{(0)} | \psi_{n\alpha}^{(1)} \rangle = -\langle \psi_{k\beta}^{(0)} | \hat{H}^{(1)} | \phi_{n\alpha}^{(0)} \rangle + E_{n\alpha}^{(1)} \langle \psi_{k\beta}^{(0)} | \phi_{n\alpha}^{(0)} \rangle$$

Applying the hermitian property of $\hat{H}^{(0)}$ and noting that $\psi_{k\beta}^{(0)}$ is orthogonal to all eigenfunctions belonging to the eigenvalue $E_n^{(0)}$, we have

$$(E_k^{(0)} - E_n^{(0)}) \langle \psi_{k\beta}^{(0)} | \psi_{n\alpha}^{(1)} \rangle = -\langle \psi_{k\beta}^{(0)} | \hat{H}^{(1)} | \phi_{n\alpha}^{(0)} \rangle \tag{9.68}$$

We next expand the first-order correction $\psi_{n\alpha}^{(1)}$ in terms of the complete set of unperturbed eigenfunctions

$$\psi_{n\alpha}^{(1)} = \sum_{j(\neq n)} \sum_{\gamma=1}^{g_j} a_{n\alpha, j\gamma} \psi_{j\gamma}^{(0)} \tag{9.69}$$

where the terms with $j = n$ are omitted for the same reason that they are omitted in equation (9.30). Substitution of equations (9.54) and (9.69) into (9.68) gives

$$(E_k^{(0)} - E_n^{(0)}) \sum_{j(\neq n)} \sum_{\gamma=1}^{g_j} a_{n\alpha, j\gamma} \langle \psi_{k\beta}^{(0)} | \psi_{j\gamma}^{(0)} \rangle = -\sum_{\gamma=1}^{g_n} c_{\alpha\gamma} \hat{H}_{k\beta, n\gamma}^{(1)}$$

In view of the orthonormality relations, the summation on the left-hand side may be simplified as follows

$$\sum_{j(\neq n)} \sum_{\gamma=1}^{g_j} a_{n\alpha, j\gamma} \langle \psi_{k\beta}^{(0)} | \psi_{j\gamma}^{(0)} \rangle = \sum_{j(\neq n)} \sum_{\gamma=1}^{g_j} a_{n\alpha, j\gamma} \delta_{kj} \delta_{\beta\gamma} = a_{n\alpha, k\beta}$$

Therefore, we have

$$a_{n\alpha,k\beta} = \frac{-\displaystyle\sum_{\gamma=1}^{g_n} c_{\alpha\gamma}\,\hat{H}^{(1)}_{k\beta,n\gamma}}{(E_k^{(0)} - E_n^{(0)})} \tag{9.70}$$

The eigenfunctions $\psi_{n\alpha}$ for the perturbed system to first order are obtained by combining equations (9.61), (9.69), and (9.70)

$$\psi_{n\alpha} = \phi_{n\alpha}^{(0)} - \lambda \sum_{k(\neq n)}^{g_k} \sum_{\beta=1}^{g_k} \frac{\displaystyle\sum_{\gamma=1}^{g_n} c_{\alpha\gamma}\,\hat{H}^{(1)}_{k\beta,n\gamma}}{(E_k^{(0)} - E_n^{(0)})} \psi_{k\beta}^{(1)} \tag{9.71}$$

Example: hydrogen atom in an electric field

As an illustration of the application of degenerate perturbation theory, we consider the influence, known as the *Stark effect*, of an externally applied electric field $\mathscr{E}$ on the energy levels of a hydrogen atom. The unperturbed Hamiltonian operator $\hat{H}^{(0)}$ for the hydrogen atom is given by equation (6.14), and its eigenfunctions and eigenvalues are given by equations (6.56) and (6.57), respectively. In this example, we label the eigenfunctions and eigenvalues of $\hat{H}^{(0)}$ with an index starting at 1 rather than at 0 to correspond to the principal quantum number n. The perturbation $\hat{H}'$ is the potential energy for the interaction between the atomic electron with charge $-e$ and an electric field $\mathscr{E}$ directed along the positive z-axis

$$\hat{H}' = \hat{H}^{(1)} = e\mathscr{E}z = e\mathscr{E}r\cos\theta \tag{9.72}$$

If spin effects are neglected, the ground-state unperturbed energy level $E_1^{(0)}$ is non-degenerate and its first-order perturbation correction $E_1^{(1)}$ is given by equation (9.24) as

$$E_1^{(1)} = e\mathscr{E}\langle 1s|z|1s\rangle = 0$$

This integral vanishes because the unperturbed ground state of the hydrogen atom, the 1s state, has even parity and z has odd parity.

The next lowest unperturbed energy level $E_2^{(0)}$, however, is four-fold degenerate and, consequently, degenerate perturbation theory must be used to determine its perturbation corrections. For simplicity of notation, in the quantities $\psi_{n\alpha}^{(0)}$, $\phi_{n\alpha}^{(0)}$, and $\hat{H}^{(1)}_{n\alpha,n\beta}$ we drop the index n, which has the value $n = 2$ throughout. As the initial set of eigenfunctions for the unperturbed system, we select the 2s, 2p$_0$, 2p$_1$, and 2p$_{-1}$ atomic orbitals as given by equations (6.59) and (6.60), so that

$$\psi_1^{(0)} = |2s\rangle, \qquad \psi_2^{(0)} = |2p_0\rangle$$

$$\psi_3^{(0)} = |2p_1\rangle, \qquad \psi_4^{(0)} = |2p_{-1}\rangle \tag{9.73}$$

The 'correct' set of unperturbed eigenfunction $\phi_\alpha^{(0)}$ are, then

$$\phi_\alpha^{(0)} = \sum_{\beta=1}^4 c_{\alpha\beta}\psi_\beta^{(0)} = c_{\alpha 1}|2s\rangle + c_{\alpha 2}|2p_0\rangle + c_{\alpha 3}|2p_1\rangle + c_{\alpha 4}|2p_{-1}\rangle,$$

$$\alpha = 1, 2, 3, 4 \tag{9.74}$$

The matrix elements $\hat{H}_{\alpha\beta}^{(1)}$ in this example are

$$\hat{H}_{\alpha\beta}^{(1)} = e\mathscr{E}\langle\psi_\alpha^{(0)}|z|\psi_\beta^{(0)}\rangle = e\mathscr{E}\langle\psi_\alpha^{(0)}|r\cos\theta|\psi_\beta^{(0)}\rangle$$

$$= e\mathscr{E}\int_0^{2\pi}\int_0^\pi\int_0^\infty \psi_\alpha^{(0)*}\psi_\beta^{(0)} r\cos\theta\, r^2\sin\theta\, dr\, d\theta\, d\varphi \tag{9.75}$$

These matrix elements vanish unless $\Delta m = 0$ and $\Delta l = 1$. Thus, only the matrix element $\hat{H}_{12}^{(1)}$, which equals $\hat{H}_{21}^{(1)}$, is non-zero.

To evaluate the matrix element $\hat{H}_{12}^{(1)}$, we substitute the 2s wave function from equation (6.59) and the 2p$_0$ wave function from equation (6.60a) into (9.75)

$$\hat{H}_{12}^{(1)} = \hat{H}_{21}^{(1)} = e\mathscr{E}[\pi(2a_0)^4]^{-1}\int_0^\infty r^4\left(1 - \frac{r}{2a_0}\right)e^{-r/a_0}\, dr\int_0^\pi \cos^2\theta\sin\theta\, d\theta\int_0^{2\pi}d\varphi$$

$$= -3e\mathscr{E}a_0$$

where equations (A.26) and (A.28) are used.

The secular determinant (9.65) is

$$\begin{vmatrix} -E_2^{(1)} & -3e\mathscr{E}a_0 & 0 & 0 \\ -3e\mathscr{E}a_0 & -E_2^{(1)} & 0 & 0 \\ 0 & 0 & -E_2^{(1)} & 0 \\ 0 & 0 & 0 & -E_2^{(1)} \end{vmatrix} = 0$$

which expands to

$$[(E_2^{(1)})^2 - (3e\mathscr{E}a_0)^2](E_2^{(1)})^2 = 0$$

The four roots are $E_2^{(1)} = -3e\mathscr{E}a_0,\, 3e\mathscr{E}a_0,\, 0,\, 0$, so that to first order the perturbed energy levels are

$$E_{21} = \frac{-e'^2}{8a_0} - 3e\mathscr{E}a_0, \qquad E_{23} = \frac{-e'^2}{8a_0}$$

$$E_{22} = \frac{-e'^2}{8a_0} + 3e\mathscr{E}a_0, \qquad E_{24} = \frac{-e'^2}{8a_0} \tag{9.76}$$

The four linear homogeneous simultaneous equations (9.64) are

$$-c_{\alpha 1} E_\alpha^{(1)} + c_{\alpha 2} \hat{H}_{12}^{(1)} = 0$$

$$c_{\alpha 1} \hat{H}_{12}^{(1)} - c_{\alpha 2} E_\alpha^{(1)} = 0$$

$$-c_{\alpha 3} E_\alpha^{(1)} = 0 \qquad (9.77)$$

$$-c_{\alpha 4} E_\alpha^{(1)} = 0$$

To find the 'correct' set of unperturbed eigenfunctions $\phi_\alpha^{(0)}$, we substitute first $E_2^{(1)} = -3e\mathscr{E}a_0$, then successively $E_2^{(1)} = 3e\mathscr{E}a_0$, 0, 0 into the set of equations (9.77). The results are as follows

for $E_2^{(1)} = \hat{H}_{12}^{(1)} = -3e\mathscr{E}a_0$: $c_1 = c_2$; $c_3 = c_4 = 0$

for $E_2^{(1)} = -\hat{H}_{12}^{(1)} = 3e\mathscr{E}a_0$: $c_1 = -c_2$; $c_3 = c_4 = 0$

for $E_2^{(1)} = 0$: $c_1 = c_2 = 0$; c_3 and c_4 undetermined

Thus, the 'correct' unperturbed eigenfunctions are

$$\phi_1^{(0)} = 2^{-1/2}(|2s\rangle + |2p_0\rangle)$$

$$\phi_2^{(0)} = 2^{-1/2}(|2s\rangle - |2p_0\rangle)$$

$$\phi_3^{(0)} = |2p_1\rangle \qquad (9.78)$$

$$\phi_4^{(0)} = |2p_{-1}\rangle$$

The factor $2^{-1/2}$ is needed to normalize the 'correct' eigenfunctions.

9.6 Ground state of the helium atom

In this section we examine the ground-state energy of the helium atom by means of both perturbation theory and the variation method. We may then compare the accuracy of the two procedures.

The potential energy V for a system consisting of two electrons, each with mass m_e and charge $-e$, and a nucleus with atomic number Z and charge $+Ze$ is

$$V = -\frac{Ze'^2}{r_1} - \frac{Ze'^2}{r_2} + \frac{e'^2}{r_{12}}$$

where r_1 and r_2 are the distances of electrons 1 and 2 from the nucleus, r_{12} is the distance between the two electrons, and $e' = e$ for CGS units or $e' = e/(4\pi\varepsilon_0)^{1/2}$ for SI units. If we assume that the nucleus is fixed in space, then the Hamiltonian operator for the two electrons is

$$\hat{H} = -\frac{\hbar^2}{2m_e}(\nabla_1^2 + \nabla_2^2) - \frac{Ze'^2}{r_1} - \frac{Ze'^2}{r_2} + \frac{e'^2}{r_{12}} \qquad (9.79)$$

The operator $\hat{H}$ applies to He for $Z = 2$, Li$^+$ for $Z = 3$, Be^{2+} for $Z = 4$, and so forth.

Perturbation theory treatment

We regard the term e'^2/r_{12} in the Hamiltonian operator as a perturbation, so that

$$\hat{H}' = \hat{H}^{(1)} = \frac{e'^2}{r_{12}} \tag{9.80}$$

In reality, this term is not small in comparison with the other terms so we should not expect the perturbation technique to give accurate results. With this choice for the perturbation, the Schrödinger equation for the unperturbed Hamiltonian operator may be solved exactly.

The unperturbed Hamiltonian operator is the sum of two hydrogen-like Hamiltonian operators, one for each electron

$$\hat{H}^{(0)} = \hat{H}_1^{(0)} + \hat{H}_2^{(0)}$$

where

$$\hat{H}_1^{(0)} = -\frac{\hbar^2}{2m_e} \nabla_1^2 - \frac{Ze'^2}{r_1}$$

$$\hat{H}_2^{(0)} = -\frac{\hbar^2}{2m_e} \nabla_2^2 - \frac{Ze'^2}{r_2}$$

If the unperturbed wave function $\psi^{(0)}$ is written as the product

$$\psi^{(0)}(\mathbf{r}_1, \mathbf{r}_2) = \psi_1^{(0)}(\mathbf{r}_1)\psi_2^{(0)}(\mathbf{r}_2)$$

and the unperturbed energy $E^{(0)}$ is written as the sum

$$E^{(0)} = E_1^{(0)} + E_2^{(0)}$$

then the Schrödinger equation for the two-electron unperturbed system

$$\hat{H}^{(0)}\psi^{(0)}(\mathbf{r}_1, \mathbf{r}_2) = E\psi^{(0)}(\mathbf{r}_1, \mathbf{r}_2)$$

separates into two independent equations,

$$H_i^{(0)}\psi_i^{(0)} = E_i^{(0)}\psi_i^{(0)}, \qquad i = 1, 2$$

which are identical except that one refers to electron 1 and the other to electron 2. The solutions are those of the hydrogen-like atom, as discussed in Chapter 6. The ground-state energy for the unperturbed two-electron system is, then, twice the ground-state energy of a hydrogen-like atom

$$E^{(0)} = -2\left(\frac{Z^2 e'^2}{2a_0}\right) = -\frac{Z^2 e'^2}{a_0} \tag{9.81}$$

The ground-state wave function for the unperturbed two-electron system is the product of two 1s hydrogen-like atomic orbitals

Table 9.1. *Ground-state energy of the helium*
atom

Method	Energy (eV)	% error
Exact	-79.0	
Perturbation theory:		
$\quad E^{(0)}$	-108.8	-37.7
$\quad E^{(0)} + E^{(1)}$	-74.8	$+5.3$
Variation theorem ($\mathscr{E}$)	-77.5	$+1.9$

$$\psi^{(0)}(\mathbf{r}_1, \mathbf{r}_2) = \frac{1}{\pi}\left(\frac{Z}{a_0}\right)^3 e^{-Zr_1/a_0} e^{-Zr_2/a_0} = \frac{1}{\pi}\left(\frac{Z}{a_0}\right)^3 e^{-(\rho_1+\rho_2)/2} \qquad (9.82)$$

where we have defined

$$\rho_i \equiv \frac{2Zr_i}{a_0}, \qquad i = 1, 2 \qquad (9.83)$$

The first-order perturbation correction $E^{(1)}$ to the ground-state energy is obtained by evaluating equation (9.24) with (9.80) as the perturbation and (9.82) as the unperturbed eigenfunction

$$E^{(1)} = \left\langle \psi^{(0)} \left| \frac{e'^2}{r_{12}} \right| \psi^{(0)} \right\rangle = \frac{2Z}{a_0}\left\langle \psi^{(0)} \left| \frac{e'^2}{\rho_{12}} \right| \psi^{(0)} \right\rangle = \frac{Ze'^2}{2^5\pi^2 a_0} I \qquad (9.84)$$

where $\rho_{12} = |\rho_2 - \rho_1|$ and where

$$I = \int \cdots \int \frac{e^{-(\rho_1+\rho_2)}}{\rho_{12}} \rho_1^2 \rho_2^2 \sin\theta_1 \sin\theta_2 \, d\rho_1 \, d\theta_1 \, d\varphi_1 \, d\rho_2 \, d\theta_2 \, d\varphi_2 \qquad (9.85)$$

This six-fold integral I is evaluated in Appendix J and is equal to $20\pi^2$. Thus, we have

$$E^{(1)} = \frac{5Ze'^2}{8a_0} = -\frac{5}{8Z}E^{(0)} \qquad (9.86)$$

The ground-state energy of the perturbed system to first order is, then

$$E = E^{(0)} + E^{(1)} = -\left(Z^2 - \frac{5Z}{8}\right)\frac{e'^2}{a_0} \qquad (9.87)$$

Numerical values of $E^{(0)}$ and $E^{(0)} + E^{(1)}$ for the helium atom ($Z = 2$) are given in Table 9.1 along with the exact value. The unperturbed energy value $E^{(0)}$ has a 37.7% error when compared with the exact value. This large inaccuracy is expected because the perturbation $\hat{H}'$ in equation (9.80) is not small. When the first-order perturbation correction is included, the calculated energy has a 5.3% error, which is still large.

Variation method treatment

As a normalized trial function ϕ for the determination of the ground-state energy by the variation method, we select the unperturbed eigenfunction $\psi^{(0)}(\mathbf{r}_1, \mathbf{r}_2)$ of the perturbation treatment, except that we replace the atomic number Z by a parameter Z'

$$\phi = \phi_1 \phi_2$$

$$\phi_1 = \frac{1}{\pi^{1/2}} \left(\frac{Z'}{a_0}\right)^{3/2} e^{-Z' r_1/a_0} \tag{9.88}$$

$$\phi_2 = \frac{1}{\pi^{1/2}} \left(\frac{Z'}{a_0}\right)^{3/2} e^{-Z' r_2/a_0}$$

The parameter Z' is an effective atomic number whose value is determined by the minimization of $\mathscr{E}$ in equation (9.2). Since the hydrogen-like wave functions ϕ_1 and ϕ_2 are normalized, we have

$$\langle \phi_1 | \phi_1 \rangle = \langle \phi_2 | \phi_2 \rangle = 1 \tag{9.89}$$

The quantity $\mathscr{E}$ is obtained by combining equations (9.2), (9.79), (9.88), and (9.89) to give

$$\mathscr{E} = \left\langle \phi_1 \left| -\frac{\hbar^2}{2m_e} \nabla_1^2 - \frac{Ze'^2}{r_1} \right| \phi_1 \right\rangle + \left\langle \phi_2 \left| -\frac{\hbar^2}{2m_e} \nabla_2^2 - \frac{Ze'^2}{r_2} \right| \phi_2 \right\rangle$$

$$+ \left\langle \phi_1 \phi_2 \left| \frac{e'^2}{r_{12}} \right| \phi_1 \phi_2 \right\rangle \tag{9.90}$$

Note that while the trial function $\phi = \phi_1 \phi_2$ depends on the parameter Z', the Hamiltonian operator contains the true atomic number Z. Therefore, we rewrite equation (9.90) in the form

$$\mathscr{E} = \left\langle \phi_1 \left| -\frac{\hbar^2}{2m_e} \nabla_1^2 - \frac{Z'e'^2}{r_1} \right| \phi_1 \right\rangle + \left\langle \phi_1 \left| \frac{(Z' - Z)e'^2}{r_1} \right| \phi_1 \right\rangle$$

$$+ \left\langle \phi_2 \left| -\frac{\hbar^2}{2m_e} \nabla_2^2 - \frac{Z'e'^2}{r_2} \right| \phi_2 \right\rangle + \left\langle \phi_2 \left| \frac{(Z' - Z)e'^2}{r_2} \right| \phi_2 \right\rangle$$

$$+ \left\langle \phi_1 \phi_2 \left| \frac{e'^2}{r_{12}} \right| \phi_1 \phi_2 \right\rangle \tag{9.91}$$

The first term on the right-hand side is just the energy of a hydrogen-like atom with nuclear charge Z', namely, $-Z'^2 e'^2/2a_0$. The third term has the same value as the first. The second term is evaluated as follows

$$\left\langle \phi_1 \left| \frac{(Z' - Z)e'^2}{r_1} \right| \phi_1 \right\rangle = (Z' - Z)e'^2 \frac{1}{\pi} \left(\frac{Z'}{a_0}\right)^3 \int_0^\infty r_1^{-1} \, e^{-2Z'r_1/a_0} 4\pi r_1^2 \, dr_1$$

$$= (Z' - Z)e'^2 \frac{Z'}{a_0}$$

where equations (A.26) and (A.28) have been used. The fourth term equals the second. The fifth term is identical to $E^{(1)}$ of the perturbation treatment given by equation (9.86) except that Z is replaced by Z' and therefore this term equals $5Z'e'^2/8a_0$. Thus, the quantity $\mathscr{E}$ in equation (9.91) is

$$\mathscr{E} = 2\left[-\frac{Z'^2 e'^2}{2a_0}\right] + 2\left[\frac{Z'(Z'-Z)e'^2}{a_0}\right] + \frac{5Z'e'^2}{8a_0} = [Z'^2 - 2(Z - \tfrac{5}{16})Z']\frac{e'^2}{a_0}$$

$$(9.92)$$

We next minimize $\mathscr{E}$ with respect to the parameter Z'

$$\frac{d\mathscr{E}}{dZ'} = 2[Z' - (Z - \tfrac{5}{16})]\frac{e'^2}{a_0} = 0$$

so that

$$Z' = Z - \tfrac{5}{16}$$

Substituting this result into equation (9.92) gives

$$\mathscr{E} = -(Z - \tfrac{5}{16})^2 \frac{e'^2}{a_0} \geqslant E_0 \tag{9.93}$$

as an upper bound for the ground-state energy E_0.

When applied to the helium atom ($Z = 2$), this upper bound is

$$\mathscr{E} = -\left(\frac{27}{16}\right)^2 \frac{e'^2}{a_0} = -2.85 \frac{e'^2}{a_0} \tag{9.94}$$

The numerical value of $\mathscr{E}$ is listed in Table 9.1. The simple variation function (9.88) gives an upper bound to the energy with a 1.9% error in comparison with the exact value. Thus, the variation theorem leads to a more accurate result than the perturbation treatment. Moreover, a more complex trial function with more parameters should be expected to give an even more accurate estimate.

Problems

9.1 The Hamiltonian operator for a hydrogen atom in a uniform external electric field E along the *z*-coordinate axis is

$$\hat{H} = \frac{-\hbar^2}{2\mu}\nabla^2 - \frac{e'^2}{r} - e\mathsf{E}z$$

Use the variation trial function $\phi = \psi_{1s}(1 + \lambda z)$, where λ is the variation parameter, to estimate the ground-state energy for this system.

9.2 Apply the gaussian function

$$\phi = e^{-\lambda r^2 / a_0^2}$$

where λ is a parameter, as the variation trial function to estimate the energy of the ground state of the hydrogen atom. What is the percent error?

9.3 Apply the variation trial function $\phi(x) = x(a - x)(a - 2x)$ to estimate the energy of a particle in a box with $V(x) = 0$ for $0 \le x \le a$, $V(x) = \infty$ for $x < 0$, $x > a$. To which energy level does this estimate apply?

9.4 Consider a particle in a one-dimensional potential well such that

$$V(x) = (b\hbar^2 / ma^4)x(x - a), \qquad 0 \le x \le a$$

$$= \infty, \qquad\qquad\qquad x < 0, x > a$$

where b is a dimensionless parameter. Using the particle in a box with $V(x) = 0$ for $0 \le x \le a$, $V(x) = \infty$ for $x < 0$, $x > a$ as the unperturbed system, calculate the first-order perturbation correction to the energy levels. (See Appendix A for the evaluation of the resulting integrals.)

9.5 Consider a particle in a one-dimensional potential well such that

$$V(x) = (b\hbar^2 / ma^3)x, \qquad 0 \le x \le a$$

$$= \infty, \qquad\qquad\quad x < 0, x > a$$

where b is a dimensionless parameter. Using the particle in a box with $V(x) = 0$ for $0 \le x \le a$, $V(x) = \infty$ for $x < 0$, $x > a$ as the unperturbed system, calculate the first-order perturbation correction to the energy levels. (See Appendix A for the evaluation of the resulting integral.)

9.6 Calculate the second-order perturbation correction to the ground-state energy for the system in problem 9.5. (Use integration by parts and see Appendix A for the evaluation of the resulting integral.)

9.7 Apply the linear variation function

$$\phi = c_1(2/a)^{1/2} \sin(\pi x/a) + c_2(2/a)^{1/2} \sin(2\pi x/a)$$

for $0 \le x \le a$ to the system in problem 9.5. Set the parameter b in the potential equal to $\pi^2/8$. Solve the secular equation to obtain estimates for the energies E_1 and E_2 of the ground state and first-excited state. Compare this estimate for E_1 with the ground-state energies obtained by first-order and second-order perturbation theory. Then determine the variation functions ϕ_1 and ϕ_2 that correspond to E_1 and E_2.

9.8 Consider a particle in a one-dimensional champagne bottle[2] for which

$$V(x) = (\pi^2 \hbar^2 / 8ma^2) \sin(\pi x/a), \qquad 0 \le x \le a$$

$$= \infty, \qquad\qquad\qquad\qquad\quad x < 0, x > a$$

[2] G. R. Miller (1979) *J. Chem. Educ.* **56**, 709.

Calculate the first-order perturbation correction to the ground-state energy level using the particle in a box with $V(x) = 0$ for $0 \leqslant x \leqslant a$, $V(x) = \infty$ for $x < 0$, $x > a$ as the unperturbed system. Then calculate the first-order perturbation correction to the ground-state wave function, terminating the expansion after the term $k = 5$. (See Appendix A for trigonometric identities and integrals.)

9.9 Using first-order perturbation theory, determine the ground-state energy of a hydrogen atom in which the nucleus is not regarded as a point charge. Instead, regard the nucleus as a sphere of radius b throughout which the charge $+e$ is evenly distributed. The potential of interaction between the nucleus and the electron is

$$V(r) = \frac{-e'^2}{2b}\left(3 - \frac{r^2}{b^2}\right), \qquad 0 \leqslant r \leqslant b$$

$$= \frac{-e'^2}{r}, \qquad\qquad r > b$$

The unperturbed system is, of course, the hydrogen atom with a point nucleus. (Inside the nuclear sphere, the exponential

$$e^{-2r/a_0} = 1 - (2r/a_0) + (2r^2/a_0^2) - \cdots$$

may be approximated by unity because r is very small in that region.)

9.10 Using first-order perturbation theory, show that the spin–orbit interaction energy for a hydrogen atom is given by

$$\tfrac{1}{2}\alpha^2 |E_n^{(0)}| \frac{1}{n(l + \tfrac{1}{2})(l + 1)} \qquad \text{for } j = l + \tfrac{1}{2},\ l \neq 0$$

$$-\tfrac{1}{2}\alpha^2 |E_n^{(0)}| \frac{1}{nl(l + \tfrac{1}{2})} \qquad \text{for } j = l - \tfrac{1}{2},\ l \neq 0$$

The Hamiltonian operator is given in equation (7.33), where $\hat{H}_0$ represents the unperturbed system and $\hat{H}_{\text{so}}$ is the perturbation. Use equations (6.74) and (6.78) to evaluate the expectation value of $\xi(r)$.

10

Molecular structure

A molecule is composed of positively charged nuclei surrounded by electrons. The stability of a molecule is due to a balance among the mutual repulsions of nuclear pairs, attractions of nuclear–electron pairs, and repulsions of electron pairs as modified by the interactions of their spins. Both the nuclei and the electrons are in constant motion relative to the center of mass of the molecule. However, the nuclear masses are much greater than the electronic mass and, as a result, the nuclei move much more slowly than the electrons. Thus, the basic molecular structure is a stable framework of nuclei undergoing rotational and vibrational motions surrounded by a cloud of electrons described by the electronic probability density.

In this chapter we present in detail the separation of the nuclear and electronic motions, the nuclear motion within a molecule, and the coupling between nuclear and electronic motion.

10.1 Nuclear structure and motion

We consider a molecule with Ω nuclei, each with atomic number Z_α and mass M_α ($\alpha = 1, 2, \ldots, \Omega$), and N electrons, each of mass m_e. We denote by $\mathbf{Q}$ the set of all nuclear coordinates and by $\mathbf{r}$ the set of all electronic coordinates. The positions of the nuclei and electrons are specified relative to an external set of coordinate axes which are fixed in space.

The Hamiltonian operator $\hat{H}$ for this system of $\Omega + N$ particles may be written in the form

$$\hat{H} = \hat{T}_Q + V_Q + \hat{H}_e \tag{10.1}$$

where $\hat{T}_Q$ is the kinetic energy operator for the nuclei

$$\hat{T}_Q = -\hbar^2 \sum_{\alpha=1}^{\Omega} \frac{\nabla_\alpha^2}{2M_\alpha} \tag{10.2}$$

V_Q is the potential energy of interaction between nuclear pairs

$$V_Q = \sum_{\alpha<\beta=1}^{\Omega} \frac{Z_\alpha Z_\beta e'^2}{r_{\alpha\beta}} \tag{10.3}$$

and $\hat{H}_e$ is the electronic Hamiltonian operator

$$\hat{H}_e = -\frac{\hbar^2}{2m_e} \sum_{i=1}^{N} \nabla_i^2 - \sum_{\alpha=1}^{\Omega}\sum_{i=1}^{N} \frac{Z_\alpha e'^2}{r_{\alpha i}} + \sum_{i<j=1}^{N} \frac{e'^2}{r_{ij}} \tag{10.4}$$

The symbols ∇_α^2 and ∇_i^2 are, respectively, the laplacian operators for a single nucleus and a single electron. The variable $r_{\alpha\beta}$ is the distance between nuclei α and β, $r_{\alpha i}$ the distance between nucleus α and electron i, and r_{ij} the distance between electrons i and j. The summations are taken over each pair of particles. The quantity e' is equal to the magnitude of the electronic charge e in CGS units and to $e/(4\pi\varepsilon_0)^{1/2}$ in SI units, where ε_0 is the permittivity of free space.

The Schrödinger equation for the molecule is

$$\hat{H}\Psi(\mathbf{r}, \mathbf{Q}) = E\Psi(\mathbf{r}, \mathbf{Q}) \tag{10.5}$$

where $\Psi(\mathbf{r}, \mathbf{Q})$ is an eigenfunction and E the corresponding eigenvalue. The differential equation (10.5) cannot be solved as it stands because there are too many variables. However, approximate, but very accurate, solutions may be found if the equation is simplified by recognizing that the nuclei and the electrons differ greatly in mass and, as a result, differ greatly in their relative speeds of motion.

Born–Oppenheimer approximation

The simplest approximate method for solving the Schrödinger equation (10.5) uses the so-called *Born–Oppenheimer approximation*. This method is a two-step process. The first step is to recognize that the nuclei are much heavier than an electron and, consequently, move very slowly in comparison with the electronic motion. Thus, the electronic part of the Schrödinger equation may be solved under the condition that the nuclei are motionless. The resulting electronic energy may then be determined for many different fixed nuclear configurations. In the second step, the nuclear part of the Schrödinger equation is solved by regarding the motion of the nuclei as taking place in the average potential field created by the fast-moving electrons.

In the first step of the Born–Oppenheimer approximation, the nuclei are all held at fixed equilibrium positions. Thus, the coordinates $\mathbf{Q}$ do not change with time and the kinetic energy operator $\hat{T}_Q$ in equation (10.2) vanishes. The Schrödinger equation (10.5) under this condition becomes

$$(\hat{H}_e + V_Q)\psi_\kappa(\mathbf{r}, \mathbf{Q}) = \varepsilon_\kappa(\mathbf{Q})\psi_\kappa(\mathbf{r}, \mathbf{Q}) \tag{10.6}$$

where the coordinates $\mathbf{Q}$ are no longer variables, but rather are constant parameters. For each electronic state κ, the electronic energy $\varepsilon_\kappa(\mathbf{Q})$ of the molecule and the eigenfunction $\psi_\kappa(\mathbf{r}, \mathbf{Q})$ depend parametrically on the fixed values of the coordinates $\mathbf{Q}$. The nuclear–nuclear interaction potential V_Q is now a constant and its value is included in $\varepsilon_\kappa(\mathbf{Q})$.

We assume in this section and in Section 10.2 that equation (10.6) has been solved and that the eigenfunctions $\psi_\kappa(\mathbf{r}, \mathbf{Q})$ and eigenvalues $\varepsilon_\kappa(\mathbf{Q})$ are known for any arbitrary set of values for the parameters $\mathbf{Q}$. Further, we assume that the eigenfunctions form a complete orthonormal set, so that

$$\int \psi_\kappa^*(\mathbf{r}, \mathbf{Q})\psi_\lambda(\mathbf{r}, \mathbf{Q})\,d\mathbf{r} = \delta_{\kappa\lambda} \tag{10.7}$$

In the second step of the Born–Oppenheimer approximation, the energy $\varepsilon_\kappa(\mathbf{Q})$ is used as a potential energy function to treat the nuclear motion. In this case, equation (10.5) becomes

$$[\hat{T}_Q + \varepsilon_\kappa(\mathbf{Q})]\chi_{\kappa\nu}(\mathbf{Q}) = E_{\kappa\nu}\chi_{\kappa\nu}(\mathbf{Q}) \tag{10.8}$$

where the nuclear wave function $\chi_{\kappa\nu}(\mathbf{Q})$ depends on the nuclear coordinates $\mathbf{Q}$ and on the electronic state κ. Each electronic state κ gives rise to a series of nuclear states, indexed by ν. Thus, for each electronic state κ, the eigenfunctions of $[\hat{T}_Q + \varepsilon_\kappa(\mathbf{Q})]$ are $\chi_{\kappa\nu}(\mathbf{Q})$ with eigenvalues $E_{\kappa\nu}$. In practice, the nuclear states are differentiated by several quantum numbers; the index ν represents, then, a set of these quantum numbers. In the solution of the differential equation (10.8), the nuclear coordinates $\mathbf{Q}$ in $\varepsilon_\kappa(\mathbf{Q})$ are treated as variables. The nuclear energy $E_{\kappa\nu}$, of course, does not depend on any parameters. Most applications of equation (10.8) are to molecules in their electronic ground states ($\kappa = 0$).

In the original mathematical treatment[1] of nuclear and electronic motion, M. Born and J. R. Oppenheimer (1927) applied perturbation theory to equation (10.5) using the kinetic energy operator $\hat{T}_Q$ for the nuclei as the perturbation. The proper choice for the expansion parameter is $\lambda = (m_e/\overline{M})^{1/4}$, where $\overline{M}$ is the mean nuclear mass

$$\overline{M} = \frac{1}{\Omega}\sum_{\alpha=1}^{\Omega} M_\alpha$$

When terms up to λ^2 are retained, the exact total energy of the molecule is

---

[1] M. Born and J. R. Oppenheimer (1927) *Ann. Physik* **84**, 457.

approximated by the energy $E_{\kappa\nu}$ of equation (10.8) and the eigenfunction $\Psi(\mathbf{Q}, \mathbf{r})$ is approximated by the product

$$\Psi(\mathbf{Q}, \mathbf{r}) \approx \chi_{\kappa\nu}(\mathbf{Q})\psi_\kappa(\mathbf{Q}, \mathbf{r}) \qquad (10.9)$$

Perturbation terms in the Hamiltonian operator up to λ^4 still lead to the uncoupling of the nuclear and electronic motions, but change the form of the electronic potential energy function in the equation for the nuclear motion. Rather than present the details of the Born–Oppenheimer perturbation expansion, we follow instead the equivalent, but more elegant procedure[2] of M. Born and K. Huang (1954).

Born–Huang treatment

Under the assumption that the Schrödinger equation (10.6) has been solved for the complete set of orthonormal eigenfunctions $\psi_\kappa(\mathbf{r}, \mathbf{Q})$, we may expand the eigenfunction $\Psi(\mathbf{r}, \mathbf{Q})$ of equation (10.5) in terms of $\psi_\kappa(\mathbf{r}, \mathbf{Q})$

$$\Psi(\mathbf{r}, \mathbf{Q}) = \sum_\lambda \chi_\lambda(\mathbf{Q})\psi_\lambda(\mathbf{r}, \mathbf{Q}) \qquad (10.10)$$

where $\chi_\lambda(\mathbf{Q})$ are the expansion coefficients. Substitution of equation (10.10) into (10.5) using (10.1) gives

$$\sum_\lambda (\hat{T}_Q + V_Q + \hat{H}_e - E)\chi_\lambda(\mathbf{Q})\psi_\lambda(\mathbf{r}, \mathbf{Q}) = 0 \qquad (10.11)$$

where the operators have been placed inside the summation. Since the operator $\hat{H}_e$ commutes with the function $\chi_\lambda(\mathbf{Q})$, we may substitute equation (10.6) into (10.11) to obtain

$$\sum_\lambda [\hat{T}_Q + \varepsilon_\lambda(\mathbf{Q}) - E]\chi_\lambda(\mathbf{Q})\psi_\lambda(\mathbf{r}, \mathbf{Q}) = 0 \qquad (10.12)$$

We next multiply equation (10.12) by $\psi_\kappa^*(\mathbf{r}, \mathbf{Q})$ and integrate over the set of electronic coordinates $\mathbf{r}$, giving

$$\sum_\lambda \int \psi_\kappa^*(\mathbf{r}, \mathbf{Q})\hat{T}_Q[\chi_\lambda(\mathbf{Q})\psi_\lambda(\mathbf{r}, \mathbf{Q})]\, d\mathbf{r} + [\varepsilon_\kappa(\mathbf{Q}) - E]\chi_\kappa(\mathbf{Q}) = 0 \qquad (10.13)$$

where we have used the orthonormal property (equation (10.7)). The operator $\hat{T}_Q$ acts on both functions in the product $\chi_\lambda(\mathbf{Q})\psi_\lambda(\mathbf{r}, \mathbf{Q})$ and involves the second derivative with respect to the nuclear coordinates $\mathbf{Q}$. To expand the expression $\hat{T}_Q[\chi_\lambda(\mathbf{Q})\psi_\lambda(\mathbf{r}, \mathbf{Q})]$, we note that

---

[2] M. Born and K. Huang (1954) *Dynamical Theory of Crystal Lattices* (Oxford University Press, Oxford), pp. 406–7.

$$\nabla_a \chi \psi = \psi \nabla_a \chi + \chi \nabla_a \psi$$

$$\nabla_a^2 \chi \psi = \psi \nabla_a^2 \chi + \chi \nabla_a^2 \psi + 2 \nabla_a \chi \cdot \nabla_a \psi$$

Therefore, we obtain

$$\hat{T}_Q[\chi_\lambda(\mathbf{Q})\psi_\lambda(\mathbf{r},\,\mathbf{Q})] = -\hbar^2 \sum_{a=1}^{\Omega} \frac{1}{2M_a} \nabla_a^2[\chi_\lambda(\mathbf{Q})\psi_\lambda(\mathbf{r},\,\mathbf{Q})]$$

$$= \psi_\lambda(\mathbf{r},\,\mathbf{Q})\hat{T}_Q\chi_\lambda(\mathbf{Q}) + \chi_\lambda(\mathbf{Q})\hat{T}_Q\psi_\lambda(\mathbf{r},\,\mathbf{Q})$$

$$- \hbar^2 \sum_{a=1}^{\Omega} \frac{1}{M_a} \nabla_a \chi_\lambda(\mathbf{Q}) \cdot \nabla_a \psi_\lambda(\mathbf{r},\,\mathbf{Q}) \qquad (10.14)$$

Substitution of equation (10.14) into (10.13) yields

$$[\hat{T}_Q + \varepsilon_\kappa(\mathbf{Q}) - E]\chi_\kappa(\mathbf{Q}) + \sum_\lambda (c_{\kappa\lambda} + \hat{\Lambda}_{\kappa\lambda})\chi_\lambda(\mathbf{Q}) = 0 \qquad (10.15)$$

where the coefficients $c_{\kappa\lambda}(\mathbf{Q})$ and the operators $\hat{\Lambda}_{\kappa\lambda}$ are defined by

$$c_{\kappa\lambda}(\mathbf{Q}) \equiv \int \psi_\kappa^*(\mathbf{r},\,\mathbf{Q})\hat{T}_Q\psi_\lambda(\mathbf{r},\,\mathbf{Q})\,d\mathbf{r} \qquad (10.16)$$

$$\ddot{\Lambda}_{\kappa\lambda} \equiv -\hbar^2 \sum_{a=1}^{\Omega} \frac{1}{M_a} \int \psi_\kappa^*(\mathbf{r},\,\mathbf{Q})\nabla_a\psi_\lambda(\mathbf{r},\,\mathbf{Q})\,d\mathbf{r} \cdot \nabla_a \qquad (10.17)$$

and equation (10.7) has been used. Since we have assumed that the electronic eigenfunctions $\psi_\kappa(\mathbf{r},\,\mathbf{Q})$ are known for all values of the parameters $\mathbf{Q}$, the coefficients $c_{\kappa\lambda}(\mathbf{Q})$ and the operators $\hat{\Lambda}_{\kappa\lambda}$ may be determined. The set of coupled equations (10.15) for the functions $\chi_\kappa(\mathbf{Q})$ is exact.

The integral I contained in the operator $\hat{\Lambda}_{\kappa\kappa}$ is

$$I \equiv \int \psi_\kappa^*(\mathbf{r},\,\mathbf{Q})\nabla_a\psi_\kappa(\mathbf{r},\,\mathbf{Q})\,d\mathbf{r}$$

For stationary states, the eigenfunctions $\psi_\kappa(\mathbf{r},\,\mathbf{Q})$ may be chosen to be real functions, so that this integral can also be written as

$$I = \tfrac{1}{2}\nabla_a \int [\psi_\kappa(\mathbf{r},\,\mathbf{Q})]^2 \, d\mathbf{r}$$

According to equation (10.7), the integral I vanishes and, therefore, we have $\hat{\Lambda}_{\kappa\kappa} = 0$.

We now write equation (10.15) as

$$[\hat{T}_Q + U_\kappa(\mathbf{Q}) - E]\chi_\kappa(\mathbf{Q}) + \sum_{\lambda(\neq\kappa)} (c_{\kappa\lambda} + \hat{\Lambda}_{\kappa\lambda})\chi_\lambda(\mathbf{Q}) = 0 \qquad (10.18)$$

where $U_\kappa(\mathbf{Q})$ is defined by

$$U_\kappa(\mathbf{Q}) = \varepsilon_\kappa(\mathbf{Q}) + c_{\kappa\kappa}(\mathbf{Q}) \tag{10.19}$$

The first term on the left-hand side of equation (10.18) has the form of a Schrödinger equation for nuclear motion, so that we may identify the expansion coefficient $\chi_\kappa(\mathbf{Q})$ as a nuclear wave function for the electronic state κ. The second term couples the influence of all the other electronic states to the nuclear motion for a molecule in the electronic state κ.

If the coefficients $c_{\kappa\kappa}(\mathbf{Q})$ and $c_{\kappa\lambda}(\mathbf{Q})$ and the operators $\hat{\Lambda}_{\kappa\lambda}$ are sufficiently small, the summation on the left-hand side of equation (10.18) and $c_{\kappa\kappa}(\mathbf{Q})$ in (10.19) may be neglected, giving a zeroth-order equation for the nuclear motion

$$[\hat{T}_Q + \varepsilon_\kappa(\mathbf{Q}) - E^{(0)}_{\kappa\nu}]\chi^{(0)}_{\kappa\nu}(\mathbf{Q}) = 0 \tag{10.20}$$

where $\chi^{(0)}_{\kappa\nu}(\mathbf{Q})$ and $E^{(0)}_{\kappa\nu}$ are the zeroth-order approximations to the nuclear wave functions and energy levels. The index ν represents a set of quantum numbers which determine the nuclear state. The neglect of these coefficients and operators is the Born–Oppenheimer approximation and equation (10.20) is identical to (10.8). Furthermore, the molecular wave function $\Psi(\mathbf{r}, \mathbf{Q})$ in equation (10.10) reduces to the product of a nuclear and an electronic wave function as shown in equation (10.9).

When the coupling coefficients $c_{\kappa\lambda}$ for $\kappa \neq \lambda$ and the coupling operators $\hat{\Lambda}_{\kappa\lambda}$ are neglected, but the coefficient $c_{\kappa\kappa}(\mathbf{Q})$ is retained, equation (10.18) becomes

$$[\hat{T}_Q + U_\kappa(\mathbf{Q}) - E^{(1)}_{\kappa\nu}]\chi^{(1)}_{\kappa\nu}(\mathbf{Q}) = 0 \tag{10.21}$$

where $\chi^{(1)}_{\kappa\nu}(\mathbf{Q})$ and $E^{(1)}_{\kappa\nu}$ are the first-order approximations to the nuclear wave functions and energy levels. Since the term $c_{\kappa\kappa}(\mathbf{Q})$ is added to $\varepsilon_\kappa(\mathbf{Q})$ in this approximation, equation (10.21) is different from (10.20) and, therefore, $\chi^{(1)}_{\kappa\nu}(\mathbf{Q})$ and $E^{(1)}_{\kappa\nu}$ differ from $\chi^{(0)}_{\kappa\nu}(\mathbf{Q})$ and $E^{(0)}_{\kappa\nu}$. In this first-order approximation, the molecular wave function $\Psi(\mathbf{r}, \mathbf{Q})$ in equation (10.10) also takes the form of (10.9). The factor $\chi^{(1)}_{\kappa\nu}(\mathbf{Q})$ describes the nuclear motion, which takes place in a potential field $U_\kappa(\mathbf{Q})$ determined by the electrons moving as though the nuclei were fixed in their instantaneous positions. Thus, the electronic state of the molecule changes in a continuous manner as the nuclei move slowly in comparison with the electronic motion. In this situation, the electrons are said to follow the nuclei *adiabatically* and this first-order approximation is known as the *adiabatic approximation*. This adiabatic feature does not occur in higher-order approximations, in which coupling terms appear.

An analysis using perturbation theory shows that the influence of the coupling terms with $c_{\kappa\lambda}(\mathbf{Q})$ and $\hat{\Lambda}_{\kappa\lambda}$ is small when the electronic energy levels $\varepsilon_\kappa(\mathbf{Q})$ and $\varepsilon_\lambda(\mathbf{Q})$ are not closely spaced. Since the ground-state electronic energy of a molecule is usually widely separated from the first-excited electronic energy level, the Born–Oppenheimer approximation and especially the adiabatic approximation are quite accurate for the electronic ground state. The influence of the coupling terms for the first few excited electronic energy levels may then be calculated using perturbation theory.

10.2 Nuclear motion in diatomic molecules

The application of the Born–Oppenheimer and the adiabatic approximations to separate nuclear and electronic motions is best illustrated by treating the simplest example, a diatomic molecule in its electronic ground state. The diatomic molecule is sufficiently simple that we can also introduce center-of-mass coordinates and show explicitly how the translational motion of the molecule as a whole is separated from the internal motion of the nuclei and electrons.

Center-of-mass coordinates

The total number of spatial coordinates for a molecule with Ω nuclei and N electrons is $3(\Omega + N)$, because each particle requires three cartesian coordinates to specify its location. However, if the motion of each particle is referred to the center of mass of the molecule rather than to the external spaced-fixed coordinate axes, then the three translational coordinates that specify the location of the center of mass relative to the external axes may be separated out and eliminated from consideration. For a diatomic molecule ($\Omega = 2$) we are left with only three *relative* nuclear coordinates and with $3N$ *relative* electronic coordinates. For mathematical convenience, we select the center of mass of the nuclei as the reference point rather than the center of mass of the nuclei and electrons together. The difference is negligibly small. We designate the two nuclei as A and B, and introduce a new set of nuclear coordinates defined by

$$\mathbf{X} = \frac{M_A}{M}\mathbf{Q}_A + \frac{M_B}{M}\mathbf{Q}_B \tag{10.22a}$$

$$\mathbf{R} = \mathbf{Q}_B - \mathbf{Q}_A \tag{10.22b}$$

where $\mathbf{X}$ locates the center of mass of the nuclei in the external coordinate system, $\mathbf{R}$ is the vector distance between the two nuclei, and M is the sum of the nuclear masses

$$M = M_A + M_B$$

The kinetic energy operator $\hat{T}_Q$ for the two nuclei, as given by equation (10.2), is

$$\hat{T}_Q = \frac{-\hbar^2}{2}\left(\frac{\nabla_A^2}{M_A} + \frac{\nabla_B^2}{M_B}\right) \tag{10.23}$$

The laplacian operators in equation (10.23) refer to the spaced-fixed coordinates $\mathbf{Q}_\alpha$ with components $Q_{x\alpha}$, $Q_{y\alpha}$, $Q_{z\alpha}$, so that

$$\nabla_\alpha^2 = \frac{\partial^2}{\partial Q_{x\alpha}^2} + \frac{\partial^2}{\partial Q_{y\alpha}^2} + \frac{\partial^2}{\partial Q_{z\alpha}^2}, \qquad \alpha = A, B$$

However, these operators change their form when the reference coordinate system is transformed from space fixed to center of mass.

To transform these laplacian operators to the coordinates $\mathbf{X}$ and $\mathbf{R}$, with components X_x, X_y, X_z and R_x, R_y, R_z, respectively, we note that

$$\frac{\partial}{\partial Q_{xA}} = \frac{\partial X_x}{\partial Q_{xA}}\frac{\partial}{\partial X_x} + \frac{\partial R_x}{\partial Q_{xA}}\frac{\partial}{\partial R_x} = \frac{M_A}{M}\frac{\partial}{\partial X_x} - \frac{\partial}{\partial R_x}$$

$$\frac{\partial}{\partial Q_{xB}} = \frac{\partial X_x}{\partial Q_{xB}}\frac{\partial}{\partial X_x} + \frac{\partial R_x}{\partial Q_{xB}}\frac{\partial}{\partial R_x} = \frac{M_B}{M}\frac{\partial}{\partial X_x} + \frac{\partial}{\partial R_x}$$

from which it follows that

$$\frac{\partial^2}{\partial Q_{xA}^2} = \left(\frac{M_A}{M}\right)^2\frac{\partial^2}{\partial X_x^2} + \frac{\partial^2}{\partial R_x^2} - \frac{2M_A}{M}\frac{\partial^2}{\partial X_x \partial R_x}$$

$$\frac{\partial^2}{\partial Q_{xB}^2} = \left(\frac{M_B}{M}\right)^2\frac{\partial^2}{\partial X_x^2} + \frac{\partial^2}{\partial R_x^2} + \frac{2M_B}{M}\frac{\partial^2}{\partial X_x \partial R_x}$$

Analogous expressions apply for Q_{yA}, Q_{yB}, Q_{zA}, and Q_{zB}. Therefore, in terms of the coordinates $\mathbf{X}$ and $\mathbf{R}$, the operators ∇_A^2 and ∇_B^2 are

$$\nabla_A^2 = \left(\frac{M_A}{M}\right)^2\nabla_X^2 + \nabla_R^2 - \frac{2M_A}{M}\boldsymbol{\nabla}_X\cdot\boldsymbol{\nabla}_R \tag{10.24a}$$

$$\nabla_B^2 = \left(\frac{M_B}{M}\right)^2\nabla_X^2 + \nabla_R^2 + \frac{2M_B}{M}\boldsymbol{\nabla}_X\cdot\boldsymbol{\nabla}_R \tag{10.24b}$$

where $\nabla_X^2 = \boldsymbol{\nabla}_X\cdot\boldsymbol{\nabla}_X$ and $\nabla_R^2 = \boldsymbol{\nabla}_R\cdot\boldsymbol{\nabla}_R$ are the laplacian operators for the vectors $\mathbf{X}$ and $\mathbf{R}$ and where $\boldsymbol{\nabla}_X$ and $\boldsymbol{\nabla}_R$ are the gradient operators. When the transformations (10.24) are substituted into (10.23), the operator $\hat{T}_Q$ becomes

$$\hat{T}_Q = \frac{-\hbar^2}{2}\left(\frac{1}{M}\nabla_X^2 + \frac{1}{\mu}\nabla_R^2\right) \tag{10.25}$$

where μ is the *reduced mass* of the two nuclei

$$\frac{1}{\mu} = \frac{1}{M_A} + \frac{1}{M_B}$$

or

$$\mu = \frac{M_A M_B}{M_A + M_B} \tag{10.26}$$

The cross terms in $\nabla_X \cdot \nabla_R$ cancel each other.

For the diatomic molecule, equations (10.1), (10.3), (10.5), and (10.25) combine to give

$$\left[-\frac{\hbar^2}{2M} \nabla_X^2 - \frac{\hbar^2}{2\mu} \nabla_R^2 + \frac{Z_A Z_B e'^2}{R} + \hat{H}_e - E_{\text{tot}} \right] \Psi_{\text{tot}} = 0 \tag{10.27}$$

where $R \equiv r_{AB}$ is the magnitude of the vector $\mathbf{R}$ and where now the laplacian operator ∇_i^2 in $\hat{H}_e$ of equation (10.4) refers to the position of electron i relative to the center of mass. The interparticle distances $r_{AB} = R$, r_{Ai}, r_{Bi}, and r_{ij} are independent of the choice of reference coordinate system and do not change as a result of the transformation from external to internal coordinates. If we write Ψ_{tot} as the product

$$\Psi_{\text{tot}} = \Phi(\mathbf{X}) \Psi(\mathbf{R}, \mathbf{r})$$

and E_{tot} as the sum

$$E_{\text{tot}} = E_{\text{cm}} + E$$

then the differential equation (10.27) separates into two independent differential equations

$$-\frac{\hbar^2}{2M} \nabla_X^2 \Phi(\mathbf{X}) - E_{\text{cm}} \Phi(\mathbf{X}) \tag{10.28a}$$

and

$$\left[-\frac{\hbar^2}{2\mu} \nabla_R^2 + \frac{Z_A Z_B e'^2}{R} + \hat{H}_e - E \right] \Psi(\mathbf{R}, \mathbf{r}) = 0 \tag{10.28b}$$

Equation (10.28b) describes the internal motions of the two nuclei and the electrons relative to the center of mass. Our next goal is to solve this equation using the method described in Section 10.1. Equation (10.28a), on the other hand, describes the translational motion of the center of mass of the molecule and is not considered any further here.

Electronic motion and the nuclear potential function

The first step in the solution of equation (10.28b) is to hold the two nuclei fixed in space, so that the operator ∇_R^2 drops out. Equation (10.28b) then takes the form of (10.6). Since the diatomic molecule has axial symmetry, the eigenfunctions and eigenvalues of $\hat{H}_e$ in equation (10.6) depend only on the fixed value R of the internuclear distance, so that we may write them as $\psi_\kappa(\mathbf{r}, R)$ and $\varepsilon_\kappa(R)$. If equation (10.6) is solved repeatedly to obtain the ground-state energy $\varepsilon_0(R)$ for many values of the parameter R, then a curve of the general form shown in Figure 10.1 is obtained. The value of R for which $\varepsilon_0(R)$ is a minimum represents the equilibrium or most stable nuclear configuration for the molecule. As the parameter R increases or decreases, the molecular energy $\varepsilon_0(R)$ increases. As R becomes small, the nuclear repulsion term V_Q becomes very large and $\varepsilon_0(R)$ rapidly approaches infinity. As R becomes very large $(R \rightarrow \infty)$, the molecule dissociates into its two constituent atoms. We assume that equation (10.6) has been solved for the ground-state wave function $\psi_0(\mathbf{r}, R)$ and ground-state energy $\varepsilon_0(R)$ for all values of the parameter R from zero to infinity.

The potential energy function $U_0(R)$ for the ground electronic state is given by equations (10.19) and (10.16) with $\hat{T}_Q = (-\hbar^2/2\mu)\nabla_R^2$ as

$$U_0(R) = \varepsilon_0(R) + c_{00}(R) = \varepsilon_0(R) - \frac{\hbar^2}{2} \int \psi_0^*(\mathbf{r}, R)\nabla_R^2\psi_0(\mathbf{r}, R)\,d\mathbf{r}$$

Within the adiabatic approximation, the term $c_{00}(R)$ evaluates the coupling between the ground-state motion of the electrons and the motion of the nuclei. The magnitude of this term at distances R near the minimum of $\varepsilon_0(R)$ is not negligible[3] for the lightweight hydrogen molecule (all isotopes), the hydrogen-molecule ion (all isotopes), and the system He_2. However, the general shape of the function $U_0(R)$ for these systems does not differ appreciably from the schematic shape of $\varepsilon_0(R)$ shown in Figure 10.1. For heavier nuclei, the term $c_{00}(R)$ is small and may be neglected. For these molecules the Born–

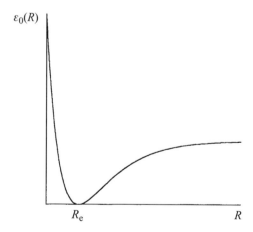

Figure 10.1 The internuclear potential energy for the ground state of a diatomic molecule.

[3] See J. O. Hirschfelder and W. J. Meath (1967) *Advances in Chemical Physics*, Vol. XII (John Wiley and Sons, New York), p. 23 and references cited therein.

Oppenheimer and the adiabatic approximations are essentially identical. Since we are interested here in only the ground electronic state, we drop the subscript on $U_0(R)$ from this point on for the sake of simplicity.

The functional form of $U(R)$ differs from one diatomic molecule to another. Accordingly, we wish to find a general form which can be used for all molecules. Under the assumption that the internuclear distance R does not fluctuate very much from its equilibrium value R_e so that $U(R)$ does not deviate greatly from its minimum value, we may expand the potential $U(R)$ in a Taylor's series about the equilibrium distance R_e

$$U = U(R_e) + U^{(1)}(R_e)(R - R_e) + \frac{1}{2!}U^{(2)}(R_e)(R - R_e)^2$$

$$+ \frac{1}{3!}U^{(3)}(R_e)(R - R_e)^3 + \frac{1}{4!}U^{(4)}(R_e)(R - R_e)^4 + \cdots$$

where

$$U^{(l)}(R_e) \equiv \frac{d^l U(R)}{dR^l}\bigg|_{R=R_e}, \qquad l = 1, 2, \ldots$$

The first derivative $U^{(1)}(R_e)$ vanishes because the potential $U(R)$ is a minimum at the distance R_e. The second derivative $U^{(2)}(R_e)$ is called the *force constant* for the diatomic molecule (see Section 4.1) and is given the symbol k. We also introduce the relative distance variable q, defined as

$$q \equiv R - R_e \tag{10.29}$$

With these substitutions, the potential takes the form

$$U(q) = U(0) + \tfrac{1}{2}kq^2 + \tfrac{1}{6}U^{(3)}(0)q^3 + \tfrac{1}{24}U^{(4)}(0)q^4 + \tag{10.30}$$

Nuclear motion

The nuclear equation (10.21) when applied to the ground electronic state of a diatomic molecule is

$$[\hat{T}_Q + U(R)]\chi_\nu(\mathbf{R}) = E_\nu\chi_\nu(\mathbf{R}) \tag{10.31}$$

where the superscript and one subscript on $\chi_{0\nu}^{(1)}(\mathbf{R})$ and on $E_{0\nu}^{(1)}$ are omitted for simplicity. In solving this differential equation, the relative coordinate vector $\mathbf{R}$ is best expressed in spherical polar coordinates R, θ, φ. The coordinate R is the magnitude of the vector $\mathbf{R}$ and is the scalar distance between the two nuclei. The angles θ and φ give the orientation of the internuclear axis relative to the external coordinate axes. The laplacian operator ∇_R^2 is then given by (A.61) as

$$\nabla_R^2 = \frac{1}{R^2}\frac{\partial}{\partial R}\left(R^2\frac{\partial}{\partial R}\right) + \frac{1}{R^2}\left[\frac{1}{\sin\theta}\frac{\partial}{\partial\theta}\left(\sin\theta\frac{\partial}{\partial\theta}\right) = \frac{1}{\sin^2\theta}\frac{\partial^2}{\partial\varphi^2}\right]$$

$$= \frac{1}{R^2}\frac{\partial}{\partial R}\left(R^2\frac{\partial}{\partial R}\right) - \frac{1}{\hbar^2 R^2}\hat{L}^2 \tag{10.32}$$

where $\hat{L}^2$ is the square of the orbital angular momentum operator given by equation (5.32). With ∇_R^2 expressed in spherical polar coordinates, equation (10.31) becomes

$$\left[-\frac{\hbar^2}{2\mu R^2}\frac{\partial}{\partial R}\left(R^2\frac{\partial}{\partial R}\right) + \frac{1}{2\mu R^2}\hat{L}^2 + U(R)\right]\chi_\nu(R,\theta,\varphi) = E_\nu\chi_\nu(R,\theta,\varphi)$$

$$\tag{10.33}$$

The operator in square brackets on the left-hand side of equation (10.33) commutes with the operator $\hat{L}^2$ and with the operator $\hat{L}_z$ in (5.31c), because $\hat{L}^2$ commutes with itself as well as with $\hat{L}_z$ and neither $\hat{L}^2$ nor $\hat{L}_z$ contain the variable R. Consequently, the three operators have simultaneous eigenfunctions. From the argument presented in Section 6.2, the nuclear wave function $\chi_\nu(R,\theta,\varphi)$ has the form

$$\chi_\nu(R,\theta,\varphi) = F(R)Y_{Jm}(\theta,\varphi) \tag{10.34}$$

where $F(R)$ is a function of only the internuclear distance R, and $Y_{Jm}(\theta,\varphi)$ are the spherical harmonics, which satisfy the eigenvalue equation

$$\hat{L}^2 Y_{Jm}(\theta,\varphi) = J(J+1)\hbar^2 Y_{Jm}(\theta,\varphi)$$

$$J = 0, 1, 2, \ldots; \qquad m = -J, -J+1, \ldots, 0, \ldots, J-1, J$$

It is customary to use the index J for the rotational quantum number. Equation (10.33) then becomes

$$\left[-\frac{\hbar^2}{2\mu R^2}\frac{d}{dR}\left(R^2\frac{d}{dR}\right) + \frac{J(J+1)\hbar^2}{2\mu R^2} + U(R) - E_\nu\right]F(R) = 0 \tag{10.35}$$

where we have divided through by $Y_{Jm}(\theta,\varphi)$.

We next replace the independent variable R in equation (10.35) by q as defined in equation (10.29). Equation (10.35) has a more useful form if we also make the substitution $S(q) \equiv RF(R)$. Since $dq/dR = 1$, we have

$$\frac{dF(R)}{dR} = \frac{1}{R}\frac{dS(q)}{dq} - \frac{1}{R^2}S(q), \qquad \frac{d}{dR}\left(R^2\frac{dF(R)}{dR}\right) = R\frac{d^2S(q)}{dq^2}$$

and equation (10.35) becomes

$$\left[-\frac{\hbar^2}{2\mu}\frac{d^2}{dq^2} + \frac{J(J+1)\hbar^2}{2\mu(R_e+q)^2} + U(q) - E_\nu\right]S(q) = 0 \tag{10.36}$$

after multiplication by the variable R.

The potential function $U(q)$ in equation (10.36) may be expanded according to (10.30). The factor $(R_e + q)^{-2}$ in the second term on the left-hand side may also be expanded in terms of the variable q as follows

$$\frac{1}{(R_e + q)^2} = \frac{1}{R_e^2 \left(1 + \dfrac{q}{R_e}\right)^2} = \frac{1}{R_e^2}\left(1 - \frac{2q}{R_e} + \frac{3q^2}{R_e^2} - \cdots\right) \tag{10.37}$$

where the expansion (A.3) is used. For small values of the ratio q/R_e, equation (10.37) gives the approximation $R \approx R_e$.

If we retain only the first two terms in the expansion (10.30) and let R be approximated by R_e, equation (10.36) becomes

$$\frac{-\hbar^2}{2\mu}\frac{d^2 S(q)}{dq^2} + (\tfrac{1}{2}kq^2 - W)S(q) = 0 \tag{10.38}$$

where

$$W \equiv E_v - U(0) - J(J + 1)B_e \tag{10.39}$$

$$B_e \equiv \hbar^2/2\mu R_e^2 = \hbar^2/2I \tag{10.40}$$

The quantity $I \ (= \mu R_e^2)$ is the moment of inertia for the diatomic molecule with the internuclear distance fixed at R_e and B_e is known as the *rotational constant* (see Section 5.4).

Equation (10.38) is recognized as the Schrödinger equation (4.13) for the one-dimensional harmonic oscillator. In order for equation (10.38) to have the same eigenfunctions and eigenvalues as equation (4.13), the function $S(q)$ must have the same asymptotic behavior as $\psi(x)$ in (4.13). As the internuclear distance R approaches infinity, the relative distance variable q also approaches infinity and the functions $F(R)$ and $S(q) = RF(R)$ must approach zero in order for the nuclear wave functions to be well-behaved. As $R \to 0$, which is equivalent to $q \to -R_e$, the potential $U(q)$ becomes infinitely large, so that $F(R)$ and $S(q)$ rapidly approach zero. Thus, the function $S(q)$ approaches zero as $q \to -R_e$ and as $R \to \infty$. The harmonic-oscillator eigenfunctions $\psi(x)$ decrease rapidly in value as $|x|$ increases from $x = 0$ and approach zero as $x \to \pm\infty$. They have essentially vanished at the value of x corresponding to $q = -R_e$. Consequently, the functions $S(q)$ in equation (10.38) and $\psi(x)$ in (4.13) have the same asymptotic behavior and the eigenfunctions and eigenvalues of (10.38) are those of the harmonic oscillator. The eigenfunctions $S_n(q)$ are the harmonic-oscillator eigenfunctions given by equation (4.41) with x replaced by q and the mass m replaced by the reduced mass μ. The eigenvalues, according to equation (4.30), are

$$W_n = (n + \tfrac{1}{2})\hbar\omega, \qquad n = 0, 1, 2, \ldots$$

where

$$\omega = \sqrt{k/\mu}$$

In this approximation, the nuclear energy levels are

$$E_{nJ} = U(0) + (n + \tfrac{1}{2})\hbar\omega + J(J+1)B_e \qquad (10.41)$$

and the nuclear wave functions are

$$\chi_{nJm}(R, \theta, \varphi) = \frac{1}{R} S_n(R - R_e) Y_{Jm}(\theta, \varphi) \qquad (10.42)$$

Higher-order approximation for nuclear motion

The next higher-order approximation to the energy levels of the diatomic molecule is obtained by retaining in equation (10.36) terms up to q^4 in the expansion (10.30) of $U(q)$ and terms up to q^2 in the expansion (10.37) of $(R_e + q)^{-2}$. Equation (10.36) then becomes

$$\frac{-\hbar^2}{2\mu} \frac{d^2 S(q)}{dq^2} + [\tfrac{1}{2}kq^2 + B_e J(J+1) + V'] S(q) = [E_v - U(0)] S(q) \qquad (10.43)$$

where

$$V' = -\frac{2B_e J(J+1)}{R_e} q + \frac{3B_e J(J+1)}{R_e^2} q^2 + \frac{1}{6} U^{(3)}(0) q^3 + \frac{1}{24} U^{(4)}(0) q^4$$

$$= b_1 q + b_2 q^2 + b_3 q^3 + b_4 q^4 \qquad (10.44)$$

For simplicity in subsequent evaluations, we have introduced in equation (10.44) the following definitions

$$b_1 \equiv -\frac{2B_e J(J+1)}{R_e} \qquad (10.45a)$$

$$b_2 \equiv \frac{3B_e J(J+1)}{R_e^2} \qquad (10.45b)$$

$$b_3 \equiv \frac{1}{6} U^{(3)}(0) \qquad (10.45c)$$

$$b_4 \equiv \frac{1}{24} U^{(4)}(0) \qquad (10.45d)$$

Since equation (10.43) with $V' = 0$ is already solved, we may treat V' as a perturbation and solve equation (10.43) using perturbation theory. The unperturbed eigenfunctions $S_n^{(0)}(q)$ are the eigenkets $|n\rangle$ for the harmonic oscillator. The first-order perturbation correction $E_{nJ}^{(1)}$ to the energy E_{nJ} as given by equation (9.24) is

$$E_{nJ}^{(1)} = \langle n|V'|n\rangle = b_1\langle n|q|n\rangle + b_2\langle n|q^2|n\rangle + b_3\langle n|q^3|n\rangle + b_4\langle n|q^4|n\rangle \tag{10.46}$$

The matrix elements $\langle n|q^k|n\rangle$ are evaluated in Section 4.4. According to equations (4.45c) and (4.50e), the first and third terms on the right-hand side of (10.46) vanish. The matrix elements in the second and fourth terms are given by equations (4.48b) and (4.51c), respectively. Thus, the first-order correction in equation (10.46) is

$$E_{nJ}^{(1)} = b_2 \frac{\hbar}{\mu\omega}\left(n + \tfrac{1}{2}\right) + b_4 \frac{3}{2}\left(\frac{\hbar}{\mu\omega}\right)^2 \left(n^2 + n + \tfrac{1}{2}\right)$$

$$= \frac{6B_e^2}{\hbar\omega}\left(n + \tfrac{1}{2}\right)J(J+1) + \frac{1}{16}\left(\frac{\hbar}{\mu\omega}\right)^2 U^{(4)}(0)\left[\left(n + \tfrac{1}{2}\right)^2 + \tfrac{1}{4}\right] \tag{10.47}$$

where equations (10.40), (10.45b), and (10.45d) have been substituted.

Since the perturbation corrections due to $b_1 q$ and $b_3 q^3$ vanish in first order, we must evaluate the second-order corrections $E_{nJ}^{(2)}$ in order to find the influence of these perturbation terms on the nuclear energy levels. According to equation (9.34), this second-order correction is

$$E_{nJ}^{(2)} = -\sum_{k(\neq n)} \frac{\langle k|b_1 q + b_3 q^3|n\rangle^2}{E_k^{(0)} - E_n^{(0)}}$$

$$= -b_1^2 \sum_{k(\neq n)} \frac{\langle k|q|n\rangle^2}{(k-n)\hbar\omega} - 2b_1 b_3 \sum_{k(\neq n)} \frac{\langle k|q|n\rangle\langle k|q^3|n\rangle}{(k-n)\hbar\omega} - b_3^2 \sum_{k(\neq n)} \frac{\langle k|q^3|n\rangle^2}{(k-n)\hbar\omega} \tag{10.48}$$

where the unperturbed energy levels are given by equation (4.30). The matrix elements in equation (10.48) are given by (4.45) and (4.50), so that $E_{nJ}^{(2)}$ becomes

$$E_{nJ}^{(2)} = -\frac{b_1^2}{\hbar\omega}\left[\frac{(n+1)\hbar}{2\mu\omega} - \frac{n\hbar}{2\mu\omega}\right]$$

$$-\frac{2b_1 b_3}{\hbar\omega}\left[\left(\frac{(n+1)\hbar}{2\mu\omega}\right)^{1/2} 3\left(\frac{(n+1)\hbar}{2\mu\omega}\right)^{3/2} - \left(\frac{n\hbar}{2\mu\omega}\right)^{1/2} 3\left(\frac{n\hbar}{2\mu\omega}\right)^{3/2}\right]$$

$$-\frac{b_3^2}{\hbar\omega}\left(\frac{\hbar}{2\mu\omega}\right)^3 \left[\frac{(n+1)(n+2)(n+3)}{3} + 9(n+1)^3\right.$$

$$\left. -9n^3 - \frac{n(n-1)(n-2)}{3}\right]$$

This equation simplifies to

$$E_{nJ}^{(2)} = -\frac{b_1^2}{2\mu\omega^2} - \frac{3b_1 b_3 \hbar}{\mu^2 \omega^3}\left(n + \tfrac{1}{2}\right) - \frac{b_3^2 \hbar^2}{8\mu^3 \omega^4}(30n^2 + 30n + 11)$$

Substitution of equations (10.40), (10.45a), and (10.45c) leads to

$$E_{nJ}^{(2)} = -\frac{4B_e^2}{\hbar^2 \omega^2}J^2(J+1)^2 + \frac{2B_e^2 R_e U^{(3)}(0)}{\mu\hbar\omega^3}\left(n+\tfrac{1}{2}\right)J(J+1)$$

$$- \frac{\hbar^2[U^{(3)}(0)]^2}{288\mu^3\omega^4}\left[30\left(n+\tfrac{1}{2}\right)^2 + \tfrac{7}{2}\right] \tag{10.49}$$

The nuclear energy levels in this higher-order approximation are given to second order in the perturbation by combining equations (10.41), (10.47), and (10.49) to give

$$E_{nJ} \approx E_{nJ}^{(0)} + E_{nJ}^{(1)} + E_{nJ}^{(2)}$$

$$= \mathscr{U}(0) + \hbar\omega\left(n+\tfrac{1}{2}\right) - \hbar\omega x_e\left(n+\tfrac{1}{2}\right)^2 + B_e J(J+1)$$

$$- DJ^2(J+1)^2 - \alpha_e\left(n+\tfrac{1}{2}\right)J(J+1) \tag{10.50}$$

where we have defined

$$x_e \equiv \frac{\hbar}{16\mu^2\omega^3}\left(\frac{5[U^{(3)}(0)]^2}{3\mu\omega^2} - U^{(4)}(0)\right) \tag{10.51a}$$

$$D \equiv \frac{4B_e^2}{\hbar^2\omega^2} \tag{10.51b}$$

$$\alpha_e \equiv \frac{-6B_e^2}{\hbar\omega}\left(1 + \frac{R_e U^{(3)}(0)}{3\mu\omega^2}\right) \tag{10.51c}$$

$$\mathscr{U}(0) \equiv U(0) + \frac{1}{64}\left(\frac{\hbar}{\mu\omega}\right)^2\left(U^{(4)}(0) - \frac{7[U^{(3)}(0)]^2}{9\mu\omega^2}\right) \tag{10.51d}$$

The approximate expression (10.50) for the nuclear energy levels E_{nJ} is observed to contain the initial terms of a power series expansion in $(n+\tfrac{1}{2})$ and $J(J+1)$. Only terms up to $(n+\tfrac{1}{2})^2$ and $[J(J+1)]^2$ and the cross term in $(n+\tfrac{1}{2})J(J+1)$ are included. Higher-order terms in the expansion may be found from higher-order perturbation corrections.

The second term on the right-hand side of equation (10.50) is the energy of a harmonic oscillator. Since the factor x_e in equation (10.51a) depends on the third and fourth derivatives of the internuclear potential at R_e, the third term in equation (10.50) gives the change in energy due to the anharmonicity of that potential. The fourth term is the energy of a rigid rotor with moment of inertia I. The fifth term is the correction to the energy due to centrifugal distortion in this non-rigid rotor. As the rotational energy increases, the internuclear distance increases, resulting in an increased moment of inertia and consequently a lower energy. Thus, this term is negative and increases as J increases. The magnitude of the centrifugal distortion is influenced by the value of the force constant k as reflected by the factor ω^{-2} in D. The last term contains both quantum numbers n and J and represents a direct coupling between the vibrational and rotational motions. This term contains two contributions: a change in vibrational energy due to the centrifugal stretching of the molecule and a change in rotational energy due to changes in the internuclear distance from anharmonic vibrations. The constant term $\mathscr{U}(0)$ merely shifts the zero-point energy of the nuclear energy levels and is usually omitted completely.

The molecular constants ω, B_e, x_e, D, and α_e for any diatomic molecule may be determined with great accuracy from an analysis of the molecule's vibrational and rotational spectra.[4] Thus, it is not necessary in practice to solve the electronic Schrödinger equation (10.28b) to obtain the ground-state energy $\varepsilon_0(R)$.

Problems

10.1 Derive equation (10.47) as outlined in the text.

10.2 Derive equation (10.49) as outlined in the text.

10.3 Derive equation (10.50) as outlined in the text.

10.4 An approximation to the potential $U(R)$ for a diatomic molecule is the Morse potential

$$U(R) = -D_e(2e^{-a(R-R_e)} - e^{-2a(R-R_e)}) = -D_e(2e^{-aq} - e^{-2aq})$$

where a is a parameter characteristic of the molecule. The Morse potential has the general form of Figure 10.2

(a) Show that $U(R_e) = -D_e$, that $U(\infty) = 0$, and that $U(0)$ is very large.

(b) If the Morse potential is expanded according to equation (10.30), relate the parameter a to μ, ω, and D_e

(c) Relate the quantities x_e, α_e, and $\mathscr{U}(0)$ in equation (10.50) to μ, ω, and D_e for the Morse potential.

10.5 Another approximate potential $U(R)$ for a diatomic molecule is the Rydberg potential

$$U(R) = -D_e[1 + b(R - R_e)]e^{-b(R-R_e)} = -D_e(1 + bq)e^{-bq}$$

where b is a parameter characteristic of the molecule.

(a) Show that $U(R_e) = -D_e$, that $U(\infty) = 0$, and that $U(0)$ is very large.

---

[4] Comprehensive tables of molecular constants for diatomic molecules may be found in K. P. Huber and G. Herzberg (1979) *Molecular Spectra and Molecular Structure: IV. Constants of Diatomic Molecules* (Van Nostrand Reinhold, New York).

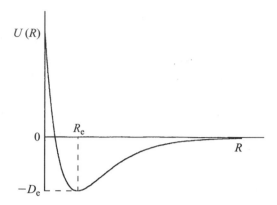

Figure 10.2 The Morse potential for the ground state of a diatomic molecule.

(b) If the Rydberg potential is expanded according to equation (10.30), relate the parameter b to μ, ω, and D_e.
(c) Relate the quantities x_e, α_e, and $\mathscr{U}(0)$ in equation (10.50) to μ, ω, and D_e for the Rydberg potential.

10.6 Consider a diatomic molecule in its ground electronic and rotational states. Its energy levels are given by equation (10.50) with $J = 0$. The value of $U(R)$ at $R = R_e$ is $-D_e$.

(a) If the anharmonic factor x_e is positive, show that the spacing of the energy levels decreases as the vibrational quantum number n increases.
(b) When the vibrational quantum number n becomes sufficiently large that the difference in energies between adjacent levels becomes zero, the molecule dissociates into its constituent atoms. By setting equal to zero the derivative of E_{n0} with respect to n, find the value of n in terms of x_e at which dissociation takes place.
(c) Relate the well depth D_e to the anharmonic factor x_e and compare with the corresponding expressions in problems 10.4 and 10.5.

Appendix A

Mathematical formulas

Useful power series expansions

$$e^{az} = \sum_{n=0}^{\infty} \frac{(az)^n}{n!}$$
(A.1)

Binomial expansions

$$(u+v)^n = \sum_{\alpha=0}^{n} \frac{n!}{\alpha!(n-\alpha)!} u^{n-\alpha} v^\alpha$$
(A.2)

$$(u+v)^{-n} = \sum_{\alpha=0}^{\infty} \frac{(-1)^\alpha (n+\alpha-1)!}{\alpha!(n-1)!} u^{-(n+\alpha)} v^\alpha$$
(A.3)

$$\frac{d^n uv}{dz^n} = \sum_{\alpha=0}^{n} \frac{n!}{\alpha!(n-\alpha)!} \frac{d^\alpha u}{dz^\alpha} \frac{d^{n-\alpha} v}{dz^{n-\alpha}}$$
(A.4)

Useful integrals

$$\int_{-\infty}^{\infty} e^{-z^2} \, dz = 2 \int_{0}^{\infty} e^{-z^2} \, dz = \pi^{1/2}$$
(A.5)

$$\int_{0}^{\infty} z e^{-z^2} \, dz = \tfrac{1}{2}$$
(A.6)

$$\int_{-\infty}^{\infty} z^2 e^{-z^2} \, dz = 2 \int_{0}^{\infty} z^2 e^{-z^2} \, dz = \tfrac{1}{2}\pi^{1/2}$$
(A.7)

$$\int_{-\infty}^{\infty} e^{-a^2 z^2 + bz} \, dz = \frac{\pi^{1/2}}{a} e^{b^2/4a^2}$$
(A.8)

$$\int_{-\infty}^{\infty} e^{-az^2} e^{c(b-z)^2} \, dz = \sqrt{\frac{\pi}{a-c}} \, e^{acb^2/(a-c)}$$
(A.9)

$$\int_{-\infty}^{\infty} \frac{\sin^2 z}{z^2} \, dz = \pi$$
(A.10)

$$\int_{-\infty}^{\infty} \frac{e^{i\rho s}}{(1+is)^{k+1}}\, ds = \frac{2\pi}{k!}\rho^k e^{-\rho} \tag{A.11}$$

$$\int_{-\pi}^{\pi} \cos n\theta\, d\theta = 2\pi, \qquad n = 0$$

$$= 0, \qquad n = 1, 2, 3, \ldots \tag{A.12}$$

$$\int_{-\pi}^{\pi} \sin n\theta\, d\theta = 0, \qquad n = 0, 1, 2, \ldots \tag{A.13}$$

$$\int_{-\pi}^{\pi} \cos m\theta \cos n\theta\, d\theta = 2\int_{0}^{\pi} \cos m\theta \cos n\theta\, d\theta = \pi\delta_{mn} \tag{A.14}$$

$$\int_{-\pi}^{\pi} \sin m\theta \sin n\theta\, d\theta = 2\int_{0}^{\pi} \sin m\theta \sin n\theta\, d\theta = \pi\delta_{mn} \tag{A.15}$$

$$\int_{-\pi}^{\pi} \cos m\theta \sin n\theta\, d\theta = 0 \tag{A.16}$$

$$\int \sin^2\theta\, d\theta = \tfrac{1}{2}\theta - \tfrac{1}{4}\sin 2\theta = \tfrac{1}{2}(\theta - \sin\theta\cos\theta) \tag{A.17}$$

$$\int \theta \sin^2\theta\, d\theta = \tfrac{1}{4}\theta^2 - \tfrac{1}{4}\theta\sin 2\theta - \tfrac{1}{8}\cos 2\theta \tag{A.18}$$

$$\int \theta^2 \sin^2\theta\, d\theta = \tfrac{1}{6}\theta^3 - \tfrac{1}{4}(\theta^2 - \tfrac{1}{2})\sin 2\theta - \tfrac{1}{4}\theta\cos 2\theta \tag{A.19}$$

$$\int \sin^3\theta\, d\theta = \tfrac{1}{3}\cos^3\theta - \cos\theta = -\tfrac{3}{4}\cos\theta + \tfrac{1}{12}\cos 3\theta \tag{A.20}$$

$$\int \sin^5\theta\, d\theta = -\tfrac{5}{8}\cos\theta + \tfrac{5}{48}\cos 3\theta - \tfrac{1}{80}\cos 5\theta \tag{A.21}$$

$$\int \sin^7\theta\, d\theta = -\tfrac{35}{64}\cos\theta + \tfrac{7}{64}\cos 3\theta - \tfrac{7}{320}\cos 5\theta + \tfrac{1}{448}\cos 7\theta \tag{A.22}$$

$$\int \sin k\theta \sin n\theta\, d\theta = \frac{\sin(k-n)\theta}{2(k-n)} - \frac{\sin(k+n)\theta}{2(k+n)}, \qquad (k^2 \neq n^2) \tag{A.23}$$

Integration by parts

$$\int u\, dv = uv - \int v\, du \tag{A.24}$$

$$\int u(x)\frac{dv(x)}{dx}\, dx = u(x)v(x) - \int v(x)\frac{du(x)}{dx}\, dx \tag{A.25}$$

Gamma function

$$\Gamma(n) = \int_0^\infty z^{n-1} e^{-z}\, dz \tag{A.26}$$

$$\Gamma(n+1) = n\Gamma(n) \tag{A.27}$$

$$\Gamma(n) = (n-1)!, \qquad n = \text{integer} \tag{A.28}$$

$$\Gamma(\tfrac{1}{2}) = \pi^{1/2} \tag{A.29}$$

$$\int_{-\infty}^{\infty} x^{2n} e^{-x^2}\, dx = \int_0^{\infty} z^{(2n-1)/2} e^{-z}\, dz = \Gamma\left(\frac{2n+1}{2}\right) \tag{A.30}$$

Trigonometric functions

$$e^{i\theta} = \cos\theta + i\sin\theta \tag{A.31}$$

$$\cos\theta = \tfrac{1}{2}(e^{i\theta} + e^{-i\theta}) \tag{A.32}$$

$$\sin\theta = \tfrac{1}{2i}(e^{i\theta} - e^{-i\theta}) \tag{A.33}$$

$$\sin^2\theta + \cos^2\theta = 1 \tag{A.34}$$

$$\cos(\theta + \varphi) = \cos\theta\cos\varphi - \sin\theta\sin\varphi \tag{A.35}$$

$$\sin(\theta + \varphi) = \sin\theta\cos\varphi + \cos\theta\sin\varphi \tag{A.36}$$

$$\cos 2\theta = \cos^2\theta - \sin^2\theta \tag{A.37}$$

$$\sin 2\theta = 2\sin\theta\cos\theta \tag{A.38}$$

$$\sin 3\theta = 3\sin\theta - 4\sin^3\theta \tag{A.39}$$

$$\sin 5\theta = 5\sin\theta - 20\sin^3\theta + 16\sin^5\theta \tag{A.40}$$

$$\sin 7\theta = 7\sin\theta - 56\sin^3\theta + 112\sin^5\theta - 64\sin^7\theta \tag{A.41}$$

$$\frac{d}{d\theta}\cos\theta = -\sin\theta \tag{A.42}$$

$$\frac{d}{d\theta}\sin\theta = \cos\theta \tag{A.43}$$

$$\frac{d}{dz}\sin^{-1}z = \frac{d}{dz}\arcsin z = (1 - z^2)^{-1/2} \tag{A.44}$$

Hyperbolic functions

$$\cosh\theta = \tfrac{1}{2}(e^{\theta} + e^{-\theta}) \tag{A.45}$$

$$\sinh\theta = \tfrac{1}{2}(e^{\theta} - e^{-\theta}) \tag{A.46}$$

$$\cosh i\theta = \cos\theta \tag{A.47}$$

$$\sinh i\theta = i\sin\theta \tag{A.48}$$

$$\cosh^2\theta - \sinh^2\theta = 1 \tag{A.49}$$

$$\tanh\theta = \frac{\sinh\theta}{\cosh\theta} \tag{A.50}$$

$$\sinh 2\theta = 2\sinh\theta\cosh\theta \tag{A.51}$$

$$\frac{d\cosh\theta}{d\theta} = \sinh\theta \tag{A.52}$$

$$\frac{d\sinh\theta}{d\theta} = \cosh\theta \tag{A.53}$$

$$\frac{d\tanh\theta}{d\theta} = \frac{1}{\cosh^2\theta} \tag{A.54}$$

Schwarz's inequality

$$\int |a(x)|^2\, dx \int |b(x)|^2\, dx \geq \left| \int a^*(x) b(x)\, dx \right|^2 \tag{A.55}$$

For $z = x + iy$, $|z|^2 \geq |\text{Im}\, z|^2$; since $\text{Im}\, z = z - z^*/2i$, $|z|^2 \geq \frac{1}{4}|z - z^*|^2$

$$\int |a(x)|^2\, dx \cdot \int |b(x)|^2\, dx \geq \frac{1}{4} \left| \int [a^*(x)b(x) - a(x)b^*(x)]\, dx \right|^2 \tag{A.56}$$

Vector relations

$$\mathbf{A} \cdot \mathbf{B} = AB \cos\theta \tag{A.57}$$
$$|\mathbf{A} \times \mathbf{B}| = AB \sin\theta \tag{A.58}$$

θ is the angle between $\mathbf{A}$ and $\mathbf{B}$

Spherical coordinates (r, θ, φ)

$$x = r \sin\theta \cos\varphi, \qquad y = r \sin\theta \sin\varphi, \qquad z = r \cos\theta \tag{A.59}$$
$$d\tau = r^2 \sin\theta\, dr\, d\theta\, d\varphi \tag{A.60}$$
$$\nabla^2 \psi = \frac{1}{r^2}\frac{\partial}{\partial r}\left(r^2 \frac{\partial \psi}{\partial r}\right) + \frac{1}{r^2 \sin\theta}\frac{\partial}{\partial \theta}\left(\sin\theta \frac{\partial \psi}{\partial \theta}\right) + \frac{1}{r^2 \sin^2\theta}\frac{\partial^2 \psi}{\partial \varphi^2} \tag{A.61}$$

Plane polar coordinates (r, φ)

$$x = r \cos\varphi, \qquad y = r \sin\varphi \tag{A.62}$$
$$\nabla^2 \psi = \frac{1}{r}\frac{\partial}{\partial r}\left(r \frac{\partial \psi}{\partial r}\right) + \frac{1}{r^2}\frac{\partial^2 \psi}{\partial \varphi^2} \tag{A.63}$$

Appendix B

Fourier series and Fourier integral

Fourier series

An arbitrary function $f(\theta)$ which satisfies the *Dirichlet conditions* can be expanded as

$$f(\theta) = \frac{a_0}{2} + \sum_{n=1}^{\infty}(a_n \cos n\theta + b_n \sin n\theta) \tag{B.1}$$

where θ is a real variable, n is a positive integer, and the coefficients a_n and b_n are constants. The Dirichlet conditions specify that $f(\theta)$ is single-valued, is continuous except for a finite number of finite discontinuities, and has a finite number of maxima and minima. The series expansion (B.1) of the function $f(\theta)$ is known as a *Fourier series*.

We note that

$$\cos n(\theta + 2\pi) = \cos n\theta$$
$$\sin n(\theta + 2\pi) = \sin n\theta$$

so that each term in equation (B.1) repeats itself in intervals of 2π. Thus, the function $f(\theta)$ on the left-hand side of equation (B.1) has the property

$$f(\theta + 2\pi) = f(\theta)$$

which is to say, $f(\theta)$ is periodic with period 2π. For convenience, we select the range $-\pi \leqslant \theta \leqslant \pi$ for the period, although any other range of width 2π is acceptable. If a function $F(\varphi)$ has period p, then it may be converted into a function $f(\theta)$ with period 2π by introducing the new variable θ defined by $\theta \equiv 2\pi\varphi/p$, so that $f(\theta) = F(2\pi\varphi/p)$. If a non-periodic function $F(\theta)$ is expanded in a Fourier series, the function $f(\theta)$ obtained from equation (B.1) is identical with $F(\theta)$ over the range $-\pi \leqslant \theta \leqslant \pi$, but outside that range the two functions do not agree.

To find the coefficients a_n and b_n in the Fourier series, we first multiply both sides of equation (B.1) by $\cos m\theta$ and integrate from $-\pi$ to π. The resulting integrals are evaluated in equations (A.12), (A.14), and (A.16). For $n = 0$, all the integrals on the right-hand side vanish except the first, so that

$$\int_{-\pi}^{\pi} f(\theta)\,d\theta = \frac{a_0}{2} \times 2\pi = \pi a_0$$

For $m > 0$, all the integrals on the right-hand side vanish except for the one in which $n = m$, giving

$$\int_{-\pi}^{\pi} f(\theta)\cos m\theta \, d\theta = \pi a_m$$

If we multiply both sides of equation (B.1) by $\sin m\theta$, integrate from $-\pi$ to π, and apply equations (A.13), (A.15), and (A.16), we find

$$\int_{-\pi}^{\pi} f(\theta)\sin m\theta \, d\theta = \pi b_m$$

Thus, the coefficients in the Fourier series are given by

$$a_n = \frac{1}{\pi} \int_{-\pi}^{\pi} f(\theta)\cos n\theta \, d\theta, \qquad n = 0, 1, 2, \ldots \tag{B.2a}$$

$$b_n = \frac{1}{\pi} \int_{-\pi}^{\pi} f(\theta)\sin n\theta \, d\theta, \qquad n = 1, 2, \ldots \tag{B.2b}$$

In deriving these expressions for a_n and b_n, we assumed that $f(\theta)$ is continuous. If $f(\theta)$ has a finite discontinuity at some angle θ_0 where $-\pi < \theta_0 < \pi$, then the expression for a_n in equation (B.2a) becomes

$$a_n = \frac{1}{\pi} \int_{-\pi}^{\theta_0} f(\theta)\cos n\theta \, d\theta + \frac{1}{\pi} \int_{\theta_0}^{\pi} f(\theta)\cos n\theta \, d\theta$$

A similar expression applies for b_n. The generalization for a function $f(\theta)$ with a finite number of finite discontinuities is straightforward. At an angle θ_0 of discontinuity, the Fourier series converges to a value of $f(\theta)$ mid-way between the left and right values of $f(\theta)$ at θ_0; i.e., it converges to

$$\lim_{\varepsilon \to 0} \frac{1}{2}[f(\theta_0 - \varepsilon) + f(\theta_0 + \varepsilon)]$$

The Fourier expansion (B.1) may also be expressed as a cosine series or as a sine series by the introduction of phase angles α_n

$$f(\theta) = \frac{a_0}{2} + \sum_{n=1}^{\infty} c_n \cos(n\theta + \alpha_n) \tag{B.3a}$$

$$= \sum_{n=0}^{\infty} c'_n \sin(n\theta + \alpha'_n) \tag{B.3b}$$

where c_n, c'_n, α_n, α'_n are constants. Using equation (A.35), we may write

$$c_n \cos(n\theta + \alpha_n) = c_n \cos n\theta \cos \alpha_n - c_n \sin n\theta \sin \alpha_n$$

If we let

$$a_n = c_n \cos \alpha_n$$

$$b_n = -c_n \sin \alpha_n$$

then equations (B.1) and (B.3a) are seen to be equivalent. Using equation (A.36), we have

$$c'_n \sin(n\theta + \alpha'_n) = c'_n \sin n\theta \cos \alpha'_n + c'_n \cos n\theta \sin \alpha'_n$$

Letting

$$a_0 = 2c'_0 \sin \alpha'_0$$

$$a_n = c'_n \sin \alpha'_n, \qquad n > 0$$

$$b_n = c'_n \cos \alpha'_n, \qquad n > 0$$

we see that equations (B.1) and (B.3b) are identical.

Other variables

The Fourier series (B.1) and (B.3) are expressed in terms of an angle θ. However, in many applications the variable may be a distance x or the time t. If the Fourier series is to represent a function $f(x)$ of the distance x in a range $-l \leqslant x \leqslant l$, we make the substitution

$$\theta = \frac{\pi x}{l}$$

in equation (B.1) to give

$$f(x) = \frac{a_0}{2} + \sum_{n=1}^{\infty} \left(a_n \cos \frac{n\pi x}{l} + b_n \sin \frac{n\pi x}{l} \right) \tag{B.4}$$

with a_n and b_n given by

$$a_n = \frac{1}{l} \int_{-l}^{l} f(x)\cos \frac{n\pi x}{l} \, dx, \quad n = 0, 1, 2, \dots \tag{B.5a}$$

$$b_n = \frac{1}{l} \int_{-l}^{l} f(x)\sin \frac{n\pi x}{l} \, dx, \quad n = 1, 2, \dots \tag{B.5b}$$

If time is the variable, then we may make either of the substitutions

$$\theta = \frac{2\pi t}{p} = \omega t$$

where p is the period of the function $f(t)$ and ω is the angular frequency, so that equation (B.1) becomes

$$f(t) = \frac{a_0}{2} + \sum_{n=1}^{\infty} \left(a_n \cos \frac{2\pi nt}{p} + b_n \sin \frac{2\pi nt}{p} \right) = \frac{a_0}{2} + \sum_{n=1}^{\infty} (a_n \cos n\omega t + b_n \sin n\omega t) \tag{B.6}$$

The constants a_n and b_n in equations (B.2) for the variable t are

$$a_n = \frac{2}{p} \int_{-p/2}^{p/2} f(t)\cos \frac{2\pi nt}{p} \, dt = \frac{\omega}{\pi} \int_{-\pi/\omega}^{\pi/\omega} f(t)\cos n\omega t \, dt \tag{B.7a}$$

$$b_n = \frac{2}{p} \int_{p/2}^{p/2} f(t)\sin \frac{2\pi nt}{p} \, dt = \frac{\omega}{\pi} \int_{-\pi/\omega}^{\pi/\omega} f(t)\sin n\omega t \, dt \tag{B.7b}$$

Complex form

The Fourier series (B.1) can also be written in complex form by the substitution of equations (A.32) and (A.33) for $\cos n\theta$ and $\sin n\theta$, respectively, to yield

$$f(\theta) = \sum_{n=-\infty}^{\infty} c_n e^{in\theta} \tag{B.8}$$

where

$$c_n \equiv \frac{a_n - ib_n}{2}, \quad n > 0$$

$$c_{-n} \equiv \frac{a_n + ib_n}{2}, \quad n > 0 \tag{B.9}$$

$$c_0 \equiv \frac{a_0}{2}$$

The coefficients c_n in equation (B.8) may be obtained from (B.9) with a_n and b_n given by (B.2). The result is

$$c_n = \frac{1}{2\pi} \int_{-\pi}^{\pi} f(\theta) e^{-in\theta} \, d\theta \tag{B.10}$$

which applies to all values of n, positive and negative, including $n = 0$. We note in passing that c_{-n} is the complex conjugate c_n^* of c_n.

In terms of the distance variable x, equations (B.8) and (B.10) become

$$f(x) = \sum_{n=-\infty}^{\infty} c_n e^{in\pi x/l} \tag{B.11}$$

$$c_n = \frac{1}{2l} \int_{-l}^{l} f(x) e^{-in\pi x/l} \, dx \tag{B.12}$$

while in terms of the time t, they take the form

$$f(t) = \sum_{n=-\infty}^{\infty} c_n e^{in\omega t} \tag{B.13}$$

$$c_n = \frac{\omega}{2\pi} \int_{-\pi/\omega}^{\pi/\omega} f(t) e^{-in\omega t} \, dt \tag{B.14}$$

Parseval's theorem

We now investigate the relation between the average of the square of $f(\theta)$ and the coefficients in the Fourier series for $f(\theta)$. For this purpose we select the Fourier series (B.8), although any of the other expansions would serve as well. In this case the average of $|f(\theta)|^2$ over the interval $-\pi \leqslant \theta \leqslant \pi$ is

$$\frac{1}{2\pi} \int_{-\pi}^{\pi} |f(\theta)|^2 \, d\theta$$

The square of the absolute value of $f(\theta)$ in equation (B.8) is

$$|f(\theta)|^2 = \left| \sum_{n=-\infty}^{\infty} c_n e^{in\theta} \right|^2 = \sum_{m=-\infty}^{\infty} \sum_{n=-\infty}^{\infty} c_m^* c_n e^{i(n-m)\theta} \tag{B.15}$$

where the two independent summations have been assigned different dummy indices. Integration of both sides of equation (B.15) over θ from $-\pi$ to π gives

$$\int_{-\pi}^{\pi} |f(\theta)|^2 \, d\theta = \sum_{m=-\infty}^{\infty} \sum_{n=-\infty}^{\infty} c_m^* c_n \int_{-\pi}^{\pi} e^{i(n-m)\theta} \, d\theta$$

Since m and n are integers, the integral on the right-hand side vanishes except when $m = n$, so that we have

$$\int_{-\pi}^{\pi} e^{i(n-m)\theta} \, d\theta = 2\pi\delta_{mn}$$

The final result is

$$\frac{1}{2\pi}\int_{-\pi}^{\pi} |f(\theta)|^2 \, d\theta = \sum_{n=-\infty}^{\infty} |c_n|^2 \tag{B.16}$$

which is one form of *Parseval's theorem*. Other forms of Parseval's theorem are obtained using the various alternative Fourier expansions.

Parseval's theorem is also known as the completeness relation and may be used to verify that the set of functions $e^{in\theta}$ for $-\infty \leqslant n \leqslant \infty$ are *complete*, as discussed in Section 3.4. If some of the terms in the Fourier series are missing, so that the set of basis functions in the expansion is incomplete, then the corresponding coefficients on the right-hand side of equation (B.16) will also be missing and the equality will not hold.

Fourier integral

The Fourier series expansions of a function $f(x)$ of the variable x over the range $-l \leqslant x \leqslant l$ may be generalized to the case where the range is infinite, i.e., where $-\infty \leqslant x \leqslant \infty$. By a suitable limiting process in which $l \to \infty$, equations (B.11) and (B.12) may be extended to the form

$$f(x) = \frac{1}{\sqrt{2\pi}}\int_{-\infty}^{\infty} g(k)e^{ikx} \, dk \tag{B.17}$$

$$g(k) = \frac{1}{\sqrt{2\pi}}\int_{-\infty}^{\infty} f(x)e^{-ikx} \, dx \tag{B.18}$$

Equation (B.17) is the *Fourier integral* representation of $f(x)$. The function $g(k)$ is the *Fourier transform* of $f(x)$, which in turn is the *inverse Fourier transform* of $g(k)$. For any function $f(x)$ which satisfies the Dirichlet conditions over the range $-\infty \leqslant x \leqslant \infty$ and for which the integral

$$\int_{\infty}^{\infty} |f(x)|^2 \, dx$$

converges, the Fourier integral in equation (B.17) converges to $f(x)$ wherever $f(x)$ is continuous and to the mean value of $f(x)$ at any point of discontinuity.

In some applications a function $f(x, t)$, where x is a distance variable and t is the time, is represented as a Fourier integral of the form

$$f(x, t) = \frac{1}{\sqrt{2\pi}}\int_{-\infty}^{\infty} G(k)e^{i[kx-\omega(k)t]} \, dk \tag{B.19}$$

where the frequency $\omega(k)$ depends on the variable k. In this case the Fourier transform $g(k)$ takes the form

$$g(k) = G(k)e^{-i\omega(k)t}$$

and equation (B.18) may be written as

$$G(k) = \frac{1}{\sqrt{2\pi}}\int_{-\infty}^{\infty} f(x, t)e^{-i[kx-\omega(k)t]} \, dx \tag{B.20}$$

The functions $f(x, t)$ and $G(k)$ are, then, a generalized form of Fourier transforms.

Another generalized form may be obtained by exchanging the roles of x and t in equations (B.19) and (B.20), so that

$$f(x, t) = \frac{1}{\sqrt{2\pi}} \int_{-\infty}^{\infty} G(\omega) e^{i[k(\omega)x - \omega t]} \, d\omega \tag{B.21}$$

$$G(\omega) = \frac{1}{\sqrt{2\pi}} \int_{-\infty}^{\infty} f(x, t) e^{-i[k(\omega)x - \omega t]} \, dt \tag{B.22}$$

Fourier integral in three dimensions

The Fourier integral may be readily extended to functions of more than one variable. We now derive the result for a function $f(x, y, z)$ of the three spatial variables x, y, z. If we consider $f(x, y, z)$ as a function only of x, with y and z as parameters, then we have

$$f(x, y, z) = \frac{1}{\sqrt{2\pi}} \int_{-\infty}^{\infty} g_1(k_x, y, z) e^{ik_x x} \, dk_x \tag{B.23a}$$

$$g_1(k_x, y, z) = \frac{1}{\sqrt{2\pi}} \int_{-\infty}^{\infty} f(x, y, z) e^{-ik_x x} \, dx \tag{B.23b}$$

We next regard $g_1(k_x, y, z)$ as a function only of y with k_x and z as parameters and express $g_1(k_x, y, z)$ as a Fourier integral

$$g_1(k_x, y, z) = \frac{1}{\sqrt{2\pi}} \int_{-\infty}^{\infty} g_2(k_x, k_y, z) e^{ik_y y} \, dk_y \tag{B.24a}$$

$$g_2(k_x, k_y, z) = \frac{1}{\sqrt{2\pi}} \int_{-\infty}^{\infty} g_1(k_x, y, z) e^{-ik_y y} \, dy \tag{B.24b}$$

Considering $g_2(k_x, k_y, z)$ as a function only of z, we have

$$g_2(k_x, k_y, z) = \frac{1}{\sqrt{2\pi}} \int_{-\infty}^{\infty} g(k_x, k_y, k_z) e^{ik_z z} \, dk_z \tag{B.25a}$$

$$g(k_x, k_y, k_z) = \frac{1}{\sqrt{2\pi}} \int_{-\infty}^{\infty} g_2(k_x, k_y, z) e^{-ik_z z} \, dz \tag{B.25b}$$

Combining equations (B.23a), (B.24a), and (B.25a), we obtain

$$f(x, y, z) = \frac{1}{(2\pi)^{3/2}} \iiint_{-\infty}^{\infty} g(k_x, k_y, k_k) e^{i(k_x x + k_y y + k_z z)} \, dk_x \, dk_y \, dk_z \tag{B.26a}$$

Combining equations (B.23b), (B.24b), and (B.25b), we have

$$g(k_x, k_y, k_k) = \frac{1}{(2\pi)^{3/2}} \iiint_{-\infty}^{\infty} f(x, y, z) e^{-i(k_x x + k_y y + k_z z)} \, dx \, dy \, dz \tag{B.26b}$$

If we define the vector $\mathbf{r}$ with components x, y, z and the vector $\mathbf{k}$ with components k_x, k_y, k_z and write the volume elements as

$$d\mathbf{r} = dx \, dy \, dz$$

$$d\mathbf{k} = dk_x \, dk_y \, dk_z$$

then equations (B.26) become

$$f(\mathbf{r}) = \frac{1}{(2\pi)^{3/2}} \int g(\mathbf{k}) e^{i\mathbf{k}\cdot\mathbf{r}} \, d\mathbf{k} \tag{B.27a}$$

$$g(\mathbf{k}) = \frac{1}{(2\pi)^{3/2}} \int f(\mathbf{r}) e^{-i\mathbf{k}\cdot\mathbf{r}} \, d\mathbf{r} \tag{B.27b}$$

Parseval's theorem

To obtain Parseval's theorem for the function $f(x)$ in equation (B.17), we first take the complex conjugate of $f(x)$

$$f^*(x) = \frac{1}{\sqrt{2\pi}} \int_{-\infty}^{\infty} g^*(k') e^{-ik'x} \, dk'$$

where we have used a different dummy variable of integration. The integral of the square of the absolute value of $f(x)$ is then given by

$$\int_{-\infty}^{\infty} |f(x)|^2 \, dx = \int_{-\infty}^{\infty} f^*(x) f(x) \, dx = \frac{1}{2\pi} \iint_{-\infty}^{\infty} g^*(k') g(k) e^{i(k-k')x} \, dk \, dk' \, dx$$

The order of integration on the right-hand side may be interchanged. If we integrate over x while noting that according to equation (C.6)

$$\int_{-\infty}^{\infty} e^{i(k-k')x} \, dx = 2\pi\delta(k-k')$$

we obtain

$$\int_{-\infty}^{\infty} |f(x)|^2 \, dx = \iint_{\infty}^{\infty} g^*(k') g(k) \delta(k-k') \, dk \, dk'$$

Finally, integration over the variable k' yields Parseval's theorem for the Fourier integral,

$$\int_{-\infty}^{\infty} |f(x)|^2 \, dx = \int_{-\infty}^{\infty} |g(k)|^2 \, dk \tag{B.28}$$

Parseval's theorem for the functions $f(\mathbf{r})$ and $g(\mathbf{k})$ in equations (B.27) is

$$\int |f(\mathbf{r})|^2 \, d\mathbf{r} = \int |g(\mathbf{k})|^2 \, d\mathbf{k} \tag{B.29}$$

This relation may be obtained by the same derivation as that leading to equation (B.28), using the integral representation (C.7) for the three-dimensional Dirac delta function.

Appendix C

Dirac delta function

The Dirac delta function $\delta(x)$ is defined by the conditions

$$\delta(x) = 0, \qquad \text{for } x \neq 0$$
$$= \infty, \qquad \text{for } x = 0 \tag{C.1}$$

such that

$$\int_{-\infty}^{\infty} \delta(x)\, dx = 1 \tag{C.2}$$

As a consequence of this definition, if $f(x)$ is an arbitrary function which is well-defined at $x = 0$, then integration of $f(x)$ with the delta function selects out the value of $f(x)$ at the origin

$$\int f(x)\delta(x)\, dx = f(0) \tag{C.3}$$

The integration is taken over the range of x for which $f(x)$ is defined, provided that the range includes the origin. It also follows that

$$\int f(x)\delta(x - x_0)\, dx = f(x_0) \tag{C.4}$$

since $\delta(x - x_0) = 0$ except when $x = x_0$. The range of integration in equation (C.4) must include the point $x = x_0$.

The following properties of the Dirac delta function can be demonstrated by multiplying both sides of each expression by $f(x)$ and observing that, on integration, each side gives the same result

$$\delta(-x) = \delta(x) \tag{C.5a}$$

$$\delta(cx) = \frac{1}{|c|}\delta(x), \qquad c \text{ real} \tag{C.5b}$$

$$x\delta(x - x_0) = x_0\delta(x - x_0) \tag{C.5c}$$

$$x\delta(x) = 0 \tag{C.5d}$$

$$f(x)\delta(x - x_0) = f(x_0)\delta(x - x_0) \tag{C.5e}$$

As defined above, the delta function by itself lacks mathematical rigor and has no meaning. Only when it appears in an integral does it have an operational meaning. That two integrals are equal does not imply that the integrands are equal. However, for the sake of convenience we often write mathematical expressions involving $\delta(x)$ such as those in equations (C.5a–e). Thus, the expressions (C.5a–e) and similar ones involving $\delta(x)$ are not to be taken as mathematical identities, but rather as operational identities. One side can replace the other within an integral that includes the origin, for $\delta(0)$, or the point x_0 for $\delta(x - x_0)$.

The concept of the Dirac delta function can be made more mathematically rigorous by regarding $\delta(x)$ as the limit of a function which becomes successively more peaked at the origin when a parameter approaches zero. One such function is

$$\delta(x) = \lim_{\varepsilon \to 0} \frac{1}{\pi^{1/2}\varepsilon} e^{-x^2/\varepsilon^2}$$

since

$$\frac{1}{\pi^{1/2}\varepsilon} \int_{-\infty}^{\infty} e^{-x^2/\varepsilon^2} \, dx = 1$$

and

$$\frac{1}{\varepsilon} e^{-x^2/c^2} \to \infty \quad \text{as} \quad x \to 0, \, \varepsilon \to 0$$

$$\to 0 \quad \text{as} \quad x \to \pm\infty$$

Equation (C.3) then becomes

$$\lim_{\varepsilon \to 0} \frac{1}{\pi^{1/2}\varepsilon} \int_{-\infty}^{\infty} f(x)e^{-x^2/\varepsilon^2} \, dx = f(0)$$

Other expressions which can be used to define $\delta(x)$ include

$$\lim_{\varepsilon \to 0} \frac{1}{\pi} \frac{\varepsilon}{x^2 + c^2}$$

and

$$\lim_{\varepsilon \to 0} \frac{1}{2\varepsilon} e^{-|x|/\varepsilon}$$

The delta function is the derivative of the Heaviside unit step function $H(x)$, defined as the limit as $\varepsilon \to 0$ of $H(x, \varepsilon)$ (see Figure C.1)

$$H(x, \varepsilon) = 0 \qquad \text{for} \quad x < \frac{-\varepsilon}{2}$$

$$= \frac{x}{\varepsilon} + \frac{1}{2} \quad \text{for} \quad \frac{-\varepsilon}{2} \leqslant x \leqslant \frac{\varepsilon}{2}$$

$$= 1 \qquad \text{for} \quad x > \frac{\varepsilon}{2}$$

Thus, in the limit we have

$$H(x) = 0 \quad \text{for} \quad x < 0$$

$$= \tfrac{1}{2} \quad \text{for} \quad x = 0$$

$$= 1 \quad \text{for} \quad x > 0$$

and dH/dx, which equals $\delta(x)$, satisfies equation (C.1). The differential $dH(x, \varepsilon)$ equals dx/ε for x between $-\varepsilon/2$ and $\varepsilon/2$ and is zero otherwise. If we take the integral of $\delta(x)$ from $-\infty$ to ∞, we have

$$\int_{-\infty}^{\infty} \delta(x) \, dx = \int_{-\infty}^{\infty} dH = \lim_{\varepsilon \to 0} \int_{-\infty}^{\infty} dH(x, \varepsilon) = \int_{-\varepsilon/2}^{\varepsilon/2} \frac{1}{\varepsilon} \, dx = \frac{1}{\varepsilon}\left(\frac{\varepsilon}{2} + \frac{\varepsilon}{2}\right) = 1$$

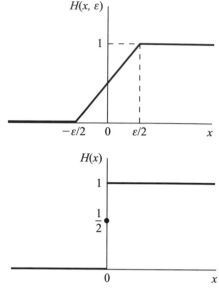

Figure C.1 The Heaviside unit step function $H(x)$, defined as the limit as $\varepsilon \to 0$ of $H(x, \varepsilon)$.

and condition (C.2) is satisfied.

We next assume that the derivative $\delta'(x)$ of $\delta(x)$ with respect to x exists. If we integrate the product $f(x)\delta'(x)$ by parts and note that the integrated part vanishes, we obtain

$$\int_{-\infty}^{\infty} f(x)\delta'(x)\,dx = -\int_{-\infty}^{\infty} f'(x)\delta(x)\,dx = -f'(0)$$

where $f'(x)$ is the derivative of $f(x)$. From equations (C.5a) and (C.5d), it follows that

$$\delta'(-x) = \delta'(x)$$

$$x\delta'(x) = -\delta(x)$$

We may also evaluate the Fourier transform $\overline{\delta}(k)$ of the Dirac delta function $\delta(x - x_0)$

$$\overline{\delta}(k) = \frac{1}{\sqrt{2\pi}}\int_{-\infty}^{\infty} \delta(x - x_0)e^{-ikx}\,dx = \frac{1}{\sqrt{2\pi}}e^{-ikx_0}$$

The inverse Fourier transform then gives an integral representation of the delta function

$$\delta(x - x_0) = \frac{1}{\sqrt{2\pi}}\int_{-\infty}^{\infty} \overline{\delta}(k)e^{ikx}\,dk = \frac{1}{2\pi}\int_{-\infty}^{\infty} e^{ik(x-x_0)}\,dk \qquad (C.6)$$

The Dirac delta function may be readily generalized to three-dimensional space. If $\mathbf{r}$ represents the position vector with components x, y, and z, then the three-dimensional delta function is

$$\delta(\mathbf{r} - \mathbf{r}_0) = \delta(x - x_0)\delta(y - y_0)\delta(z - z_0)$$

and possesses the property that

$$\int f(\mathbf{r})\delta(\mathbf{r} - \mathbf{r}_0)\,d\mathbf{r} = f(\mathbf{r}_0)$$

or, equivalently

$$\int\int\int f(x,\,y,\,z)\delta(x - x_0)\delta(y - y_0)\delta(z - z_0)\,dx\,dy\,dz = f(x_0,\,y_0,\,z_0)$$

where the range of integration includes the points x_0, y_0, and z_0. The integral representation is

$$\delta(\mathbf{r} - \mathbf{r}_0) = \frac{1}{(2\pi)^3}\int_{-\infty}^{\infty} e^{ik(\mathbf{r} - \mathbf{r}_0)}\,d\mathbf{k} \tag{C.7}$$

where $\mathbf{k}$ is a vector with components k_x, k_y, and k_z and where

$$d\mathbf{k} = dk_x\,dk_y\,dk_z$$

Appendix D

Hermite polynomials

The Hermite polynomials $H_n(\xi)$ are defined by means of an infinite series expansion of the generating function $g(\xi, s)$,

$$g(\xi, s) \equiv e^{2\xi s - s^2} = e^{\xi^2 - (s-\xi)^2} \equiv \sum_{n=0}^{\infty} H_n(\xi) \frac{s^n}{n!} \qquad \text{(D.1)}$$

where $-\infty \leq \xi \leq \infty$ and where $|s| < 1$ in order for the Taylor series expansion to converge. The coefficients $H_n(\xi)$ of the Taylor expansion are given by

$$H_n(\xi) = \left. \frac{\partial^n g(\xi, s)}{\partial s^n} \right|_{s=0} = e^{\xi^2} \left. \frac{\partial^n}{\partial s^n} (e^{-(s-\xi)^2}) \right|_{s=0} \qquad \text{(D.2)}$$

For a function $f(x + y)$ of the sum of two variables x and y, we note that

$$\left(\frac{\partial f}{\partial x} \right)_y = \left(\frac{\partial f}{\partial y} \right)_x$$

Applying this property with $x = s$ and $y = -\xi$ to the nth-order partial derivative in equation (D.2), we obtain

$$\left. \frac{\partial^n}{\partial s^n} (e^{-(s-\xi)^2}) \right|_{s=0} = (-1)^n \left. \frac{\partial^n}{\partial \xi^n} (e^{-(s-\xi)^2}) \right|_{s=0} = (-1)^n \frac{d^n}{d\xi^n} e^{-\xi^2}$$

and equation (D.2) becomes

$$H_n(\xi) = (-1)^n e^{\xi^2} \frac{d^n}{d\xi^n} e^{-\xi^2} \qquad \text{(D.3)}$$

Another expression for the Hermite polynomials may be obtained by expanding $g(\xi, s)$ using equation (A.1)

$$g(\xi, s) = e^{s(2\xi - s)} = \sum_{k=0}^{\infty} \frac{s^k (2\xi - s)^k}{k!}$$

Applying the binomial expansion (A.2) to the factor $(2\xi - s)^k$, we obtain

$$(2\xi - s)^k = \sum_{\alpha=0}^{k} \frac{(-1)^\alpha k!}{\alpha! (k - \alpha)!} (2\xi)^{k-\alpha} s^\alpha$$

and $g(\xi, s)$ takes the form

$$g(\xi, s) = \sum_{k=0}^{\infty} \sum_{\alpha=0}^{k} \frac{(-1)^\alpha 2^{k-\alpha} \xi^{k-\alpha} s^{k+\alpha}}{\alpha!(k-\alpha)!}$$

We next collect all the coefficients of s^n for an arbitrary n, so that $k + \alpha = n$, and replace the summation over k by a summation over n. When $k = n$, the index α equals zero; when $k = n - 1$, the index α equals one; when $k = n - 2$, the index α equals two; and so on until we have $k = n - M$ and $\alpha = M$. Since the index α runs from 0 to k so that $\alpha \leqslant k$, this final term gives $M \leqslant n - M$ or $M \leqslant n/2$. Thus, for $k + \alpha = n$, the summation over α terminates at $\alpha = M$ with $M = n/2$ for n even and $M = (n-1)/2$ for n odd. The result of this resummation is

$$g(\xi, s) = \sum_{n=0}^{\infty} \sum_{\alpha=0}^{M} \frac{(-1)^\alpha 2^{n-2\alpha} \xi^{n-2\alpha}}{\alpha!(n-2\alpha)!} s^n$$

Since the Hermite polynomial $H_n(\xi)$ divided by $n!$ is the coefficient of s^n in the expansion (D.1) of $g(\xi, s)$, we have

$$H_n(\xi) = 2^n n! \sum_{\alpha=0}^{M} \frac{(-1)^\alpha}{2^{2\alpha} \alpha!(n-2\alpha)!} \xi^{n-2\alpha} \tag{D.4}$$

We note that $H_n(\xi)$ is an odd or even polynomial in ξ according to whether n is odd or even and that the coefficient of the highest power of ξ in $H_n(\xi)$ is 2^n.

Expression (D.4) is useful for obtaining the series of Hermite polynomials, the first few of which are

$$H_0(\xi) = 1 \qquad\qquad H_3(\xi) = 8\xi^3 - 12\xi$$

$$H_1(\xi) = 2\xi \qquad\qquad H_4(\xi) = 16\xi^4 - 48\xi^2 + 12$$

$$H_2(\xi) = 4\xi^2 - 2 \qquad H_5(\xi) = 32\xi^5 - 160\xi^3 + 120\xi$$

Recurrence relations

We next derive some recurrence relations for the Hermite polynomials. If we differentiate equation (D.1) with respect to s, we obtain

$$2(\xi - s)e^{2\xi s - s^2} = \sum_{n=1}^{\infty} H_n(\xi) \frac{s^{n-1}}{(n-1)!}$$

The first term ($n = 0$) in the summation on the right-hand side vanishes because it is the derivative of a constant. The exponential on the left-hand side is the generating function $g(\xi, s)$, for which equation (D.1) may be used to give

$$2(\xi - s) \sum_{n=0}^{\infty} H_n(\xi) \frac{s^n}{n!} = \sum_{n=1}^{\infty} H_n(\xi) \frac{s^{n-1}}{(n-1)!}$$

Since this equation is valid for all values of s with $|s| < 1$, we may collect terms corresponding to the same power of s, for example s^n, and obtain

$$\frac{2\xi H_n(\xi)}{n!} - \frac{2 H_{n-1}(\xi)}{(n-1)!} = \frac{H_{n+1}(\xi)}{n!}$$

or

$$H_{n+1}(\xi) - 2\xi H_n(\xi) + 2n H_{n-1}(\xi) = 0 \tag{D.5}$$

This recurrence relation may be used to obtain a Hermite polynomial when the two preceding polynomials are known.

Another recurrence relation may be obtained by differentiating equation (D.1) with respect to ξ to obtain

$$2s e^{2\xi s - s^2} = \sum_{n=0}^{\infty} \frac{\mathrm{d}H_n}{\mathrm{d}\xi} \frac{s^n}{n!}$$

Replacing the exponential on the left-hand side using equation (D.1) gives

$$2s \sum_{n=0}^{\infty} H_n(\xi) \frac{s^n}{n!} = \sum_{n=0}^{\infty} \frac{\mathrm{d}H_n}{\mathrm{d}\xi} \frac{s^n}{n!}$$

If we then equate the coefficients of s^n, we obtain the desired result

$$\frac{\mathrm{d}H_n}{\mathrm{d}\xi} = 2n H_{n-1}(\xi) \tag{D.6}$$

The relations (D.5) and (D.6) may be combined to give a third recurrence relation. Addition of the two equations gives

$$H_{n+1}(\xi) = \left(2\xi - \frac{\mathrm{d}}{\mathrm{d}\xi} \right) H_n(\xi) \tag{D.7}$$

With this recurrence relation, a Hermite polynomial may be obtained from the preceding polynomial. By applying the relation (D.7) to $H_n(\xi)$ k times, we have

$$H_{n+k}(\xi) = \left(2\xi - \frac{\mathrm{d}}{\mathrm{d}\xi} \right)^k H_n(\xi) \tag{D.8}$$

Differential equation

To find the differential equation that is satisfied by the Hermite polynomials, we first differentiate the second recurrence relation (D.6) and then substitute (D.6) with n replaced by $n-1$ to eliminate the first derivative of $H_{n-1}(\xi)$

$$\frac{\mathrm{d}^2 H_n}{\mathrm{d}\xi^2} = 2n \frac{\mathrm{d}H_{n-1}}{\mathrm{d}\xi} = 4n(n-1) H_{n-2}(\xi) \tag{D.9}$$

Replacing n by $n-1$ in the first recurrence relation (D.5), we have

$$H_n(\xi) - 2\xi H_{n-1}(\xi) + 2(n-1) H_{n-2}(\xi) = 0$$

which may be used to eliminate $H_{n-2}(\xi)$ in equation (D.9), giving

$$\frac{\mathrm{d}^2 H_n}{\mathrm{d}\xi^2} + 2n H_n(\xi) - 4n\xi H_{n-1}(\xi) = 0$$

Application of equation (D.6) again to eliminate $H_{n-1}(\xi)$ yields

$$\frac{\mathrm{d}^2 H_n}{\mathrm{d}\xi^2} - 2\xi \frac{\mathrm{d}H_n}{\mathrm{d}\xi} + 2n H_n(\xi) = 0 \tag{D.10}$$

which is the Hermite differential equation.

Integral relations

To obtain the orthogonality and normalization relations for the Hermite polynomials, we multiply together the generating functions $g(\xi, s)$ and $g(\xi, t)$, both obtained from equation (D.1), and the factor $e^{-\xi^2}$ and then integrate over ξ

$$I \equiv \int_{-\infty}^{\infty} e^{-\xi^2} g(\xi, s) g(\xi, t) \,\mathrm{d}\xi = \sum_{n=0}^{\infty} \sum_{m=0}^{\infty} \frac{s^n t^m}{n! m!} \int_{-\infty}^{\infty} e^{-\xi^2} H_n(\xi) H_m(\xi) \,\mathrm{d}\xi \tag{D.11}$$

For convenience, we have abbreviated the integral with the symbol I. To evaluate the left integral, we substitute the analytical forms for the generating functions from equation (D.1) to give

$$I = \int_{-\infty}^{\infty} e^{-\xi^2} e^{2\xi s - s^2} e^{2\xi t - t^2} \, d\xi = e^{2st} \int_{-\infty}^{\infty} e^{-(\xi - s - t)^2} \, d(\xi - s - t) = \pi^{1/2} e^{2st}$$

where equation (A.5) has been used. We next expand e^{2st} in the power series (A.1) to obtain

$$I = \pi^{1/2} \sum_{n=0}^{\infty} \frac{2^n s^n t^n}{n!}$$

Substitution of this expression for I into equation (D.11) gives

$$\pi^{1/2} \sum_{n=0}^{\infty} \frac{2^n (st)^n}{n!} = \sum_{n=0}^{\infty} \sum_{m=0}^{\infty} \frac{s^n t^m}{n! m!} \int_{-\infty}^{\infty} e^{-\xi^2} H_n(\xi) H_m(\xi) \, d\xi \qquad \text{(D.12)}$$

On the left-hand side, we see that there are no terms for which the power of s is not equal to the power of t. Therefore, terms on the right-hand side with $n \neq m$ must vanish, giving

$$\int_{-\infty}^{\infty} e^{-\xi^2} H_n(\xi) H_m(\xi) \, d\xi = 0, \qquad n \neq m \qquad \text{(D.13)}$$

The Hermite polynomials $H_n(\xi)$ form an orthogonal set over the range $-\infty \leqslant \xi \leqslant \infty$ with a weighting factor $e^{-\xi^2}$. If we equate coefficients of $(st)^n$ on each side of equation (D.12), we obtain

$$\int_{-\infty}^{\infty} e^{-\xi^2} [H_n(\xi)]^2 \, d\xi = 2^n n! \pi^{1/2}$$

which may be combined with equation (D.13) to give

$$\int_{-\infty}^{\infty} e^{-\xi^2} H_n(\xi) H_m(\xi) \, d\xi = 2^n n! \pi^{1/2} \delta_{nm} \qquad \text{(D.14)}$$

Completeness

If we define the set of functions $\psi_n(\xi)$ as

$$\phi_n(\xi) \equiv (2^n n!)^{-1/2} \pi^{-1/4} e^{-\xi^2/2} H_n(\xi) \qquad \text{(D.15)}$$

then equation (D.14) shows that the members of this set are orthonormal with weighting factor unity. We can also demonstrate[1] that this set is complete.

We begin with the integral formula (A.8) which, with suitable definitions for the parameters, may be written as

$$\int_{-\infty}^{\infty} e^{-(s^2/4) + i\xi s} \, ds = 2\pi^{1/2} e^{-\xi^2} \qquad \text{(D.16)}$$

If we replace $e^{-\xi^2}$ in equation (D.3) by the integral in (D.16), we obtain for $H_n(\xi)$

$$H_n(\xi) = \frac{(-1)^n}{2\pi^{1/2}} e^{\xi^2} \frac{\partial^n}{\partial \xi^n} \int_{-\infty}^{\infty} e^{-(s^2/4) + i\xi s} \, ds = \frac{(-1)^n}{2\pi^{1/2}} e^{\xi^2} \int_{-\infty}^{\infty} e^{-s^2/4} \frac{\partial^n}{\partial \xi^n} e^{i\xi s} \, ds$$

$$= \frac{(-i)^n}{2\pi^{1/2}} e^{\xi^2} \int_{-\infty}^{\infty} e^{-(s^2/4) + i\xi s} s^n \, ds$$

---

[1] See D. Park (1992) *Introduction to the Quantum Theory*, 3rd edition (McGraw-Hill, New York), p. 565.

The function $\phi_n(\xi)$ as defined by equation (D.15) then becomes

$$\phi_n(\xi) = \frac{(-\mathrm{i})^n}{2(2^n \pi n!)^{1/2}\pi^{1/4}} \mathrm{e}^{\xi^2/2} \int_{-\infty}^{\infty} \mathrm{e}^{-(s^2/4)+\mathrm{i}\xi s} s^n \, \mathrm{d}s \qquad (D.17)$$

We now evaluate the summation

$$\sum_{n=0}^{\infty} \phi_n(\xi)\phi_n(\xi')$$

by substituting equation (D.17) twice, once with the dummy variable of integration s and once with s replaced by t. Since the functions $\phi_n(\xi)$ are real, they equal their complex conjugates. These substitutions give

$$\sum_{n=0}^{\infty} \phi_n(\xi)\phi_n(\xi') = \frac{1}{4\pi^{3/2}} \mathrm{e}^{(\xi^2+\xi'^2)/2} \int_{-\infty}^{\infty}\int_{-\infty}^{\infty} \mathrm{e}^{-[(s^2+t^2)/4]+\mathrm{i}(\xi s+\xi' t)} \sum_{n=0}^{\infty} \frac{(-1)^n}{2^n n!}(st)^n \, \mathrm{d}s \, \mathrm{d}t$$

since $(-\mathrm{i})^{2n} = (-1)^n$. The summation on the right-hand side is easily evaluated using equation (A.1)

$$\sum_{n=0}^{\infty} \frac{(-1)^n}{n!}\left(\frac{st}{2}\right)^n = \mathrm{e}^{-st/2}$$

Noting that

$$\frac{s^2+t^2}{4} + \frac{st}{2} = \frac{(s+t)^2}{4}$$

we have

$$\sum_{n=0}^{\infty} \phi_n(\xi)\phi_n(\xi') = \frac{1}{4\pi^{3/2}} \mathrm{e}^{(\xi^2+\xi'^2)/2} \int_{-\infty}^{\infty}\int_{-\infty}^{\infty} \mathrm{e}^{-[(s+t)^2/4]+\mathrm{i}(\xi s+\xi' t)} \, \mathrm{d}s \, \mathrm{d}t \qquad (D.18)$$

The double integral may be evaluated by introducing the new variables u and v

$$u = \frac{s+t}{2}, \qquad v = \frac{s-t}{2} \qquad \text{or} \qquad s = u+v, \qquad t = u-v$$

$$\mathrm{d}s \, \mathrm{d}t = 2 \, \mathrm{d}u \, \mathrm{d}v$$

The double integral is thereby factored into

$$2\int_{-\infty}^{\infty} \mathrm{e}^{-u^2+\mathrm{i}(\xi+\xi')u} \, \mathrm{d}u \int_{-\infty}^{\infty} \mathrm{e}^{\mathrm{i}(\xi-\xi')v} \, \mathrm{d}v = 2 \times \pi^{1/2}\mathrm{e}^{-(\xi+\xi')^2/4} \times 2\pi\delta(\xi-\xi')$$

where the first integral is evaluated by equation (A.8) and the second by (C.6). Equation (D.18) now becomes

$$\sum_{n=0}^{\infty} \phi_n(\xi)\phi_n(\xi') = \mathrm{e}^{[(\xi^2+\xi'^2)/2]-[(\xi+\xi')^2/4]}\delta(\xi-\xi') = \mathrm{e}^{(\xi-\xi')^2/4}\delta(\xi-\xi')$$

Applying equation (C.5e), we obtain the completeness relation for the functions $\phi_n(\xi)$

$$\sum_{n=0}^{\infty} \phi_n(\xi)\phi_n(\xi') = \delta(\xi-\xi') \qquad (D.19)$$

demonstrating, according to equation (3.31), that the set $\phi_n(\xi)$ is a complete set.

Appendix E

Legendre and associated Legendre polynomials

Legendre polynomials

The Legendre polynomials $P_l(\mu)$ may be defined as the coefficients of s^l in an infinite series expansion of a generating function $g(\mu, s)$

$$g(\mu, s) \equiv (1 - 2\mu s + s^2)^{-1/2} \equiv \sum_{l=0}^{\infty} P_l(\mu)s^l \tag{E.1}$$

where $-1 \leqslant \mu \leqslant 1$ and $|s| < 1$ in order for the infinite series to converge.

We may also expand $g(\mu, s)$ by applying the standard formula

$$f(z) = (1 - z)^{1/2} = \sum_{n=0}^{\infty} \frac{z^n}{n!} \left(\frac{d^n f}{dz^n}\right)_{z=0} = \sum_{n=0}^{\infty} \frac{z^n}{n!} \frac{1 \cdot 3 \cdot 5 \cdot \cdots \cdot (2n-1)}{2^n}$$

$$- \sum_{n=0}^{\infty} \frac{(2n)!}{2^{2n}(n!)^2} z^n$$

If we set $z = s(2\mu - s)$, then $g(\mu, s)$ becomes

$$g(\mu, s) = \sum_{n=0}^{\infty} \frac{(2n)!}{2^{2n}(n!)^2} s^n (2\mu - s)^n$$

With the use of the binomial expansion (A.2), the factor $(2\mu - s)^n$ can be further expanded as

$$(2\mu - s)^n - \sum_{\alpha=0}^{n} \frac{(-1)^\alpha n!}{\alpha!(n-\alpha)!} (2\mu)^{n-\alpha} s^\alpha$$

so that

$$g(\mu, s) = \sum_{n=0}^{\infty} \sum_{\alpha=0}^{n} \frac{(-1)^\alpha (2n)! \mu^{n-\alpha}}{2^{n+\alpha} n! \alpha! (n-\alpha)!} s^{n+\alpha}$$

We next collect all the coefficients of s^l for some arbitrary l and replace the summation over n with a summation over l. Since $n + \alpha = l$, when $n = l$, we have $\alpha = 0$; when $n = l - 1$, we have $\alpha = 1$; and so on until $n = l - M$, $\alpha = M$, where $M \leqslant l - M$ or $M \leqslant l/2$. The summation over α terminates at $\alpha = M$, with $M = l/2$ for l even and $M = (l-1)/2$ for l odd, because α cannot be greater than n. The result is

$$g(\mu, s) = \sum_{l=0}^{\infty} \sum_{\alpha=0}^{M} \frac{(-1)^{\alpha}(2l - 2\alpha)! \mu^{l-2\alpha}}{2^l \alpha! (l - \alpha)! (l - 2\alpha)!} s^l$$

Since the Legendre polynomials are the coefficients of s^l in the expansion (E.1) of $g(\mu, s)$, we have

$$P_l(\mu) = \sum_{\alpha=0}^{M} \frac{(-1)^{\alpha}(2l - 2\alpha)!}{2^l \alpha! (l - \alpha)! (l - 2\alpha)!} \mu^{l-2\alpha} \tag{E.2}$$

We see from equation (E.2) that $P_l(\mu)$ for even l is a polynomial with only even powers of μ, while for odd l only odd powers of μ are present.

The first few Legendre polynomials may be readily obtained from equation (E.2) and are

$$P_0(\mu) = 1 \qquad\qquad P_3(\mu) = \tfrac{1}{2}(5\mu^3 - 3\mu)$$

$$P_1(\mu) = \mu \qquad\qquad P_4(\mu) = \tfrac{1}{8}(35\mu^4 - 30\mu^2 + 3)$$

$$P_2(\mu) = \tfrac{1}{2}(3\mu^2 - 1) \qquad P_5(\mu) = \tfrac{1}{8}(63\mu^5 - 70\mu^3 + 15\mu)$$

We observe that $P_l(1) = 1$, which can be shown rigorously by setting $\mu = 1$ in equation (E.1) and noting that

$$g(1, s) = (1 - s)^{-1} = \sum_{l=0}^{\infty} s^l = \sum_{l=0}^{\infty} P_l(1)s^l$$

Since $P_l(\mu)$ is either even or odd in μ, it follows that $P_l(-1) = (-1)^l$ and that $P_l(0) = 0$ for l odd.

Recurrence relations

We next derive some recurrence relations for the Legendre polynomials. Differentiation of the generating function $g(\mu, s)$ with respect to s gives

$$\frac{\partial g}{\partial s} = \frac{\mu - s}{(1 - 2\mu + s^2)^{3/2}} = \frac{(\mu - s)g}{1 - 2\mu + s^2} = \sum_{l=1}^{\infty} l P_l(\mu) s^{l-1} \tag{E.3}$$

The term with $l = 0$ in the summation vanishes, so that the summation now begins with the $l = 1$ term. We may write equation (E.3) as

$$(\mu - s) \sum_{l=0}^{\infty} P_l(\mu) s^l = (1 - 2\mu s + s^2) \sum_{l=1}^{\infty} l P_l(\mu) s^{l-1}$$

If we equate coefficients of s^{l-1} on each side of the equation, we obtain

$$\mu P_{l-1}(\mu) - P_{l-2}(\mu) = l P_l(\mu) - 2(l - 1)\mu P_{l-1}(\mu) + (l - 2)P_{l-2}(\mu)$$

or

$$l P_l(\mu) - (2l - 1)\mu P_{l-1}(\mu) + (l - 1)P_{l-2}(\mu) = 0 \tag{E.4}$$

The recurrence relation (E.4) is useful for evaluating $P_l(\mu)$ when the two preceding polynomials are known.

Differentiation of the generating function $g(\mu, s)$ in equation (E.1) with respect to μ yields

$$\frac{\partial g}{\partial \mu} = \frac{sg}{1 - 2\mu s + s^2}$$

which may be combined with equation (E.3) to give

$$s\frac{\partial g}{\partial s} = (\mu - s)\frac{\partial g}{\partial \mu}$$

so that

$$\sum_{l=1}^{\infty} lP_l(\mu)s^l = (\mu - s)\sum_{l=0}^{\infty} \frac{dP_l}{d\mu}s^l$$

Equating coefficients of s^l on each side of this equation yields a second recurrence relation

$$\mu\frac{dP_l}{d\mu} - \frac{dP_{l-1}}{d\mu} - lP_l(\mu) = 0 \tag{E.5}$$

A third recurrence relation may be obtained by differentiating equation (E.4) to give

$$l\frac{dP_l}{d\mu} - (2l-1)\mu\frac{dP_{l-1}}{d\mu} - (2l-1)P_{l-1}(\mu) + (l-1)\frac{dP_{l-2}}{d\mu} = 0$$

and then eliminating $dP_{l-2}/d\mu$ by the substitution of equation (E.5) with l replaced by $l-1$. The result is

$$\frac{dP_l}{d\mu} - \mu\frac{dP_{l-1}}{d\mu} - lP_{l-1}(\mu) = 0 \tag{E.6}$$

Differential equation

To find the differential equation satisfied by the polynomials $P_l(\mu)$, we first multiply equation (E.5) by $-\mu$ and add the result to equation (E.6) to give

$$(1-\mu^2)\frac{dP_l}{d\mu} + l\mu P_l(\mu) - lP_{l-1}(\mu) = 0$$

We then differentiate to obtain

$$(1-\mu^2)\frac{d^2P_l}{d\mu^2} - 2\mu\frac{dP_l}{d\mu} + l\mu\frac{dP_l}{d\mu} + lP_l(\mu) - l\frac{dP_{l-1}}{d\mu} = 0$$

The third and last terms on the left-hand side may be eliminated by means of equation (E.5) to give Legendre's differential equation

$$(1-\mu^2)\frac{d^2P_l}{d\mu^2} - 2\mu\frac{dP_l}{d\mu} + l(l+1)P_l(\mu) = 0 \tag{E.7}$$

Rodrigues' formula

Rodrigues' formula for the Legendre polynomials may be derived as follows. Consider the expression

$$v = (\mu^2 - 1)^l$$

The derivative of v is

$$\frac{dv}{d\mu} = 2l\mu(\mu^2-1)^{l-1} = 2l\mu v(\mu^2-1)^{-1}$$

which is just the differential equation

$$(1-\mu^2)\frac{dv}{d\mu} + 2l\mu v = 0$$

If we differentiate this equation, we obtain

$$(1 - \mu^2)\frac{d^2v}{d\mu^2} + 2(l-1)\mu\frac{dv}{d\mu} + 2lv = 0$$

We now differentiate r times more and obtain

$$(1 - \mu^2)\frac{d^{r+2}v}{d\mu^{r+2}} + 2(l-r-1)\mu\frac{d^{r+1}v}{d\mu^{r+1}} + (r+1)(2l-r)\frac{d^r v}{d\mu^r} = 0 \qquad (E.8)$$

If we let $r = l$ and define w as

$$w \equiv \frac{d^l v}{d\mu^l} = \frac{d^l}{d\mu^l}(\mu^2 - 1)^l$$

then equation (E.8) reduces to

$$(1 - \mu^2)\frac{d^2w}{d\mu^2} - 2\mu\frac{dw}{d\mu} + l(l+1)w = 0$$

which is just Legendre's differential equation (E.7). Since the polynomials $P_l(\mu)$ represent all of the solutions of equation (E.7), these polynomials must be multiples of w, so that

$$P_l(\mu) = c_l\frac{d^l}{d\mu^l}(\mu^2 - 1)^l$$

The proportionality constants c_l may be evaluated by setting the term in μ^l, namely

$$c_l\frac{d^l}{d\mu^l}\mu^{2l} = c_l\frac{(2l)!}{l!}\mu^l$$

equal to the term in μ^l in equation (E.2), i.e.,

$$\frac{(2l)!}{2^l(l!)^2}\mu^l$$

Thus, we have

$$c_l = \frac{1}{2^l l!}$$

and

$$P_l(\mu) = \frac{1}{2^l l!}\frac{d^l}{d\mu^l}(\mu^2 - 1)^l \qquad (E.9)$$

This expression (equation (E.9)) is Rodrigues' formula.

Associated Legendre polynomials

The associated Legendre polynomials $P_l^m(\mu)$ are defined in terms of the Legendre polynomials $P_l(\mu)$ by

$$P_l^m(\mu) \equiv (1 - \mu^2)^{m/2}\frac{d^m P_l(\mu)}{d\mu^m} \qquad (E.10)$$

where m is a positive integer, $m = 0, 1, 2, \ldots, l$. If $m = 0$, then the corresponding associated Legendre polynomial is just the Legendre polynomial of degree l. If $m > l$, then the corresponding associated Legendre polynomial vanishes.

The generating functions $g^{(m)}(\mu, s)$ for the associated Legendre polynomials may be found from equation (E.1) by letting

$$g^{(m)}(\mu, s) = (1 - \mu^2)^{m/2} \frac{d^m g(\mu, s)}{d\mu^m}$$

Since

$$\frac{d^m g(\mu, s)}{d\mu^m} = 3 \cdot 5 \cdot \cdots \cdot (2m - 1)s^m (1 - 2\mu s + s^2)^{-(m+\frac{1}{2})}$$

$$= \frac{(2m)!}{2^m m!} s^m (1 - 2\mu s + s^2)^{-(m+\frac{1}{2})}$$

we have

$$g^{(m)}(\mu, s) = \sum_{l=m}^{\infty} P_l^m(\mu) s^l = \frac{(2m)!(1 - \mu^2)^{m/2} s^m}{2^m m!(1 - 2\mu s + s^2)^{m+\frac{1}{2}}} \qquad (E.11)$$

We can also write an explicit series for $P_l^m(\mu)$ by differentiating equation (E.2) m times

$$P_l^m(\mu) = (1 - \mu^2)^{m/2} \sum_{\alpha=0}^{M'} \frac{(-1)^\alpha (2l - 2\alpha)! \mu^{l-m-2\alpha}}{2^l \alpha!(l - \alpha)!(l - m - 2\alpha)!} \qquad (E.12)$$

where $M' = (l - m)/2$ or $(l - m - 1)/2$, whichever is an integer. Furthermore, combining equation (E.10) with Rodrigues' formula (E.9), we see that

$$P_l^m(\mu) = \frac{1}{2^l l!}(1 - \mu^2)^{m/2} \frac{d^{l+m}}{d\mu^{l+m}}(\mu^2 - 1)^l \qquad (E.13)$$

The first few associated Legendre polynomials are

$$P_0^0(\mu) = P_0(\mu) = 1$$

$$P_1^0(\mu) = \mu$$

$$P_1^1(\mu) = (1 - \mu^2)^{1/2}$$

$$P_2^0(\mu) = P_2(\mu) = \tfrac{1}{2}(3\mu^2 - 1)$$

$$P_2^1(\mu) = 3\mu(1 - \mu^2)^{1/2}$$

$$P_2^2(\mu) = 3(1 - \mu^2)$$

$$\vdots$$

Differential equation

The differential equation satisfied by the polynomials $P_l^m(\mu)$ may be obtained as follows. Let $r = l + m$ in equation (E.8) and define w_m as

$$w_m \equiv \frac{d^{l+m}}{d\mu^{l+m}}(\mu^2 - 1)^l = 2^l l! \frac{d^m P_l}{d\mu^m} \qquad (E.14)$$

so that

$$w_m = 2^l l!(1 - \mu^2)^{-m/2} P_l^m(\mu) \qquad (E.15)$$

Equation (E.8) then becomes

$$(1 - \mu^2)\frac{d^2 w_m}{d\mu^2} - 2(m + 1)\mu \frac{dw_m}{d\mu} + [l(l + 1) - m(m + 1)]w_m = 0 \qquad (E.16)$$

We then substitute equation (E.15) for w_m and take the first and second derivatives as indicated to obtain

$$(1 - \mu^2)\frac{d^2 P_l^m}{d\mu^2} - 2\mu\frac{dP_l^m}{d\mu} + \left[l(l+1) - \frac{m^2}{1 - \mu^2}\right]P_l^m(\mu) = 0 \qquad (E.17)$$

Equation (E.17) is the associated Legendre differential equation.

Orthogonality
Equation (E.17) as satisfied by $P_l^m(\mu)$ and by $P_{l'}^m(\mu)$ may be written as

$$\frac{d}{d\mu}\left[(1 - \mu^2)\frac{dP_l^m}{d\mu}\right] + \left[l(l+1) - \frac{m^2}{1 - \mu^2}\right]P_l^m(\mu) = 0$$

and

$$\frac{d}{d\mu}\left[(1 - \mu^2)\frac{dP_{l'}^m}{d\mu}\right] + \left[l'(l'+1) - \frac{m^2}{1 - \mu^2}\right]P_{l'}^m(\mu) = 0$$

If we multiply the first by $P_{l'}^m(\mu)$ and the second by $P_l^m(\mu)$ and then subtract, we have

$$P_{l'}^m\frac{d}{d\mu}\left[(1 - \mu^2)\frac{dP_l^m}{d\mu}\right] - P_l^m\frac{d}{d\mu}\left[(1 - \mu^2)\frac{dP_{l'}^m}{d\mu}\right] = [l'(l'+1) - l(l+1)]P_l^m P_{l'}^m$$

We then add to and subtract from the left-hand side the term

$$(1 - \mu^2)\frac{dP_l^m}{d\mu}\frac{dP_{l'}^m}{d\mu}$$

so as to obtain

$$\frac{d}{d\mu}\left[(1 - \mu^2)\left(P_{l'}^m\frac{dP_l^m}{d\mu} - P_l^m\frac{dP_{l'}^m}{d\mu}\right)\right] = [l'(l'+1) - l(l+1)]P_l^m P_{l'}^m$$

We next integrate with respect to μ from -1 to $+1$ and note that

$$\left[(1 - \mu^2)\left(P_{l'}^m\frac{dP_l^m}{d\mu} - P_l^m\frac{dP_{l'}^m}{d\mu}\right)\right]_{-1}^{1} = 0$$

giving

$$[l'(l'+1) - l(l+1)]\int_{-1}^{1} P_l^m P_{l'}^m \, d\mu = 0$$

If $l' \neq l$, then the integral must vanish

$$\int_{-1}^{1} P_l^m(\mu)P_{l'}^m(\mu) \, d\mu = 0 \qquad (E.18)$$

so that the associated Legendre polynomials $P_l^m(\mu)$ with fixed m form an orthogonal set of functions. Since equation (E.18) is valid for $m = 0$, the Legendre polynomials $P_l(\mu)$ are also an orthogonal set.

Normalization
We next wish to evaluate the integral I_{lm}

$$I_{lm} \equiv \int_{-1}^{1} [P_l^m(\mu)]^2 \, d\mu$$

As a first step, we evaluate I_{10}

$$I_{l0} = \int_{-1}^{1} [P_l(\mu)]^2 \, d\mu$$

We solve the recurrence relation (E.4) for $P_l(\mu)$, multiply both sides by $P_l(\mu)$, integrate with respect to μ from -1 to $+1$, and note that one of the integrals vanishes according to the orthogonality relation (E.18), so that

$$\int_{-1}^{1} [P_l(\mu)]^2 \, d\mu = \frac{2l-1}{l} \int_{-1}^{1} \mu P_l(\mu) P_{l-1}(\mu) \, d\mu$$

Replacing l by $l+1$ in equation (E.4), we can substitute for $\mu P_l(\mu)$ on the right-hand side. Again applying equation (E.18), we find that

$$\int_{-1}^{1} [P_l(\mu)]^2 \, d\mu = \frac{2l-1}{2l+1} \int_{-1}^{1} [P_{l-1}(\mu)]^2 \, d\mu$$

This relationship can then be applied successively to obtain

$$\int_{-1}^{1} [P_l(\mu)]^2 \, d\mu = \frac{(2l-1)(2l-3)}{(2l+1)(2l-1)} \int_{-1}^{1} [P_{l-2}(\mu)]^2 \, d\mu$$

$$\vdots$$

$$= \frac{(2l-1)(2l-3)\cdots 1}{(2l+1)(2l-1)(2l-3)\cdots 3} \int_{-1}^{1} [P_0(\mu)]^2 \, d\mu$$

$$= \frac{1}{2l+1} \int_{-1}^{1} [P_0(\mu)]^2 \, d\mu$$

Since $P_0(1) = 1$, the desired result is

$$\int_{-1}^{1} [P_l(\mu)]^2 \, d\mu = \frac{1}{2l+1} \int_{-1}^{1} d\mu = \frac{2}{2l+1} \tag{E.19}$$

We are now ready to evaluate I_{lm}. From equation (E.10) we have

$$I_{lm} = \int_{-1}^{1} (1-\mu^2)^m \left(\frac{d^m P_l}{d\mu^m} \right)^2 d\mu = \int_{-1}^{1} (1-\mu^2)^m \frac{d^m P_l}{d\mu^m} \frac{d}{d\mu} \left(\frac{d^{m-1} P_l}{d\mu^{m-1}} \right) d\mu$$

Integration by parts gives

$$I_{lm} = (1-\mu^2)^m \frac{d^m P_l}{d\mu^m} \frac{d^{m-1} P_l}{d\mu^{m-1}} \bigg|_{-1}^{1} - \int_{-1}^{1} \frac{d^{m-1} P_l}{d\mu^{m-1}} \frac{d}{d\mu} \left[(1-\mu^2)^m \frac{d^m P_l}{d\mu^m} \right] d\mu \tag{E.20}$$

The integrated part vanishes because $(1-\mu^2) = 0$ at $\mu = \pm 1$.

To evaluate the integral on the right-hand side of equation (E.20), we replace m by $m-1$ in (E.16) and multiply by $(1-\mu^2)^{m-1}$ to obtain

$$(1-\mu^2)^m \frac{d^2 w_{m-1}}{d\mu^2} - 2m\mu(1-\mu^2)^{m-1} \frac{dw_{m-1}}{d\mu}$$

$$+ [l(l+1) - m(m-1)](1-\mu^2)^{m-1} w_{m-1} = 0$$

which can be rewritten as

$$\frac{d}{d\mu} \left[(1-\mu^2)^m \frac{dw_{m-1}}{d\mu} \right] = -(l+m)(l-m+1)(1-\mu^2)^{m-1} w_{m-1} = 0$$

From equation (E.14) we see that

$$w_{m-1} = 2^l l! \frac{d^{m-1} P_l}{d\mu^{m-1}}$$

and

$$\frac{dw_{m-1}}{d\mu} = w_m = 2^l l! \frac{d^m P_l}{d\mu^m}$$

so that

$$\frac{d}{d\mu}\left[(1-\mu^2)^m \frac{d^m P_l}{d\mu^m}\right] = -(l+m)(l-m+1)(1-\mu^2)^{m-1}\frac{d^{m-1} P_l}{d\mu^{m-1}}$$

Thus, equation (E.20) takes the form

$$I_{lm} = (l+m)(l-m+1)\int_{-1}^{1}(1-\mu^2)^{m-1}\left[\frac{d^{m-1} P_l}{d\mu^{m-1}}\right]^2 d\mu$$

Using equation (E.10) to introduce $P_l^{m-1}(\mu)$, we have

$$I_{lm} = (l+m)(l-m+1)\int_{-1}^{1}[P_l^{m-1}(\mu)]^2\, d\mu$$

$$= (l+m)(l-m+1)I_{l,m-1}$$

which relates I_{lm} to $I_{l,m-1}$. This process can be repeated until I_{l0} is obtained

$$I_{lm} = [(l+m)(l+m-1)][(l-m+1)(l-m+2)]I_{l,m-2}$$

$$\vdots$$

$$= [(l+m)(l+m-1)\cdots(l+1)][(l-m+1)(l-m+2)\cdots l]I_{l0}$$

$$= \frac{(l+m)!}{l!}\frac{l!}{(l-m)!}I_{l0}$$

so that

$$\int_{-1}^{1}[P_l^m(\mu)]^2\, d\mu = \frac{2(l+m)!}{(2l+1)(l-m)!}$$

Completeness

The set of associated Legendre polynomials $P_l^m(\mu)$ with m fixed and $l = m$, $m+1, \ldots$, form a complete orthogonal set[1] in the range $-1 \leqslant \mu \leqslant 1$. Thus, an arbitrary function $f(\mu)$ can be expanded in the series

$$f(\mu) = \sum_{l=m}^{\infty} a_{lm} P_l^m(\mu)$$

with the expansion coefficients given by

---

[1] The proof of completeness may be found in W. Kaplan (1991) *Advanced Calculus*, 4th edition (Addison-Wesley, Reading, MA) p. 537 and in G. Birkhoff and G.-C. Rota (1989) *Ordinary Differential Equations*, 4th edition (John Wiley & Sons, New York), pp. 350–4.

$$a_{lm} = \frac{2l+1}{2} \frac{(l-m)!}{(l+m)!} \int_{-1}^{1} P_l^m(\mu) f(\mu) \, d\mu$$

The completeness relation for the polynomials $P_l^m(\mu)$ is

$$\sum_{l=m}^{\infty} \frac{2l+1}{2} \frac{(l-m)!}{(l+m)!} P_l^m(\mu) P_l^m(\mu') = \delta(\mu - \mu')$$

Appendix F

Laguerre and associated Laguerre polynomials

Laguerre polynomials

The Laguerre polynomials $L_k(\rho)$ are defined by means of the generating function $g(\rho, s)$

$$g(\rho, s) \equiv \frac{e^{-\rho s/(1-s)}}{1-s} \equiv \sum_{k=0}^{\infty} L_k(\rho) \frac{s^k}{k!} \tag{F.1}$$

where $0 \leqslant \rho \leqslant \infty$ and where $|s| < 1$ in order to ensure convergence of the infinite series. Since the right-hand term is a Taylor series expansion of $g(\rho, s)$, the Laguerre polynomials are given by

$$L_k(\rho) = \frac{\partial^k g(\rho, s)}{\partial s^k}\bigg|_{s=0} = \frac{\partial^k}{\partial s^k}\left(\frac{e^{-\rho s/(1-s)}}{1-s}\right)\bigg|_{s=0} \tag{F.2}$$

To evaluate $L_k(\rho)$ from equation (F.2), we first factor out e^ρ in the generating function and expand the remaining exponential function in a Taylor series

$$g(\rho, s) = \frac{e^\rho}{1-s} e^{-\rho/(1-s)} = \frac{e^\rho}{1-s} \sum_{\alpha=0}^{\infty} \frac{(-1)^\alpha}{\alpha!}\left(\frac{\rho}{1-s}\right)^\alpha = e^\rho \sum_{\alpha=0}^{\infty} \frac{(-1)^\alpha \rho^\alpha}{\alpha!}(1-s)^{-(\alpha+1)}$$

We then take k successive derivatives of $g(\rho, s)$ with respect to s

$$\frac{\partial g(\rho, s)}{\partial s} = e^\rho \sum_{\alpha=0}^{\infty} \frac{(-1)^\alpha \rho^\alpha}{\alpha!}(\alpha+1)(1-s)^{-(\alpha+2)}$$

$$\frac{\partial^2 g(\rho, s)}{\partial s^2} = e^\rho \sum_{\alpha=0}^{\infty} \frac{(-1)^\alpha \rho^\alpha}{\alpha!}(\alpha+1)(\alpha+2)(1-s)^{-(\alpha+3)}$$

$$\vdots$$

$$\frac{\partial^k g(\rho, s)}{\partial s^k} = e^\rho \sum_{\alpha=0}^{\infty} \frac{(-1)^\alpha \rho^\alpha}{\alpha!} \frac{(\alpha+k)!}{\alpha!}(1-s)^{-(\alpha+k+1)}$$

When the kth derivative is evaluated at $s = 0$, we have

$$L_k(\rho) - e^\rho \sum_{\alpha=0}^{\infty} \frac{(-1)^\alpha (\alpha+k)!}{(\alpha!)^2} \rho^\alpha \tag{F.3}$$

310

Using equation (A.1) we note that

$$\frac{d^k}{d\rho^k}(\rho^k e^{-\rho}) = \frac{d^k}{d\rho^k}\sum_{\alpha=0}^{\infty}\frac{(-1)^\alpha}{\alpha!}\rho^{\alpha+k} = \sum_{\alpha=0}^{\infty}\frac{(-1)^\alpha(\alpha+k)!}{(\alpha!)^2}\rho^\alpha \qquad (F.4)$$

Combining equations (F.3) and (F.4), we obtain the formula for the Laguerre polynomials

$$L_k(\rho) = e^\rho \frac{d^k}{d\rho^k}(\rho^k e^{-\rho}) \qquad (F.5)$$

Another relationship for the polynomials $L_k(\rho)$ can be obtained by expanding the generating function $g(\rho, s)$ in equation (F.1) using (A.1)

$$g(\rho, s) = \frac{e^{-\rho s/(1-s)}}{1-s} = \sum_{\alpha=0}^{\infty}\frac{(-1)^\alpha\rho^\alpha}{\alpha!}\frac{s^\alpha}{(1-s)^{\alpha+1}}$$

The factor $(1-s)^{-(\alpha+1)}$ may be expanded in an infinite series using equation (A.3) to obtain

$$(1-s)^{-(\alpha+1)} = \sum_{\beta=0}^{\infty}\frac{(\alpha+\beta)!}{\alpha!\beta!}s^\beta$$

so that $g(\rho, s)$ becomes

$$g(\rho, s) = \sum_{\alpha=0}^{\infty}\sum_{\beta=0}^{\infty}\frac{(-1)^\alpha(\alpha+\beta)!\rho^\alpha}{(\alpha!)^2\beta!}s^{\alpha+\beta}$$

We next collect all the coefficients of s^k for an arbitrary k, so that $\alpha + \beta = k$, and replace the summation over α by a summation over k. When $\alpha = k$, the index β equals zero; when $\alpha = k - 1$, the index β equals one; and so on until we have $\alpha = 0$ and $\beta = k$. Thus, the result of the summation is

$$g(\rho, s) = k!\sum_{k=0}^{\infty}\sum_{\beta=0}^{k}\frac{(-1)^{k-\beta}\rho^{k-\beta}}{[(k-\beta)!]^2\beta!}s^k$$

Since the Laguerre polynomial $L_k(\rho)$ divided by $k!$ is the coefficient of s^k in the expansion (F.1) of the generating function, we have

$$L_k(\rho) = (k!)^2\sum_{\beta=0}^{k}\frac{(-1)^{k-\beta}}{[(k-\beta)!]^2\beta!}\rho^{k-\beta}$$

If we let $k - \beta = \gamma$ and replace the summation over β by a summation over γ, we obtain the desired result

$$L_k(\rho) = (k!)^2\sum_{\gamma=0}^{k}\frac{(-1)^\gamma}{(\gamma!)^2(k-\gamma)!}\rho^\gamma \qquad (F.6)$$

A third relationship for the polynomials $L_k(\rho)$ can be obtained by expanding the derivative in equation (F.5), using (A.4), to give

$$L_k(\rho) = e^\rho\sum_{\alpha=0}^{k}\frac{k!}{\alpha!(k-\alpha)!}\frac{d^\alpha\rho^k}{d\rho^\alpha}\frac{d^{k-\alpha}e^{-\rho}}{d\rho^{k-\alpha}} = \sum_{\alpha=0}^{k}\frac{(-1)^{k-\alpha}k!}{\alpha!(k-\alpha)!}\frac{d^\alpha\rho^k}{d\rho^\alpha}$$

We now observe that the operator $[(d/d\rho) - 1]^k$ may be expanded according to the binomial theorem (A.2) as

$$[(d/d\rho) - 1]^k = (-1)^k[1 - (d/d\rho)]^k = (-1)^k \sum_{\alpha=0}^{k} \frac{(-1)^\alpha k!}{\alpha!(k-\alpha)!} \frac{d^\alpha}{d\rho^\alpha}$$

so that

$$L_k(\rho) = [(d/d\rho) - 1]^k \rho^k \tag{F.7}$$

where we have noted that $(-1)^\alpha = (-1)^{-\alpha}$.

From equation (F.2), (F.5), (F.6), or (F.7), we observe that the polynomial $L_k(\rho)$ is of degree k and we may readily obtain the first few polynomials of the set

$$L_0(\rho) = 1$$

$$L_1(\rho) = 1 - \rho$$

$$L_2(\rho) = 2 - 4\rho + \rho^2$$

$$L_3(\rho) = 6 - 18\rho + 9\rho^2 - \rho^3$$

We also note that $L_k(0) = k!$.

Differential equation

Equation (F.5) can be used to find the differential equation satisfied by the polynomials $L_k(\rho)$. We note that the function $f(\rho)$ defined as

$$f(\rho) \equiv \rho^k e^{-\rho}$$

satisfies the relation

$$\rho \frac{df}{d\rho} + (\rho - k)f = 0$$

If we differentiate this expression $k + 1$ times, we obtain

$$\rho \frac{d^2 f^{(k)}}{d\rho^2} + (1 + \rho) \frac{df^{(k)}}{d\rho} + (k + 1)f^{(k)} = 0$$

where $f^{(k)}$ is the kth derivative of $f(\rho)$. Since from equation (F.5) we have

$$f^{(k)} = e^{-\rho} L_k(\rho)$$

the Laguerre polynomials $L_k(\rho)$ satisfy the differential equation

$$\rho \frac{d^2 L_k}{d\rho^2} + (1 - \rho) \frac{dL_k}{d\rho} + kL_k(\rho) = 0 \tag{F.8}$$

Associated Laguerre polynomials

The associated Laguerre polynomials $L_k^j(\rho)$ are defined in terms of the Laguerre polynomials by

$$L_k^j(\rho) \equiv \frac{d^j}{d\rho^j} L_k(\rho) \tag{F.9}$$

Since $L_k(\rho)$ is a polynomial of degree k, $L_k^k(\rho)$ is a constant and $L_k^j(\rho) = 0$ for $j > k$. The generating function $g(\rho, s; j)$ for the associated Laguerre polynomials with fixed j is readily obtained by differentiation of equation (F.1)

$$g(\rho, s; j) = \frac{d^j}{d\rho^j}\left(\frac{e^{-\rho s/(1-s)}}{1-s}\right) = \frac{(-1)^j s^j}{(1-s)^{j+1}}e^{-\rho s/(1-s)} = \sum_{k=j}^{\infty} L_k^j(\rho)\frac{s^k}{k!} \qquad \text{(F.10)}$$

The summation in the right-hand term begins with $k = j$, since j cannot exceed k.
We can write an explicit series for $L_k^j(\rho)$ by substituting equation (F.6) into (F.9)

$$L_k^j(\rho) = (k!)^2 \sum_{\gamma=0}^{k} \frac{(-1)^\gamma}{(\gamma!)^2(k-\gamma)!}\frac{d^j}{d\rho^j}\rho^\gamma = (k!)^2 \sum_{\gamma=j}^{k}\frac{(-1)^\gamma}{\gamma!(k-\gamma)!(\gamma-j)!}\rho^{\gamma-j}$$

The summation over γ now begins with the term $\gamma = j$ because the earlier terms
vanish in the differentiation. If we let $\gamma - j = \alpha$ and replace the summation over γ by
a summation over α, we have

$$L_k^j(\rho) = (k!)^2 \sum_{\alpha=0}^{k-j}\frac{(-1)^{j+\alpha}\rho^\alpha}{\alpha!(k-j-\alpha)!(j+\alpha)!} \qquad \text{(F.11)}$$

For the purpose of deriving some useful relationships involving the polynomials
$L_k^j(\rho)$, we define the polynomial $\Lambda_i^j(\rho)$ as

$$\Lambda_i^j(\rho) \equiv \frac{i!}{(i+j)!}L_{i+j}^j(\rho) \qquad \text{(F.12)}$$

If we replace the dummy index of summation k in equation (F.10) by i, where
$i = k - j$, then (F.10) takes the form

$$g(\rho, s; j) - \sum_{i=0}^{\infty}\frac{L_{i+j}^j(\rho)}{(i+j)!}s^{i+j} = s^j\sum_{i=0}^{\infty}\frac{\Lambda_i^j(\rho)}{i!}s^i$$

Thus, $\Lambda_i^j(\rho)$ are just the coefficients in a Taylor series expansion of the function
$s^{-j}g(\rho, s; j)$ and are, therefore, given by

$$\Lambda_i^j(\rho) = \frac{\partial^i}{\partial s^i}s^{-j}g(\rho, s; j)\bigg|_{s=0}$$

Substituting for $g(\rho, s; j)$ using equation (F.10), we obtain

$$\Lambda_i^j(\rho) = (-1)^j\frac{\partial^i}{\partial s^i}\left[\frac{e^{-\rho s/(1-s)}}{(1-s)^{j+1}}\right]\bigg|_{s=0}$$

$$= (-1)^j\frac{\partial^i}{\partial s^i}\left[\frac{e^\rho e^{-\rho/(1-s)}}{(1-s)^{j+1}}\right]\bigg|_{s=0}$$

$$= (-1)^j\frac{\partial^i}{\partial s^i}\left[e^\rho\sum_{\alpha=0}^{\infty}\frac{(-\rho)^\alpha}{\alpha!(1-s)^{\alpha+j+1}}\right]\bigg|_{s=0}$$

$$= (-1)^j e^\rho\sum_{\alpha=0}^{\infty}\frac{(-\rho)^\alpha}{\alpha!}\frac{(\alpha+j+i)!}{(\alpha+j)!}(1-s)^{-(\alpha+j+i+1)}\bigg|_{s=0}$$

$$= (-1)^j e^\rho\sum_{\alpha=0}^{\infty}\frac{(\alpha+j+i)!}{\alpha!(\alpha+j)!}(-\rho)^\alpha \qquad \text{(F.13)}$$

We next note that

$$\frac{d^i}{d\rho^i}(\rho^{i+j}e^{-\rho}) = \frac{d^i}{d\rho^i}\sum_{\alpha=0}^{\infty}\frac{(-1)^\alpha}{\alpha!}\rho^{\alpha+j+i} = \sum_{\alpha=0}^{\infty}\frac{(-1)^\alpha}{\alpha!}\frac{(\alpha+j+i)!}{(\alpha+j)!}\rho^{\alpha+j}$$

$$= \rho^j\sum_{\alpha=0}^{\infty}\frac{(\alpha+j+i)!}{\alpha!(\alpha+j)!}(-\rho)^\alpha \tag{F.14}$$

Comparison of equations (F.13) and (F.14) yields the result that

$$\Lambda_i^j(\rho) = (-1)^j\rho^{-j}e^{\rho}\frac{d^i}{d\rho^i}(\rho^{i+j}e^{-\rho})$$

From equation (F.12) we obtain

$$L_{i+j}^j(\rho) = (-1)^j\frac{(i+j)!}{i!}\rho^{-j}e^{\rho}\frac{d^i}{d\rho^i}(\rho^{i+j}e^{-\rho})$$

Finally, replacing i by the original index $k(=i+j)$, we have

$$L_k^j(\rho) = (-1)^j\frac{k!}{(k-j)!}\rho^{-j}e^{\rho}\frac{d^{k-j}}{d\rho^{k-j}}(\rho^k e^{-\rho}) \tag{F.15}$$

Equation (F.15) for the associated Laguerre polynomials is the analog of (F.5) for the Laguerre polynomials and, in fact, when $j=0$, equation (F.15) reduces to (F.5).

Differential equation
The differential equation satisfied by the associated Laguerre polynomials may be obtained by repeatedly differentiating equations (F.8) j times

$$\rho\frac{d^3 L_k}{d\rho^3} + (2-\rho)\frac{d^2 L_k}{d\rho^2} + (k-1)\frac{dL_k}{d\rho} = 0$$

$$\rho\frac{d^4 L_k}{d\rho^4} + (3-\rho)\frac{d^3 L_k}{d\rho^3} + (k-2)\frac{d^2 L_k}{d\rho^2} = 0$$

$$\vdots$$

$$\rho\frac{d^{j+2} L_k}{d\rho^{j+2}} + (j+1-\rho)\frac{d^{j+1} L_k}{d\rho^{j+1}} + (k-j)\frac{d^j L_k}{d\rho^j} = 0$$

When the polynomials $L_k^j(\rho)$ are introduced with equation (F.9), the differential equation is

$$\rho\frac{d^2 L_k^j}{d\rho^2} + (j+1-\rho)\frac{dL_k^j}{d\rho} + (k-j)L_k^j(\rho) = 0 \tag{F.16}$$

Integral relations
In order to obtain the orthogonality and normalization relations of the associate Laguerre polynomials, we make use of the generating function (F.10). We multiply together $g(\rho, s; j)$, $g(\rho, t; j)$, and the factor $\rho^{j+\nu}e^{-\rho}$ and then integrate over ρ to give an integral that we abbreviate with the symbol I

$$I \equiv \int_0^\infty \rho^{j+\nu}e^{-\rho}g(\rho, s; j)g(\rho, t; j)\,d\rho = \sum_{\alpha=j}^{\infty}\sum_{\beta=j}^{\infty}\frac{s^\alpha t^\beta}{\alpha!\beta!}\int_0^\infty \rho^{j+\nu}e^{-\rho}L_\alpha^j(\rho)L_\beta^j(\rho)\,d\rho$$

$$\tag{F.17}$$

To evaluate the left-hand integral, we substitute the analytical forms of the generating functions from equation (F.10) to give

$$I = \frac{(st)^j}{(1-s)^{j+1}(1-t)^{j+1}} \int_0^\infty \rho^{j+\nu} e^{-a\rho}\, d\rho \qquad (F.18)$$

where

$$a \equiv 1 + \frac{s}{1-s} + \frac{t}{1-t} = \frac{1-st}{(1-s)(1-t)}$$

The integral in equation (F.18) is just the gamma function (A.26), so that

$$\int_0^\infty \rho^{j+\nu} e^{-a\rho}\, d\rho = \frac{\Gamma(j+\nu+1)}{a^{j+\nu+1}} = \frac{(j+\nu)!}{a^{j+\nu+1}}, \qquad j+\nu>0$$

where we have restricted ν to integer values. Thus, I in equation (F.18) is

$$I = \frac{(j+\nu)!(st)^j(1-s)^\nu(1-t)^\nu}{(1-st)^{j+\nu+1}}$$

Applying the expansion formula (A.3), we have

$$(1-st)^{-(j+\nu+1)} = \sum_{i=0}^\infty \frac{(j+\nu+i)!}{(j+\nu)!i!}(st)^i$$

If we replace the dummy index i by α, where $\alpha = i + j$, then this expression becomes

$$(1-st)^{-(j+\nu+1)} = \sum_{\alpha=j}^\infty \frac{(\alpha+\nu)!}{(j+\nu)!(\alpha-j)!}(st)^{\alpha-j}$$

and I takes the form

$$I = (1-s)^\nu(1-t)^\nu \sum_{\alpha=j}^\infty \frac{(\alpha+\nu)!}{(\alpha-j)!}(st)^\alpha$$

Combining this result with equation (F.17), we have

$$\sum_{\alpha=j}^\infty \sum_{\beta=j}^\infty \frac{s^\alpha t^\beta}{\alpha!\beta!} \int_0^\infty \rho^{j+\nu} e^{-\rho} L_\alpha^j(\rho) L_\beta^j(\rho)\, d\rho = (1-s)^\nu(1-t)^\nu \sum_{\alpha=j}^\infty \frac{(\alpha+\nu)!}{(\alpha-j)!}(st)^\alpha \quad (F.19)$$

We now equate coefficients of like powers of s and t on each side of this equation. Since the integer ν appears as an exponent of both s and t on the right-hand side, the effect of equating coefficients depends on the value of ν. Accordingly, we shall first have to select a value for ν.

For $\nu = 0$, equation (F.19) becomes

$$\sum_{\alpha=j}^\infty \sum_{\beta=j}^\infty \frac{s^\alpha t^\beta}{\alpha!\beta!} \int_0^\infty \rho^j e^{-\rho} L_\alpha^j(\rho) L_\beta^j(\rho)\, d\rho = \sum_{\alpha=j}^\infty \frac{\alpha!}{(\alpha-j)!}(st)^\alpha$$

Since the exponent of s on the right-hand side is always the same as the exponent of t, the coefficients of $s^\alpha t^\beta$ for $\alpha \neq \beta$ on the left-hand side must vanish, i.e.

$$\int_0^\infty \rho^j e^{-\rho} L_\alpha^j(\rho) L_\beta^j(\rho)\, d\rho = 0; \qquad \alpha \neq \beta \qquad (F.20)$$

Thus, the associated Laguerre polynomials form an orthogonal set over the range $0 \leqslant \rho \leqslant \infty$ with a weighting factor $\rho^j e^{-\rho}$. For the case where s and t on the left-hand side have the same exponent, we pick out the term $\beta = \alpha$ in the summation over β, giving

$$\sum_{a=j}^{\infty} \frac{(st)^{a}}{(a!)^{2}} \int_{0}^{\infty} \rho^{j} e^{-\rho} [L_{a}^{j}(\rho)]^{2} \, d\rho = \sum_{a=j}^{\infty} \frac{a!}{(a-j)!} (st)^{a}$$

Equating coefficients of $(st)^{a}$ on each side yields

$$\int_{0}^{\infty} \rho^{j} e^{-\rho} [L_{a}^{j}(\rho)]^{2} \, d\rho = \frac{(a!)^{3}}{(a-j)!} \tag{F.21}$$

Equations (F.20) and (F.21) may be combined into a single expression

$$\int_{0}^{\infty} \rho^{j} e^{-\rho} L_{a}^{j}(\rho) L_{\beta}^{j}(\rho) \, d\rho = \frac{(a!)^{3}}{(a-j)!} \delta_{a\beta} \tag{F.22}$$

For $\nu = 1$, equation (F.19) becomes

$$\sum_{a=j}^{\infty} \sum_{\beta=j}^{\infty} \frac{s^{a} t^{\beta}}{a! \beta!} \int_{0}^{\infty} \rho^{j+1} e^{-\rho} L_{a}^{j}(\rho) L_{\beta}^{j}(\rho) \, d\rho$$

$$= \sum_{a=j}^{\infty} \frac{(a+1)!}{(a-j)!} [(st)^{a} + (st)^{a+1} - s^{a+1} t^{a} - s^{a} t^{a+1}]$$

Equating coefficients of like powers of s and t on both sides of this equation, we see that

$$\int_{0}^{\infty} \rho^{j+1} e^{-\rho} L_{a}^{j}(\rho) L_{\beta}^{j}(\rho) \, d\rho = 0; \qquad \beta \neq a, \, a \pm 1 \tag{F.23}$$

and that

$$\int_{0}^{\infty} \rho^{j+1} e^{-\rho} L_{a}^{j}(\rho) L_{a+1}^{j}(\rho) \, d\rho = -\frac{a! [(a+1)!]^{2}}{(a-j)!} \tag{F.24}$$

$$\int_{0}^{\infty} \rho^{j+1} e^{-\rho} [L_{a}^{j}(\rho)]^{2} \, d\rho = (a!)^{2} \left[\frac{(a+1)!}{(a-j)!} + \frac{a!}{(a-1-j)!} \right] = \frac{(2a-j+1)(a!)^{3}}{(a-j)!} \tag{F.25}$$

The term in which $\beta = a - 1$ is equivalent to the term in which $\beta = a + 1$ after the dummy indices a and β are interchanged. Equations (F.23), (F.24), and (F.25) are pertinent to the wave functions for the hydrogen atom.

Completeness

We define the set of functions $\chi_{kj}(\rho)$ by the relation

$$\chi_{kj}(\rho) = \left[\frac{(k-j)!}{(k!)^{3}} \rho^{j} e^{-\rho} \right]^{1/2} L_{k}^{j}(\rho) \tag{F.26}$$

According to equation (F.22), the functions $\chi_{kj}(\rho)$ constitute an orthonormal set. We now show[1] that this set is complete.

Substitution of equation (F.15) into (F.26) gives

$$\chi_{kj}(\rho) = (-1)^{j} \left[\frac{\rho^{-j} e^{\rho}}{k! (k-j)!} \right]^{1/2} \frac{d^{k-j}}{d\rho^{k-j}} (\rho^{k} e^{-\rho}) \tag{F.27}$$

If we apply equation (A.11), we may express the derivative in (F.27) as

---

[1] D. Park, personal communication. This method parallels the procedure used to demonstrate the completeness of the set of functions in equation (D.15).

$$\frac{d^{k-j}}{d\rho^{k-j}}(\rho^k e^{-\rho}) = \frac{k!}{2\pi}\frac{d^{k-j}}{d\rho^{k-j}}\int_{-\infty}^{\infty}\frac{e^{i\rho s}}{(1+is)^{k+1}}\,ds = \frac{i^{k-j}k!}{2\pi}\int_{-\infty}^{\infty}\frac{e^{i\rho s}}{(1+is)^{k+1}}s^{k-j}\,ds$$

so that $\chi_{kj}(\rho)$ in integral form is

$$\chi_{kj}(\rho) = \frac{(-1)^j i^{k-j}}{2\pi}\left[\frac{k!\rho^{-j}e^{\rho}}{(k-j)!}\right]^{1/2}\int_{-\infty}^{\infty}\frac{e^{i\rho s}}{(1+is)^{k+1}}s^{k-j}\,ds \qquad (F.28)$$

To demonstrate that the set $\chi_{kj}(\rho)$ is complete, we need to evaluate the sum

$$\sum_{k=j}^{\infty}\chi_{kj}(\rho)\chi_{kj}(\rho')$$

Expressing (F.28) in terms of the dummy variable of integration s for ρ and in terms of t for ρ', we obtain for the summation

$$\sum_{k=j}^{\infty}\chi_{kj}(\rho)\chi_{kj}(\rho')$$
$$= \frac{1}{4\pi^2}(\rho\rho')^{-j/2}e^{(\rho+\rho')/2}\int_{-\infty}^{\infty}\int_{-\infty}^{\infty}e^{i(\rho s+\rho' t)}\left[\sum_{k=j}^{\infty}\frac{(-1)^{k-j}k!}{(k-j)!}\frac{(st)^{k-j}}{[1+i(s+t)-st]^{k+1}}\right]ds\,dt$$

By letting $\alpha = k - j$, we may express the sum on the right-hand side as

$$\sum_{\alpha=0}^{\infty}\frac{(-1)^{\alpha}(\alpha+j)!}{\alpha!}[1+i(s+t)-st]^{-(\alpha+j+1)}(st)^{\alpha} = j![1+i(s+t)]^{-(j+1)}$$

where we have applied equation (A.3) to evaluate the sum over α. We now have

$$\sum_{k=j}^{\infty}\chi_{kj}(\rho)\chi_{kj}(\rho') = \frac{j!}{4\pi^2}(\rho\rho')^{-j/2}e^{(\rho+\rho')/2}\int_{-\infty}^{\infty}\int_{-\infty}^{\infty}\frac{e^{i(\rho s+\rho' t)}}{[1+i(s+t)]^{j+1}}\,ds\,dt \qquad (F.29)$$

To evaluate the double integral, we introduce the variables u and v

$$u = \frac{s+t}{2}, \quad v = \frac{s-t}{2} \quad \text{or} \quad s = u+v, \quad t = u-v$$
$$ds\,dt = 2\,du\,dv$$

The double integral then factors into

$$2\int_{-\infty}^{\infty}\frac{e^{i(\rho+\rho')u}}{(1+2iu)^{j+1}}\,du\int_{-\infty}^{\infty}e^{i(\rho-\rho')v}\,dv = \frac{2\pi}{j!}\left(\frac{\rho+\rho'}{2}\right)^{j}e^{-(\rho+\rho')/2}[2\pi\delta(\rho-\rho')]$$

where the first integral is evaluated by equation (A.11) and the second by (C.6). Equation (F.29) becomes

$$\sum_{k=j}^{\infty}\chi_{kj}(\rho)\chi_{kj}(\rho') = \left[\frac{\rho+\rho'}{2(\rho\rho')^{1/2}}\right]^{j}\delta(\rho-\rho')$$

By applying equation (C.5e), we obtain the completeness relation

$$\sum_{k=j}^{\infty}\chi_{kj}(\rho)\chi_{kj}(\rho') = \delta(\rho-\rho') \qquad (F.30)$$

demonstrating according to equation (3.31) that the set $\chi_{kj}(\rho)$ is complete.

Appendix G

Series solutions of differential equations

General procedure

The application of the time-independent Schrödinger equation to a system of chemical interest requires the solution of a linear second-order homogeneous differential equation of the general form

$$p(x)\frac{d^2 u(x)}{dx^2} + q(x)\frac{du(x)}{dx} + r(x)u(x) = 0 \tag{G.1}$$

where $p(x)$, $q(x)$, and $r(x)$ are polynomials in x and where $p(x)$ does not vanish in some interval which contains the point $x = 0$. Equation (G.1) is *linear* because each term contains u or a derivative of u to the first power only. The order of the highest derivative determines that equation (G.1) is *second-order*. In a *homogeneous* differential equation, every term contains u or one of its derivatives.

The Frobenius or series solution method for solving equation (G.1) assumes that the solution may be expressed as a power series in x

$$u = \sum_{k=0}^{\infty} a_k x^{k+s} = a_0 x^s + a_1 x^{s+1} + \cdots \tag{G.2}$$

where a_k ($k = 0, 1, 2, \ldots$) and s are constants to be determined. The constant s is chosen such that a_0 is not equal to zero. The first and second derivatives of u are then given by

$$\frac{du}{dx} = u' = \sum_{k=0}^{\infty} a_k(k+s)x^{k+s-1} = a_0 s x^{s-1} + a_1(s+1)x^s + \cdots \tag{G.3}$$

$$\frac{d^2 u}{dx^2} = u'' = \sum_{k=0}^{\infty} a_k(k+s)(k+s-1)x^{k+s-2} = a_0 s(s-1)x^{s-2} + a_1(s+1)s x^{s-1} + \cdots$$
$$\tag{G.4}$$

A second-order differential equation has two solutions of the form of equation (G.2), each with a different set of values for the constant s and the coefficients a_k.

Not all differential equations of the general form (G.1) possess solutions which can be expressed as a power series (equation (G.2)).[1] However, the differential equations encountered in quantum mechanics can be treated in this manner. Moreover, the power

---

[1] For a thorough treatment see F. B. Hildebrand (1949) *Advanced Calculus for Engineers* (Prentice-Hall, New York).

series expansion of u is valid for many differential equations in which $p(x)$, $q(x)$, and/or $r(x)$ are functions other than polynomials,[2] but such differential equations do not occur in quantum-mechanical applications.

The Frobenius procedure consists of the following steps.

1. Equations (G.2), (G.3), and (G.4) are substituted into the differential equation (G.1) to obtain a series of the form

$$\sum_{k=0}^{\infty} a_k[(k+s)(k+s-1)p(x)x^{k+s-2} + (k+s)q(x)x^{k+s-1} + r(x)x^{k+s}] = 0$$

2. The terms are arranged in order of ascending powers of x to obtain

$$\sum_{k=\alpha}^{\infty} c_k x^{k+s-2} = 0 \tag{G.5}$$

where the coefficients c_k are combinations of the constant s, the coefficients a_k, and the coefficients in the polynomials $p(x)$, $q(x)$, and $r(x)$. The lower limit α of the summation is selected such that the coefficients c_k for $k < \alpha$ are identically zero, but c_α is not.

3. Since the right-hand side of equation (G.5) is zero, the left-hand side must also equal zero for all values of x in an interval that includes $x = 0$. The only way to meet this condition is to set each of the coefficients c_k equal to zero, i.e., $c_k = 0$ for $k = \alpha, \alpha + 1, \ldots$

4. The coefficient c_α of the lowest power of x in equation (G.5) always has the form $c_\alpha = a_0 f(s)$, where $f(s)$ is quadratic in s because the differential equation is second-order. The expression $c_\alpha = a_0 f(s) = 0$ is called the *indicial equation* and has two roots, s_1 and s_2, assuming that $a_0 \neq 0$.

5. For each of the two values of s, the remaining expressions $c_k = 0$ for $k = \alpha + 1$, $\alpha + 2, \ldots$ determine successively $a_1, a_2, \ldots$ in terms of u_0. Each value of s yields a different set of values for a_k; one set is denoted here as a_k, the other as a'_k.

6. The two mathematical solutions of the differential equation are u_1 and u_2

$$u_1 = a_0 x^{s_1}[1 + (a_1/a_0)x + (a_2/a_0)x^2 + \cdots]$$

$$u_2 = a'_0 x^{s_2}[1 + (a'_1/a'_0)x + (a'_2/a'_0)x^2 + \cdots]$$

where a_0 and a'_0 are arbitrary constants. Physical solutions are obtained by applying boundary and normalization conditions to u_1 and u_2.

7. For some differential equations, the two roots s_1 and s_2 of the indicial equation differ by an integer. Under this circumstance, there are two possible outcomes: (a) steps 1 to 6 lead to two independent solutions, or (b) for the larger root s_1, steps 1 to 6 give a solution u_1, but for the root s_2 the recursion relation gives infinite values for the coefficients a_k beyond some specific value of k and therefore these steps fail to provide a second solution. For some other differential equations, the two roots of

---

[2] See for example E. T. Whittaker and G. N. Watson (1927) *A Course of Modern Analysis*, 4th edition (Cambridge University Press, Cambridge), pp. 194–8; see also the reference in footnote 1 of this Appendix.

the indicial equation are the same ($s_1 = s_2$) and therefore only one solution u_1 is obtained. In those cases where steps 1 to 6 give only one solution u_1, a second solution u_2 may be obtained[3] by a slightly more complex procedure. This second solution has the form

$$u_2 = cu_1 \ln x + c \sum_{k=0}^{\infty} b_k x^{k+s_2}$$

where c is an arbitrary constant and the coefficients b_k are related to the coefficients a_k. However, a solution containing $\ln x$ is not well-behaved and the arbitrary constant c is set equal to zero in quantum-mechanical applications.

8. The interval of convergence for each of the series solutions u_1 and u_2 may be determined by applying the *ratio test*. For convergence, the condition

$$\lim_{k \to \infty} \left| \frac{a_{k+1}}{a_k} \right| |x| < 1$$

must be satisfied. Thus, a series converges for values of x in the range

$$-\frac{1}{R} < x < \frac{1}{R}$$

where R is defined by

$$R \equiv \lim_{k \to \infty} \left| \frac{a_{k+1}}{a_k} \right|$$

For R equal to zero, the corresponding series converges for $-\infty < x < \infty$. If R equals unity, the corresponding series converges for $-1 < x < 1$.

Applications

In Chapters 4, 5, and 6 the Schrödinger equation is applied to three systems: the harmonic oscillator, the orbital angular momentum, and the hydrogen atom, respectively. The ladder operator technique is used in each case to solve the resulting differential equation. We present here the solutions of these differential equations using the Frobenius method.

Harmonic oscillator

The Schrödinger equation for the linear harmonic oscillator leads to the differential equation (4.17)

$$-\frac{d^2\phi(\xi)}{d\xi^2} + \xi^2\phi(\xi) = \frac{2E}{\hbar\omega}\phi(\xi) \tag{G.6}$$

If we define λ by the relation

$$2\lambda + 1 = \frac{2E}{\hbar\omega} \tag{G.7}$$

and introduce this expression into equation (G.6), we obtain

$$\phi'' + (2\lambda + 1 - \xi^2)\phi = 0 \tag{G.8}$$

---

[3] See footnote 1 of this Appendix.

We first investigate the asymptotic behavior of $\phi(\xi)$. For large values of ξ, the constant $2\lambda + 1$ may be neglected in comparison with ξ^2 and equation (G.8) becomes

$$\phi'' = \xi^2 \phi$$

The approximate solutions of this differential equation are

$$\phi = ce^{\pm\xi^2/2}$$

because we have

$$\phi'' = (\xi^2 \pm 1)\phi \approx \xi^2\phi \quad \text{for large } \xi$$

The function $e^{\xi^2/2}$ is not a satisfactory solution because it becomes infinite as $\xi \to \pm\infty$, but the function $e^{-\xi^2/2}$ is well-behaved. This asymptotic behavior of $\phi(\xi)$ suggests that a satisfactory solution of equation (G.8) has the form

$$\phi(\xi) = u(\xi)e^{-\xi^2/2} \tag{G.9}$$

where $u(\xi)$ is a function to be determined.

Substitution of equation (G.9) into (G.8) gives

$$u'' - 2\xi u' + 2\lambda u = 0 \tag{G.10}$$

We solve this differential equation by the series solution method. Applying equations (G.2), (G.3), and (G.4), we obtain

$$\sum_{k=0}^{\infty} a_k(k+s)(k+s-1)\xi^{k+s-2} + \sum_{k=0}^{\infty} a_k[-2(k+s) + 2\lambda]\xi^{k+s} = 0 \tag{G.11}$$

The coefficient of ξ^{s-2} gives the indicial equation

$$a_0 s(s-1) = 0 \tag{G.12}$$

with two solutions, $s = 0$ and $s = 1$. The coefficient of ξ^{s-1} gives

$$a_1(s+1)s = 0 \tag{G.13}$$

For the case $s = 0$, the coefficient a_1 has an arbitrary value; for $s = 1$, we have $a_1 = 0$.

If we omit the first two terms (they vanish according to equations (G.12) and (G.13)) in the first summation on the left-hand side of (G.11) and replace the dummy index k by $k + 2$ in that summation, we obtain

$$\sum_{k=0}^{\infty} \{a_{k+2}(k+s+2)(k+s+1) + a_k[-2(k+s) + 2\lambda]\}\xi^{k+s} = 0 \tag{G.14}$$

Setting the coefficient of each power of ξ equal to zero gives the recursion formula

$$a_{k+2} = \frac{2(k+s-\lambda)}{(k+s+2)(k+s+1)} a_k \tag{G.15}$$

For the case $s = 0$, the constants a_0 and a_1 are arbitrary and we have the following two sets of expansion constants

a_0

$a_2 = -\lambda a_0$

$a_4 = \dfrac{2(2-\lambda)}{4\cdot 3} a_2 = -\dfrac{2^2\lambda(2-\lambda)}{4!} a_0$

$a_6 = \dfrac{2(4-\lambda)}{6\cdot 5} a_4 = -\dfrac{2^3\lambda(2-\lambda)(4-\lambda)}{6!} a_0$

$\vdots$

a_1

$a_3 = \dfrac{2(1-\lambda)}{3!} a_1$

$a_5 = \dfrac{2(3-\lambda)}{5\cdot 4} a_3 = \dfrac{2^2(1-\lambda)(3-\lambda)}{5!} a_1$

$a_7 = \dfrac{2(5-\lambda)}{7\cdot 6} a_5 = \dfrac{2^3(1-\lambda)(3-\lambda)(5-\lambda)}{7!} a_1$

$\vdots$

Thus, the two solutions of the second-order differential equation (G.10) are

$$u_1 = a_0 \left[1 - \lambda\xi^2 - \frac{2^2\lambda(2-\lambda)}{4!}\xi^4 - \frac{2^3\lambda(2-\lambda)(4-\lambda)}{6!}\xi^6 - \cdots \right] \qquad \text{(G.16a)}$$

$$u_2 = a_1 \left[\xi + \frac{2(1-\lambda)}{3!}\xi^3 + \frac{2^2(1-\lambda)(3-\lambda)}{5!}\xi^5 + \frac{2^3(1-\lambda)(3-\lambda)(5-\lambda)}{7!}\xi^7 + \cdots \right] \qquad \text{(G.16b)}$$

The solution u_1 is an even function of the variable ξ and u_2 is an odd function of ξ. Accordingly, u_1 and u_2 are independent solutions. For the case $s = 1$, we again obtain the solution u_2.

The ratio of consecutive terms in either series solution u_1 or u_2 is given by the recursion formula with $s = 0$ as

$$\frac{a_{k+2}\xi^{k+2}}{a_k\xi^k} = \frac{2(k-\lambda)}{(k+2)(k+1)}\xi^2$$

In the limit as $k \to \infty$, this ratio approaches zero

$$\lim_{k\to\infty} \frac{a_{k+2}\xi^{k+2}}{a_k\xi^k} \to \lim_{k\to\infty} \frac{2}{k}\xi^2 \to 0$$

so that the series u_1 and u_2 converge for all finite values of ξ. To see what happens to u_1 and u_2 as $\xi \to \pm\infty$, we consider the Taylor series expansion of e^{ξ^2}

$$e^{\xi^2} = \sum_{n=0}^{\infty} \frac{\xi^{2n}}{n!} = 1 + \xi^2 + \frac{\xi^4}{2!} + \frac{\xi^6}{3!} + \cdots$$

The coefficient a_n is given by $a_n = 1/(n/2)!$ for n even and $a_n = 0$ for n odd, so that

$$\frac{a_{n+2}\xi^{n+2}}{a_n\xi^n} = \frac{\left(\dfrac{n}{2}\right)!}{\left(\dfrac{n+2}{2}\right)!}\xi^2 = \frac{1}{\left(\dfrac{n}{2}+1\right)}\xi^2 \approx \frac{2\xi^2}{n} \qquad \text{as } n \to \infty$$

Thus, u_1 and u_2 behave like e^{ξ^2} as $\xi \to \pm\infty$. For large $|\xi|$, the function $\phi(\xi)$ behaves like

$$\phi(\xi) = u(\xi)e^{-\xi^2/2} \approx e^{\xi^2}e^{-\xi^2/2} = e^{\xi^2/2} \to \infty \qquad \text{as } \xi \to \pm\infty$$

which is not satisfactory behavior for a wave function.

In order to obtain well-behaved solutions for the differential equation (G.8), we need to terminate the infinite power series u_1 and u_2 in (G.16) to a finite polynomial. If we let λ equal an integer n ($n = 0, 1, 2, 3, \ldots$), then we obtain well-behaved solutions $\phi(\xi)$

$$n = 0, \qquad \phi_0 = a_0 e^{-\xi^2/2}, \qquad a_1 = 0$$

$$n = 1, \qquad \phi_1 = a_1\xi e^{-\xi^2/2}, \qquad a_0 = 0$$

$$n = 2, \qquad \phi_2 = a_0(1 - 2\xi^2)e^{-\xi^2/2}, \qquad a_1 = 0$$

$$n = 3, \qquad \phi_3 = a_1\xi(1 - \tfrac{2}{3}\xi^2)e^{-\xi^2/2}, \qquad a_0 = 0$$

$$n = 4, \qquad \phi_4 = a_0(1 - 4\xi^2 + \tfrac{4}{3}\xi^4)e^{-\xi^2/2}, \qquad a_1 = 0$$

$$n = 5, \qquad \phi_5 = a_1\xi(1 - \tfrac{4}{3}\xi^2 + \tfrac{4}{15}\xi^4)e^{-\xi^2/2}, \qquad a_0 = 0$$

$$\vdots \qquad\qquad \vdots$$

Since the parameter λ is equal to a positive integer n, the energy E of the harmonic oscillator in equation (G.7) is

$$E_n = (n + \tfrac{1}{2})\hbar\omega, \qquad n = 0, 1, 2, \ldots$$

in agreement with equation (4.30). Setting λ in equation (G.10) equal to the integer n gives

$$u'' - 2\xi u' + 2nu = 0 \tag{G.17}$$

A comparison of equation (G.17) with (D.10) shows that the solutions $u(\xi)$ are the Hermite polynomials, whose properties are discussed in Appendix D. Thus, the functions $\phi_n(\xi)$ for the harmonic oscillator are

$$\phi_n(\xi) = a_n H_n(\xi) e^{-\xi^2/2}$$

where a_n are the constants which normalize $\phi_n(\xi)$. Application of equation (D.14) yields the final result

$$\phi_n(\xi) = (2^n n!)^{-1/2} \pi^{-1/4} H_n(\xi) e^{-\xi^2/2}$$

which agrees with equation (4.40).

Orbital angular momentum

We wish to solve the differential equation

$$\hat{L}^2 \psi(\theta, \varphi) = \lambda\hbar^2 \psi(\theta, \varphi) \tag{G.18}$$

where $\hat{L}^2$ is given by equation (5.32) as

$$\hat{L}^2 = -\hbar^2 \left[\frac{1}{\sin\theta} \frac{\partial}{\partial\theta} \left(\sin\theta \frac{\partial}{\partial\theta} \right) + \frac{1}{\sin^2\theta} \frac{\partial^2}{\partial\varphi^2} \right] \tag{G.19}$$

We write the function $\psi(\theta, \varphi)$ as the product of two functions, one depending only on the angle θ, the other only on φ

$$\psi(\theta, \varphi) = \Theta(\theta)\Phi(\varphi) \tag{G.20}$$

When equations (G.19) and (G.20) are substituted into (G.18), we obtain after a little rearrangement

$$\frac{\sin\theta}{\Theta} \frac{d}{d\theta} \left(\sin\theta \frac{d\Theta}{d\theta} \right) + \lambda \sin^2\theta = -\frac{1}{\Phi} \frac{d^2\Phi}{d\varphi^2} \tag{G.21}$$

The left-hand side of equation (G.21) depends only on the variable θ, while the right-hand side depends only on φ. Following the same argument used in the solution of equation (2.28), each side of equation (G.21) must be equal to a constant, which we write as m^2. Thus, equation (G.21) separates into two differential equations

$$\frac{\sin\theta}{\Theta} \frac{d}{d\theta} \left(\sin\theta \frac{d\Theta}{d\theta} \right) + \lambda \sin^2\theta = m^2 \tag{G.22}$$

and

$$\frac{d^2\Phi}{d\varphi^2} = -m^2\Phi \tag{G.23}$$

The solution of equation (G.23) is

$$\Phi = A e^{im\varphi} \tag{G.24}$$

where A is an arbitrary constant. In order for Φ to be single-valued, we require that

$$\Phi(\varphi) = \Phi(\varphi + 2\pi)$$

or

$$e^{2\pi i m} = 1$$

so that m is an integer, $m = 0, \pm 1, \pm 2, \ldots$

To solve the differential equation (G.22), we introduce a change of variable

$$\mu = \cos\theta \qquad (G.25)$$

The function $\Theta(\theta)$ then becomes a new function $F(\mu)$ of the variable μ, $\Theta(\theta) = F(\mu)$, so that

$$\frac{d\Theta}{d\theta} = \frac{dF}{d\mu}\frac{d\mu}{d\theta} = -\sin\theta\,\frac{dF}{d\mu} = -(1-\mu^2)^{1/2}\frac{dF}{d\mu} \qquad (G.26)$$

Substitution of equations (G.25) and (G.26) into (G.22) gives

$$\frac{d}{d\mu}\left[(1-\mu^2)\frac{dF}{d\mu}\right] + \left(\lambda - \frac{m^2}{1-\mu^2}\right)F = 0$$

or

$$(1-\mu^2)F'' - 2\mu F' + \left(\lambda - \frac{m^2}{1-\mu^2}\right)F = 0 \qquad (G.27)$$

A power series solution of equation (G.27) yields a recursion formula relating a_{k+4}, a_{k+2}, and a_k, which is too complicated to be practical. Accordingly, we make the further definition

$$F(\mu) = (1-\mu^2)^{|m|/2}G(\mu) \qquad (G.28)$$

from which it follows that

$$F' = (1-\mu^2)^{|m|/2}[G' - |m|\mu(1-\mu^2)^{-1}G] \qquad (G.29)$$

$$F'' = (1-\mu^2)^{|m|/2}[G'' - 2|m|\mu(1-\mu^2)^{-1}G' - |m|(1-\mu^2)^{-1}G$$

$$+ |m|(|m|-2)\mu^2(1-\mu^2)^{-2}G] \qquad (G.30)$$

Substitution of (G.28), (G.29), and (G.30) into (G.27) with division by $(1-\mu^2)^{|m|/2}$ gives

$$(1-\mu^2)G'' - 2(|m|+1)\mu G' + [\lambda - |m|(|m|+1)]G = 0 \qquad (G.31)$$

To solve this differential equation, we substitute equations (G.2), (G.3), and (G.4) for G, G', and G'' to obtain

$$\sum_{k=0}^{\infty} a_k(k+s)(k+s-1)\mu^{k+s-2} + \sum_{k=0}^{\infty} a_k[\lambda - (k+s+|m|)(k+s+|m|+1)]\xi^{k+s}$$

$$= 0 \qquad (G.32)$$

Equating the coefficient of μ^{s-2} to zero, we obtain the indicial equation

$$a_0 s(s-1) = 0 \qquad (G.33)$$

with solutions $s = 0$ and $s = 1$. Equating the coefficient of μ^{s-1} to zero gives

$$a_1 s(s+1) = 0 \qquad (G.34)$$

For the case $s = 0$, the coefficient a_1 has an arbitrary value, while for $s = 1$, the coefficient a_1 must vanish.

If we replace the dummy index k by $k+2$ in the first summation on the left-hand side of equation (G.32), that equation becomes

$$\sum_{k=0}^{\infty}\{a_{k+2}(k+s+2)(k+s+1) + a_k[\lambda - (k+s+|m|)(k+s+|m|+1)]\}\mu^{k+s} = 0$$

$$(G.35)$$

The recursion formula is obtained by setting the coefficient of each power of μ equal to zero

$$a_{k+2} = \frac{(k+s+|m|)(k+s+|m|+1) - \lambda}{(k+s+2)(k+s+1)} a_k \qquad (G.36)$$

Thus, we obtain a result analogous to the harmonic oscillator solution. The two independent solutions are infinite series, one in odd powers of μ and the other in even powers of μ. The case $s = 0$ gives both solutions, while the case $s = 1$ merely reproduces the odd series. These solutions are

$$G_1 = a_0\left\{1 + \frac{|m|(|m|+1) - \lambda}{2!}\mu^2 + \frac{[(|m|+2)(|m|+3) - \lambda][|m|(|m|+1) - \lambda]}{4!}\mu^4\right.$$
$$\left. + \cdots\right\} \qquad (G.37a)$$

$$G_2 = a_1\left\{\mu + \frac{(|m|+1)(|m|+2) - \lambda}{3!}\mu^3\right.$$
$$\left. + \frac{[(|m|+3)(|m|+4) - \lambda][(|m|+1)(|m|+2) - \lambda]}{5!}\mu^5 + \cdots\right\} \qquad (G.37b)$$

The ratio of consecutive terms in G_1 and in G_2 is given by equation (G.36) as

$$\frac{a_{k+2}\mu^{k+2}}{a_k\mu^k} = \frac{(k+|m|)(k+|m|+1) - \lambda}{(k+1)(k+2)}\mu^2$$

In the limit as $k \to \infty$, this ratio becomes

$$\lim_{k\to\infty} \frac{a_{k+2}\mu^{k+2}}{a_k\mu^k} \to \frac{k^2 - \lambda}{k^2}\mu^2 \to \mu^2$$

As long as $|\mu| < 1$, this ratio is less than unity and the series G_1 and G_2 converge. However, for $\mu = 1$ and $\mu = -1$, this ratio equals unity and neither of the infinite power series converges. For the solutions to equation (G.31) to be well-behaved, we must terminate the series G_1 and G_2 to polynomials by setting

$$\lambda = (k+|m|)(k+|m|+1) = l(l+1) \qquad (G.38)$$

where l is an integer defined as $l = |m| + k$, so that $l = |m|, |m| + 1, |m| + 2, \ldots$ We observe that $|m| \leqslant l$, so that m takes on the values $-l, -l + 1, \ldots, -1, 0, 1, \ldots, l - 1, l$.

Substitution of equation (G.38) into the differential equation (G.27) gives

$$(1 - \mu^2)F'' - 2\mu F' + \left(l(l+1) - \frac{m^2}{1 - \mu^2}\right)F = 0 \qquad (G.39)$$

which is identical to the associated Legendre differential equation (E.17). Thus, the well-behaved solutions to (G.27) are proportional to the associated Legendre polynomials $P_l^{|m|}(\mu)$ introduced in Appendix E

$$F(\mu) = cP_l^{|m|}(\mu)$$

Since we have $\Theta(\theta) = F(\mu)$, where $\mu = \cos\theta$, the functions $\Theta(\theta)$ are

$$\Theta_{lm}(\theta) = cP_l^{|m|}(\cos\theta)$$

and the eigenfunctions $\psi(\theta, \varphi)$ of $\hat{L}^2$ are

$$\psi_{lm}(\theta, \varphi) = c_{lm} P_l^{|m|}(\cos\theta)\, \mathrm{e}^{\mathrm{i}m\varphi} \qquad (G.40)$$

where c_{lm} are the normalization constants. A comparison of equation (G.40) with (5.59) shows that the functions $\psi_{lm}(\theta, \varphi)$ are the spherical harmonics $Y_{lm}(\theta, \varphi)$.

Radial equation for the hydrogen-like atom

The radial differential equation for the hydrogen-like atom is given by equation (6.24) as

$$\frac{\mathrm{d}^2 S}{\mathrm{d}\rho^2} + \frac{2}{\rho}\frac{\mathrm{d}S}{\mathrm{d}\rho} + \left[-\frac{1}{4} + \frac{\lambda}{\rho} - \frac{l(l+1)}{\rho^2}\right] S = 0 \qquad (G.41)$$

where l is a positive integer. If a power series solution is applied directly to equation (G.41), the resulting recursion relation involves a_{k+2}, a_{k+1}, and a_k. Since such a three-term recursion relation is difficult to handle, we first examine the asymptotic behavior of $S(\rho)$. For large values of ρ, the terms in ρ^{-1} and ρ^{-2} become negligible and equation (G.41) reduces to

$$\frac{\mathrm{d}^2 S}{\mathrm{d}\rho^2} = \frac{S}{4}$$

or

$$S = c\mathrm{e}^{\pm\rho/2}$$

where c is the integration constant. Since ρ, as defined in equation (6.22), is always real and positive for $E \leq 0$, the function $\mathrm{e}^{\rho/2}$ is not well-behaved, but $\mathrm{e}^{-\rho/2}$ is. Therefore, we let $S(\rho)$ take the form

$$S(\rho) = F(\rho)\mathrm{e}^{-\rho/2} \qquad (G.42)$$

Substitution of equation (G.42) into (G.41) yields

$$\rho^2 F'' + \rho(2 - \rho)F' + [(\lambda - 1)\rho - l(l+1)]F = 0 \qquad (G.43)$$

where we have multiplied through by $\rho^2 \mathrm{e}^{\rho/2}$. To solve this differential equation by the series solution method, we substitute equations (G.2), (G.3), and (G.4) for F, F', and F'' to obtain

$$\sum_{k=0}^{\infty} a_k[(k+s)(k+s+1) - l(l+1)]\rho^{k+s} + \sum_{k=0}^{\infty} a_k(\lambda - 1 - k - s)\rho^{k+s+1} = 0$$

$$(G.44)$$

The indicial equation is given by the coefficient of ρ^s as

$$a_0[s(s+1) - l(l+1)] = 0 \qquad (G.45)$$

with solutions $s = l$ and $s = -(l+1)$. For the case $s = -(l+1)$, we have

$$F(\rho) = a_0\rho^{-(l+1)} + a_1\rho^{-l} + a_2\rho^{-l+1} + \cdots \qquad (G.46)$$

which diverges at the origin.[4] Thus, the case $s = l$ is the only acceptable solution.

Omitting the vanishing first term in the first summation on the left-hand side of (G.44) and replacing k by $k+1$ in that summation, we have

---

[4] The reason for rejecting the solution $s = -(l+1)$ is actually more complicated for states with $l = 0$. I. N. Levine (1991) *Quantum Chemistry*, 4th edition (Prentice-Hall, Englewood Cliffs, NJ), p. 124, summarizes the arguments with references to more detailed discussions. The complications here strengthen the reasons for preferring the ladder operator technique used in the main text.

$$\sum_{k=0}^{\infty}\{a_{k+1}[(k+1)(k+2l+2)] + a_k(\lambda - l - 1 - k)\}\rho^{k+s+1} = 0 \qquad \text{(G.47)}$$

Since the coefficient of each power of ρ must vanish, we have for the recursion formula

$$a_{k+1} = \frac{k+l+1-\lambda}{(k+1)(k+2l+2)}a_k \qquad \text{(G.48)}$$

Thus, we obtain the following set of expansion constants

$$a_0$$

$$a_1 = \frac{l+1-\lambda}{2l+2}a_0$$

$$a_2 = \frac{l+2-\lambda}{2(2l+3)}a_1 = \frac{(l+2-\lambda)(l+1-\lambda)}{2(2l+3)(2l+2)}a_0$$

$$u_3 = \frac{l+3-\lambda}{3(2l+4)}a_2 = \frac{(l+3-\lambda)(l+2-\lambda)(l+1-\lambda)}{3!(2l+4)(2l+3)(2l+2)}a_0$$

$$\vdots$$

so that the solution of (G.43) is

$$F = a_0\rho^l\left(1 + \sum_{k=1}^{\infty}\frac{(l+k-\lambda)(l+k-1-\lambda)\cdots(l+1-\lambda)}{k!(2l+k+1)(2l+k)\cdots(2l+2)}\rho^k\right) \qquad \text{(G.49)}$$

We have already discarded the second solution, equation (G.46).

The ratio of consecutive terms in the power series expansion F is given by equation (G.48) as

$$\frac{a_{k+1}\rho^{k+l+1}}{a_k\rho^{k+l}} = \frac{k+l+1-\lambda}{(k+1)(k+2l+2)}\rho$$

In the limit as $k \to \infty$, this ratio becomes ρ/k, which approaches zero for finite ρ. Thus, the series converges for all finite values of ρ. To test the behavior of the power series as $\rho \to \infty$, we consider the Taylor series expansion of e^ρ

$$e^\rho = \sum_{k=0}^{\infty}\frac{\rho^k}{k!}$$

and note that the ratio of consecutive terms is also ρ/k. Since the behavior of F as $\rho \to \infty$ is determined by the expansion terms with large values of k ($k \to \infty$), we see that F behaves like e^ρ as $\rho \to \infty$. This behavior is not acceptable because $S(\rho)$ in equation (G.42) would take the form

$$S(\rho) \to \rho^l e^\rho e^{-\rho/2} = \rho^l e^{\rho/2} \to \infty \qquad \text{as } \rho \to \infty$$

and could not be normalized.

The only way to avoid this convergence problem is to terminate the infinite series (equation (G.49)) after a finite number of terms. If we let λ take on the successive values $l+1, l+2, \ldots$, then we obtain a series of acceptable solutions of the differential equation (G.43)

$$l + 1, \qquad F_0 = a_0 \rho^l$$

$$l + 2, \qquad F_1 = a_0 \rho^l \left(\left(1 - \frac{1}{2l + 2}\rho\right)\right)$$

$$l + 3, \qquad F_2 = a_0 \rho^l \left(1 - \frac{1}{l + 1}\rho + \frac{1}{(2l + 3)(2l + 2)}\rho^2\right)$$

$$\vdots \qquad \vdots$$

Since l is an integer with values $0, 1, 2, \ldots$, the parameter λ takes on integer values n, $n = 1, 2, 3, \ldots$, so that $n = l + 1, l + 2, \ldots$ When the *quantum number n* equals 1, the value of l is 1; when $n = 2$, we have $l = 0, 1$; when $n = 3$, we have $l = 0, 1, 2$; etc.

The energy E of the hydrogen-like atom is related to λ by equation (6.21). If we solve this equation for E and set λ equal to n, we obtain

$$E_n = -\frac{\mu Z^2 e'^4}{2\hbar^2 n^2}, \qquad n = 1, 2, 3, \ldots$$

in agreement with equation (6.48).

To identify the polynomial solutions for $F(\rho)$, we make the substitution

$$F(\rho) = \rho^l u(\rho) \tag{G.50}$$

in the differential equation (G.43) and set λ equal to n to obtain

$$\rho u'' + [2(l + 1) - \rho]u' + (n - l - 1)u = 0 \tag{G.51}$$

Since n and l are integers, equation (G.51) is identical to the associated Laguerre differential equation (F.16) with $k = n + l$ and $j = 2l + 1$. Thus, the solutions $u(\rho)$ are proportional to the associated Laguerre polynomials $L_{n+l}^{2l+1}(\rho)$, whose properties are discussed in Appendix F

$$u(\rho) = cL_{n+l}^{2l+1}(\rho) \tag{G.52}$$

Combining equations (G.42), (G.50), and (G.52), we obtain

$$S_{nl}(\rho) = c_{nl}\rho^l e^{-\rho/2} L_{n+l}^{2l+1}(\rho) \tag{G.53}$$

where c_{nl} are the normalizing constants. Equation (G.53) agrees with equation (6.53).

Appendix H

Recurrence relation for hydrogen-atom expectation values

The expectation values $\langle r^k \rangle_{nl}$ of various powers of the radial variable r for a hydrogen-like atom with quantum numbers n and l are given by equation (6.69)

$$\langle r^k \rangle_{nl} = \int_0^\infty r^k [R_{nl}(r)]^2 r^2 \, dr \tag{H.1}$$

where $R_{nl}(r)$ are the solutions of the radial differential equation (6.17). In this appendix, we show that these expectation values are related by the recurrence relation

$$\frac{k+1}{n^2} \langle r^k \rangle_{nl} - (2k+1) \frac{a_0}{Z} \langle r^{k-1} \rangle_{nl} + k \left[l(l+1) + \frac{1-k^2}{4} \right] \frac{a_0^2}{Z^2} \langle r^{k-2} \rangle_{nl} = 0 \tag{H.2}$$

To simplify the notation, we define the real function $u(r)$ by $u = rR_{nl}(r)$ and denote the first and second derivatives of $u(r)$ by u' and u''. Equation (H.1) then takes the form

$$\langle r^k \rangle_{nl} = \int_0^\infty r^k u^2 \, dr \tag{H.3}$$

Since we have

$$\frac{dR(r)}{dr} = \frac{u'}{r} - \frac{u}{r^2}, \qquad \frac{d}{dr}\left(r^2 \frac{dR(r)}{dr} \right) - ru''$$

equation (6.17) becomes

$$u'' = \left[\frac{l(l+1)}{r^2} - \frac{2Z}{a_0 r} + \frac{Z^2}{n^2 a_0^2} \right] u \tag{H.4}$$

where equation (6.57) for the energy E_n has also been introduced.

Before beginning the direct derivation of equation (H.2), we first derive a useful relationship. Consider the integral

$$\int_0^\infty r^v uu' \, dr$$

and integrate by parts

$$\int_0^\infty r^v uu' \, dr = r^v u^2 \Big|_0^\infty - \int_0^\infty u \frac{d}{dr}(r^v u) \, dr$$

The integrated part vanishes because $R(r) \to 0$ exponentially as $r \to \infty$ and $u(r) \to 0$ as $r \to 0$. Expanding the derivative within the integral on the right-hand side, we have

329

$$\int_0^\infty r^\nu uu' \, \mathrm{d}r = -\nu \int_0^\infty r^{\nu-1} u^2 \, \mathrm{d}r - \int_0^\infty r^\nu uu' \, \mathrm{d}r$$

Combining the integral on the left-hand side with the last one on the right-hand side, we obtain the desired result

$$\int_0^\infty r^\nu uu' \, \mathrm{d}r = -\frac{\nu}{2} \langle r^{\nu-1} \rangle_{nl} \tag{H.5}$$

To obtain the recurrence relation (H.2), we multiply equation (H.4) by $r^{k+1}u'$ and integrate over r

$$\int_0^\infty r^{k+1} u'u'' \, \mathrm{d}r = l(l+1) \int_0^\infty r^{k-1} uu' \, \mathrm{d}r - \frac{2Z}{a_0} \int_0^\infty r^k uu' \, \mathrm{d}r + \frac{Z^2}{n^2 a_0^2} \int_0^\infty r^{k+1} uu' \, \mathrm{d}r$$

$$= -\frac{l(l+1)(k-1)}{2} \langle r^{k-2} \rangle_{nl} + \frac{kZ}{a_0} \langle r^{k-1} \rangle_{nl} - \frac{(k+1)Z^2}{2n^2 a_0^2} \langle r^k \rangle_{nl} \tag{H.6}$$

where equation (H.5) was applied to the right-hand side. The integral on the left-hand side of (H.6) may be integrated by parts twice to give

$$\int_0^\infty r^{k+1} u'u'' \, \mathrm{d}r = -\int_0^\infty u' \frac{\mathrm{d}}{\mathrm{d}r} (r^{k+1} u') \, \mathrm{d}r$$

$$= -(k+1) \int_0^\infty r^k u'u' \, \mathrm{d}r - \int_0^\infty r^{k+1} u'u'' \, \mathrm{d}r$$

$$= (k+1) \int_0^\infty u \frac{\mathrm{d}}{\mathrm{d}r} (r^k u') \, \mathrm{d}r - \int_0^\infty r^{k+1} u'u'' \, \mathrm{d}r$$

$$= k(k+1) \int_0^\infty r^{k-1} uu' \, \mathrm{d}r + (k+1) \int_0^\infty r^k uu'' \, \mathrm{d}r - \int_0^\infty r^{k+1} u'u'' \, \mathrm{d}r$$

The integral on the left-hand side and the last integral on the right-hand side may be combined to give

$$\int_0^\infty r^{k+1} u'u'' \, \mathrm{d}r = -\frac{(k-1)k(k+1)}{4} \langle r^{k-2} \rangle_{nl} + \frac{(k+1)}{2} \int_0^\infty r^k uu'' \, \mathrm{d}r \tag{H.7}$$

where equation (H.5) has been used for the first integral on the right-hand side. Substitution of equation (H.4) for u'' in the last integral on the right-hand side of (H.7) yields

$$\int_0^\infty r^{k+1} u'u'' \, \mathrm{d}r = -\frac{(k-1)k(k+1)}{4} \langle r^{k-2} \rangle_{nl} + \frac{(k+1)l(l+1)}{2} \langle r^{k-2} \rangle_{nl}$$

$$- \frac{(k+1)Z}{a_0} \langle r^{k-1} \rangle_{nl} + \frac{(k+1)Z^2}{2n^2 a_0^2} \langle r^k \rangle_{nl} \tag{H.8}$$

Combining equations (H.6) and (H.8), we obtain the recurrence relation (H.2).

Appendix I

Matrices

An $m \times n$ matrix $\mathbf{A}$ is an ordered set of mn elements a_{ij} ($i = 1, 2, \ldots, m$; $j = 1, 2, \ldots, n$) arranged in a rectangular array of m rows and n columns,

$$\mathbf{A} = \begin{pmatrix} a_{11} & a_{12} & \cdots & a_{1n} \\ a_{21} & a_{22} & \cdots & a_{2n} \\ \cdots & \cdots & \cdots & \cdots \\ a_{m1} & a_{m2} & \cdots & a_{mn} \end{pmatrix} \tag{I.1}$$

If m equals n, the array is a *square matrix* of *order n*. If we have $m = 1$, then the matrix has only one row and is known as a *row matrix*. On the other hand, if we have $n = 1$, then the matrix consists of one column and is called a *column matrix*.

Matrix algebra

Two $m \times n$ matrices $\mathbf{A}$ and $\mathbf{B}$ are equal if and only if their corresponding elements are equal, i.e., $a_{ij} = b_{ij}$ for all values of i and j. Some of the rules of matrix algebra are defined by the following relations

$$\mathbf{A} + \mathbf{B} = \mathbf{B} + \mathbf{A} = \mathbf{C}; \qquad c_{ij} = a_{ij} + b_{ij}$$

$$\mathbf{A} \quad \mathbf{B} - \mathbf{C}, \qquad c_{ij} = a_{ij} - b_{ij} \tag{I.2}$$

$$k\mathbf{A} = \mathbf{C}; \qquad c_{ij} = ka_{ij}$$

where k is a constant. Clearly, the matrices $\mathbf{A}$, $\mathbf{B}$, and $\mathbf{C}$ in equations (I.2) must have the same dimensions $m \times n$.

Multiplication of an $m \times n$ matrix $\mathbf{A}$ and an $n \times p$ matrix $\mathbf{B}$ is defined by

$$\mathbf{AB} = \mathbf{C}; \qquad c_{ik} = \sum_{j=1}^{n} a_{ij}b_{jk} \tag{I.3}$$

The matrix $\mathbf{C}$ has dimensions $m \times p$. Two matrices may be multiplied only if they are *conformable*, i.e., only if the number of columns of the first equals the number of rows of the second. As an example, suppose $\mathbf{A}$ and $\mathbf{B}$ are

$$\mathbf{A} = \begin{pmatrix} a_{11} & a_{12} & a_{13} \\ a_{21} & a_{22} & a_{23} \\ a_{31} & a_{32} & a_{33} \end{pmatrix}; \qquad \mathbf{B} = \begin{pmatrix} b_{11} & b_{12} \\ b_{21} & b_{22} \\ b_{31} & b_{32} \end{pmatrix}$$

Then the product $\mathbf{AB}$ is

331

$$\mathbf{AB} = \begin{pmatrix} a_{11}b_{11} + a_{12}b_{21} + a_{13}b_{31} & a_{11}b_{12} + a_{12}b_{22} + a_{13}b_{32} \\ a_{21}b_{11} + a_{22}b_{21} + a_{23}b_{31} & a_{21}b_{12} + a_{22}b_{22} + a_{23}b_{32} \\ a_{31}b_{11} + a_{32}b_{21} + a_{33}b_{31} & a_{31}b_{12} + a_{32}b_{22} + a_{33}b_{32} \end{pmatrix}$$

Continued products, such as **ABC**, may be defined and evaluated if the matrices are conformable. In such cases, multiplication is *associative*, for example

$$\mathbf{ABC} = \mathbf{A}(\mathbf{BC}) = (\mathbf{AB})\mathbf{C} = \mathbf{D}; \qquad d_{il} = \sum_j \sum_k a_{ij}b_{jk}c_{kl} \tag{I.4}$$

For the *null matrix* **0**, all the matrix elements are zero

$$\mathbf{0} = \begin{pmatrix} 0 & 0 & \cdots & 0 \\ 0 & 0 & \cdots & 0 \\ \cdots & \cdots & \cdots & \cdots \\ 0 & 0 & \cdots & 0 \end{pmatrix} \tag{I.5}$$

and we have

$$\mathbf{0} + \mathbf{A} = \mathbf{A} + \mathbf{0} = \mathbf{A} \tag{I.6}$$

The product of an arbitrary $m \times n$ matrix **A** with a conformable $n \times p$ null matrix is the $m \times p$ null matrix

$$\mathbf{A0} = \mathbf{0} \tag{I.7}$$

In matrix algebra it is possible for the product of two conformable matrices, neither of which is a null matrix, to be a null matrix. For example, if **A** and **B** are

$$\mathbf{A} = \begin{pmatrix} 1 & 1 & 0 \\ -2 & -2 & 2 \\ 3 & 3 & 1 \end{pmatrix}; \qquad \mathbf{B} = \begin{pmatrix} 1 & -1 \\ -1 & 1 \\ 0 & 0 \end{pmatrix}$$

then the product **AB** is the 3×2 null matrix.

The *transpose matrix* $\mathbf{A}^{\mathrm{T}}$ of a matrix **A** is obtained by interchanging the rows and columns of **A**. If the matrix **A** is given by equation (I.1), then its transpose is

$$\mathbf{A}^{\mathrm{T}} = \begin{pmatrix} a_{11} & a_{21} & \cdots & a_{m1} \\ a_{12} & a_{22} & \cdots & a_{m2} \\ \cdots & \cdots & \cdots & \cdots \\ a_{1n} & a_{2n} & \cdots & a_{mn} \end{pmatrix} \tag{I.8}$$

Thus, the elements a_{ij}^{T} of $\mathbf{A}^{\mathrm{T}}$ are given by $a_{ij}^{\mathrm{T}} = a_{ji}$.

Let the matrix **C** be the product of matrices **A** and **B** as in equation (I.3). The elements c_{ik}^{T} of the transpose of **C** are then given by

$$c_{ik}^{\mathrm{T}} = \sum_{j=1}^n a_{kj}b_{ji} = \sum_{j=1}^n a_{jk}^{\mathrm{T}}b_{ij}^{\mathrm{T}} = \sum_{j=1}^n b_{ij}^{\mathrm{T}}a_{jk}^{\mathrm{T}} \tag{I.9}$$

where we have noted that $a_{\alpha\beta}^{\mathrm{T}} = a_{\beta\alpha}$ and $b_{\alpha\beta}^{\mathrm{T}} = b_{\beta\alpha}$. Thus, we see that

$$\mathbf{C}^{\mathrm{T}} = \mathbf{B}^{\mathrm{T}}\mathbf{A}^{\mathrm{T}}$$

or

$$(\mathbf{AB})^{\mathrm{T}} = \mathbf{B}^{\mathrm{T}}\mathbf{A}^{\mathrm{T}} \tag{I.10}$$

This result may be generalized to give

$$(\mathbf{AB} \cdots \mathbf{Q})^{\mathrm{T}} = \mathbf{Q}^{\mathrm{T}} \cdots \mathbf{B}^{\mathrm{T}}\mathbf{A}^{\mathrm{T}} \tag{I.11}$$

as long as the matrices are conformable.

If each element a_{ij} in a matrix **A** is replaced by its complex conjugate a_{ij}^*, then the resulting matrix $\mathbf{A}^*$ is called the *conjugate* of **A**. The transposed conjugate of **A** is called the *adjoint*[1] of $\mathbf{A}$ and is denoted by $\mathbf{A}^\dagger$. The elements $a_{ij}^\dagger$ of $\mathbf{A}^\dagger$ are obviously given by $a_{ij}^\dagger = a_{ji}^*$.

Square matrices

Square matrices are of particular interest because they apply to many physical situations.

A square matrix of order n is *symmetric* if $a_{ij} = a_{ji}$, $(i, j = 1, 2, \ldots, n)$, so that $\mathbf{A} = \mathbf{A}^\mathrm{T}$, and is *antisymmetric* if $a_{ij} = -a_{ji}$, $(i, j = 1, 2, \ldots, n)$, so that $\mathbf{A} = -\mathbf{A}^\mathrm{T}$. The diagonal elements of an antisymmetric matrix must all be zero. Any arbitrary square matrix $\mathbf{A}$ may be written as the sum of a symmetric matrix $\mathbf{A}^{(s)}$ and an antisymmetric matrix $\mathbf{A}^{(a)}$

$$\mathbf{A} = \mathbf{A}^{(s)} + \mathbf{A}^{(a)} \tag{I.12}$$

where

$$a_{ij}^{(s)} = \tfrac{1}{2}(a_{ij} + a_{ji}); \qquad a_{ij}^{(a)} = \tfrac{1}{2}(a_{ij} - a_{ji}) \tag{I.13}$$

A square matrix $\mathbf{A}$ is *diagonal* if $a_{ij} = 0$ for $i \neq j$. Thus, a diagonal matrix has the form

$$\mathbf{A} = \begin{pmatrix} a_1 & 0 & \cdots & 0 \\ 0 & a_2 & \cdots & 0 \\ \cdots & \cdots & \cdots & \cdots \\ 0 & 0 & \cdots & a_n \end{pmatrix} \tag{I.14}$$

A diagonal matrix is *scalar* if all the diagonal elements are equal, $a_1 = a_2 = \cdots = a_n \equiv a$, so that

$$\mathbf{A} = \begin{pmatrix} a & 0 & \cdots & 0 \\ 0 & a & \cdots & 0 \\ \cdots & \cdots & \cdots & \cdots \\ 0 & 0 & \cdots & a \end{pmatrix} \tag{I.15}$$

A special case of a scalar matrix is the *unit matrix* $\mathbf{I}$, for which a equals unity

$$\mathbf{I} = \begin{pmatrix} 1 & 0 & \cdots & 0 \\ 0 & 1 & \cdots & 0 \\ \cdots & \cdots & \cdots & \cdots \\ 0 & 0 & \cdots & 1 \end{pmatrix} \tag{I.16}$$

The elements of the unit matrix are δ_{ij}, the Kronecker delta function.

For square matrices in general, the product $\mathbf{AB}$ is not equal to the product $\mathbf{BA}$. For example, if

$$\mathbf{A} = \begin{pmatrix} 0 & 1 \\ 1 & 0 \end{pmatrix}; \qquad \mathbf{B} = \begin{pmatrix} 2 & 0 \\ 0 & 3 \end{pmatrix}$$

then we have

$$\mathbf{AB} = \begin{pmatrix} 0 & 3 \\ 2 & 0 \end{pmatrix}; \qquad \mathbf{BA} = \begin{pmatrix} 0 & 2 \\ 3 & 0 \end{pmatrix} \neq \mathbf{AB}$$

If the product $\mathbf{AB}$ equals the product $\mathbf{BA}$, then $\mathbf{A}$ and $\mathbf{B}$ *commute*. Any square matrix $\mathbf{A}$ commutes with the unit matrix of the same order

---

[1] Mathematics texts use the term *transpose conjugate* for this matrix and apply the term *adjoint* to the adjugate matrix defined in equation (I.28).

$$\mathbf{AI} = \mathbf{IA} \tag{I.17}$$

Moreover, two diagonal matrices of the same order commute

$$\mathbf{AB} = \begin{pmatrix} a_1 & 0 & \cdots & 0 \\ 0 & a_2 & \cdots & 0 \\ \cdots & \cdots & \cdots & \cdots \\ 0 & 0 & \cdots & a_n \end{pmatrix} \begin{pmatrix} b_1 & 0 & \cdots & 0 \\ 0 & b_2 & \cdots & 0 \\ \cdots & \cdots & \cdots & \cdots \\ 0 & 0 & \cdots & b_n \end{pmatrix} = \begin{pmatrix} a_1 b_1 & 0 & \cdots & 0 \\ 0 & a_2 b_2 & \cdots & 0 \\ \cdots & \cdots & \cdots & \cdots \\ 0 & 0 & \cdots & a_n b_n \end{pmatrix}$$

$$= \mathbf{BA} \tag{I.18}$$

Determinants

For a square matrix $\mathbf{A}$, there exists a number called the *determinant* of the matrix. This determinant is denoted by

$$|A| = \begin{vmatrix} a_{11} & a_{12} & \cdots & a_{1n} \\ a_{21} & a_{22} & \cdots & a_{2n} \\ \cdots & \cdots & \cdots & \cdots \\ a_{n1} & a_{n2} & \cdots & a_{nn} \end{vmatrix} \tag{I.19}$$

and is defined as the summation

$$|A| = \sum_P \delta_P a_{1i} a_{2j} \cdots a_{nq} \tag{I.20}$$

where $\delta_P = \pm 1$. The summation is taken over all possible permutations $i, j, \ldots, q$ of the sequence $1, 2, \ldots, n$. The value of δ_P is $+1$ (-1) if the order $i, j, \ldots, q$ is obtained by an even (odd) number of pair interchanges from the order $1, 2, \ldots, n$. There are $n!$ terms in the summation, half with $\delta_P = 1$ and half with $\delta_P = -1$. Thus, for a second-order determinant, we have

$$|A| = \begin{vmatrix} a_{11} & a_{12} \\ a_{21} & a_{22} \end{vmatrix} = a_{11} a_{22} - a_{12} a_{21} \tag{I.21}$$

and for a third-order determinant, we have

$$|A| = \begin{vmatrix} a_{11} & a_{12} & a_{13} \\ a_{21} & a_{22} & a_{23} \\ a_{31} & a_{32} & a_{33} \end{vmatrix}$$

$$= a_{11} a_{22} a_{33} + a_{12} a_{23} a_{31} + a_{13} a_{21} a_{32} - (a_{11} a_{23} a_{32} + a_{12} a_{21} a_{33} + a_{13} a_{22} a_{31}) \tag{I.22}$$

If the determinant $|A|$ of the matrix $\mathbf{A}$ vanishes, then the matrix $\mathbf{A}$ is said to be *singular*. Otherwise, the matrix $\mathbf{A}$ is *non-singular*.

The determinant $|A|$ has the following properties, which are easily derived from the definition (I.20).

1. The interchange of any two rows or any two columns changes the sign of the determinant.
2. Multiplication of all the elements in any row or in any column by a constant k gives a new determinant of value $k|A|$. (Note that if $\mathbf{B} = k\mathbf{A}$, then $|B| = k^n |A|$.)
3. The value of the determinant is zero if any two rows or any two columns are identical, or if each element in any row or in any column is zero. As a special case of properties 2 and 3, a determinant vanishes if any two rows or any two columns are proportional.

4. The value of a determinant is unchanged if the rows are written as columns. Thus, the determinants of a matrix $\mathbf{A}$ and its transpose matrix $\mathbf{A}^T$ are equal.
5. The value of a determinant is unchanged if, to each element of one row (column) is added a constant k times the corresponding element of another row (column). Thus, we have, for example

$$\begin{vmatrix} a_{11} & a_{12} & a_{13} \\ a_{21} & a_{22} & a_{23} \\ a_{31} & a_{32} & a_{33} \end{vmatrix} = \begin{vmatrix} a_{11} + ka_{12} & a_{12} & a_{13} \\ a_{21} + ka_{22} & a_{22} & a_{23} \\ a_{31} + ka_{32} & a_{32} & a_{33} \end{vmatrix} = \begin{vmatrix} a_{11} + ka_{31} & a_{12} + ka_{32} & a_{13} + ka_{33} \\ a_{21} & a_{22} & a_{23} \\ a_{31} & a_{32} & a_{33} \end{vmatrix}$$

Each element a_{ij} of the determinant $|A|$ in equation (I.19) has a *cofactor* C_{ij}, which is an $(n-1)$-order determinant. This cofactor C_{ij} is constructed by deleting the ith row and the jth column of $|A|$ and then multiplying by $(-1)^{i+j}$. For example, the cofactor of the element a_{12} in equation (I.22) is

$$C_{12} = -\begin{vmatrix} a_{21} & a_{23} \\ a_{31} & a_{33} \end{vmatrix} = a_{23}a_{31} - a_{21}a_{33}$$

The summation on the right-hand side of equation (I.20) may be expressed in terms of the cofactors of the first row of $|A|$, so that (I.20) becomes

$$|A| = a_{11}C_{11} + a_{12}C_{12} + \cdots + a_{1n}C_{1n} = \sum_{k=1}^{n} a_{1k}C_{1k} \tag{I.23}$$

Alternatively, the expression of $|A|$ in equation (I.20) may be expanded in terms of any row i

$$|A| = \sum_{k=1}^{n} a_{ik}C_{ik}, \qquad i = 1, 2, \ldots, n \tag{I.24}$$

or in terms of any column j

$$|A| = \sum_{k=1}^{n} a_{kj}C_{kj}, \qquad j = 1, 2, \ldots, n \tag{I.25}$$

Equations (I.20), (I.24), and (I.25) are identical; they are just expressed in different notations.

Now suppose that row 1 and row i of the determinant $|A|$ are identical. Equation (I.23) then becomes

$$|A| = a_{i1}C_{11} + a_{i2}C_{12} + \cdots + a_{in}C_{1n} = \sum_{k=1}^{n} a_{ik}C_{1k} = 0$$

where the determinant $|A|$ vanishes according to property 3. This argument applies to any identical pair of rows or any identical pair of columns, so that equations (I.24) and (I.25) may be generalized

$$\sum_{k=1}^{n} a_{ik}C_{jk} = \sum_{k=1}^{n} a_{ki}C_{kj} = |A|\delta_{ij}, \qquad i, j = 1, 2, \ldots, n \tag{I.26}$$

It can be shown[2] that the determinant of the product of two square matrices of the same order is equal to the product of the two determinants, i.e., if $\mathbf{C} = \mathbf{AB}$, then

$$|C| = |A| \cdot |B| \tag{I.27}$$

---

[2] See G. D. Arfken and H. J. Weber (1995) *Mathematical Methods for Physicists*, 4th edition (Academic Press, San Diego), p. 169.

It follows from equation (I.27) that the product of two non-singular matrices is also non-singular.

Special square matrices

The *adjugate matrix* $\hat{\mathbf{A}}$ of the square matrix $\mathbf{A}$ is defined as

$$\hat{\mathbf{A}} = \begin{pmatrix} C_{11} & C_{21} & \cdots & C_{n1} \\ C_{12} & C_{22} & \cdots & C_{n2} \\ \cdots & \cdots & \cdots & \cdots \\ C_{1n} & C_{2n} & \cdots & C_{nn} \end{pmatrix} \tag{I.28}$$

where C_{ij} are the cofactors of the elements a_{ij} of the determinant $|A|$ of $\mathbf{A}$. Note that the element $\hat{a}_{kl}$ of $\hat{\mathbf{A}}$ is the cofactor C_{lk}. The matrix product $\mathbf{A}\hat{\mathbf{A}}$ is a matrix $\mathbf{B}$ whose elements b_{ij} are given by

$$b_{ij} = \sum_{k=1}^{n} a_{ik}\hat{a}_{kj} = \sum_{k=1}^{n} a_{ik}C_{jk} = |A|\delta_{ij} \tag{I.29}$$

where equation (I.26) was introduced. Thus, we have

$$\mathbf{A}\hat{\mathbf{A}} = \mathbf{B} = |A|\mathbf{I} = \hat{\mathbf{A}}\mathbf{A} \tag{I.30}$$

where $\mathbf{I}$ is the unit matrix in equation (I.16), and we see that the matrices $\mathbf{A}$ and $\hat{\mathbf{A}}$ commute.

Any non-singular square matrix $\mathbf{A}$ possesses an *inverse matrix* $\mathbf{A}^{-1}$ defined as

$$\mathbf{A}^{-1} = \hat{\mathbf{A}}/|A| \tag{I.31}$$

From equation (I.30) we observe that

$$\mathbf{A}\mathbf{A}^{-1} = \mathbf{A}^{-1}\mathbf{A} = \mathbf{I} \tag{I.32}$$

Consider three square matrices $\mathbf{A}, \mathbf{B}, \mathbf{C}$ such that $\mathbf{A}\mathbf{B} = \mathbf{C}$. Then we have

$$\mathbf{A}^{-1}\mathbf{A}\mathbf{B} = \mathbf{A}^{-1}\mathbf{C}$$

or

$$\mathbf{B} = \mathbf{A}^{-1}\mathbf{C} \tag{I.33}$$

Thus, the inverse matrix plays the role of division in matrix algebra. Multiplication of equation (I.33) from the left by $\mathbf{B}^{-1}$ and from the right by $\mathbf{C}^{-1}$ yields

$$\mathbf{C}^{-1} = \mathbf{B}^{-1}\mathbf{A}^{-1}$$

or

$$(\mathbf{A}\mathbf{B})^{-1} = \mathbf{B}^{-1}\mathbf{A}^{-1} \tag{I.34}$$

This result may easily be generalized to show that

$$(\mathbf{A}\mathbf{B} \cdots \mathbf{Q})^{-1} = \mathbf{Q}^{-1} \cdots \mathbf{B}^{-1}\mathbf{A}^{-1} \tag{I.35}$$

A square matrix $\mathbf{A}$ is *hermitian* or *self-adjoint* if it is equal to its adjoint, i.e., if $\mathbf{A} = \mathbf{A}^{\dagger}$ or $a_{ij} = a_{ji}^{*}$. Thus, the diagonal elements of a hermitian matrix are real.

A square matrix $\mathbf{A}$ is *orthogonal* if it satisfies the relation

$$\mathbf{A}\mathbf{A}^{\mathrm{T}} = \mathbf{A}^{\mathrm{T}}\mathbf{A} = \mathbf{I}$$

If we multiply $\mathbf{A}\mathbf{A}^{\mathrm{T}} = \mathbf{I}$ from the left by $\mathbf{A}^{-1}$, then we have the equivalent definition

$$\mathbf{A}^{\mathrm{T}} = \mathbf{A}^{-1}$$

Since the determinants $|A|$ and $|A^{\mathrm{T}}|$ are equal, we have from equation (I.27)

$$|A|^2 = 1 \quad \text{or} \quad |A| = \pm 1$$

The product of two orthogonal matrices is an orthogonal matrix as shown by the following sequence

$$(\mathbf{AB})^T = \mathbf{B}^T\mathbf{A}^T = \mathbf{B}^{-1}\mathbf{A}^{-1} = (\mathbf{AB})^{-1}$$

where equations (I.10) and (I.34) were used. The inverse of an orthogonal matrix is also an orthogonal matrix as shown by taking the transpose of $\mathbf{A}^{-1}$ and noting that the order of transposition and inversion may be reversed

$$(\mathbf{A}^{-1})^T = (\mathbf{A}^T)^{-1} = (\mathbf{A}^{-1})^{-1}$$

A square matrix $\mathbf{A}$ is *unitary* if its inverse is equal to its adjoint, i.e., if $\mathbf{A}^{-1} = \mathbf{A}^\dagger$ or if $\mathbf{AA}^\dagger = \mathbf{A}^\dagger\mathbf{A} = \mathbf{I}$. For a *real* matrix, with all elements real so that $a_{ij} = a_{ij}^*$, there is no distinction between an orthogonal and a unitary matrix. In that case, we have $\mathbf{A}^\dagger = \mathbf{A}^T = \mathbf{A}^{-1}$.

Linear vector space

A vector $\mathbf{x}$ in three-dimensional cartesian space may be represented as a column matrix

$$\mathbf{x} = \begin{pmatrix} x_1 \\ x_2 \\ x_3 \end{pmatrix} \tag{I.36}$$

where x_1, x_2, x_3 are the components of $\mathbf{x}$. The adjoint of the column matrix $\mathbf{x}$ is a row matrix

$$\mathbf{x}^\dagger = (x_1^* \quad x_2^* \quad x_3^*) \tag{I.37}$$

The scalar product of the vectors $\mathbf{x}$ and $\mathbf{y}$ when expressed in matrix notation is

$$\mathbf{x}^\dagger\mathbf{y} = (x_1^* \quad x_2^* \quad x_3^*) \begin{pmatrix} y_1 \\ y_2 \\ y_3 \end{pmatrix} = x_1^* y_1 + x_2^* y_2 + x_2^* y_2 \tag{I.38}$$

Consequently, the magnitude of the vector $\mathbf{x}$ is

$$(\mathbf{x}^\dagger\mathbf{x})^{1/2} = (|x_1|^2 + |x_2|^2 + |x_3|^2)^{1/2} \tag{I.39}$$

If the vectors $\mathbf{x}$ and $\mathbf{y}$ are orthogonal, then we have $\mathbf{x}^\dagger\mathbf{y} = \mathbf{0}$. The unit vectors $\mathbf{i}$, $\mathbf{j}$, $\mathbf{k}$ in matrix notation are

$$\mathbf{i} = \begin{pmatrix} 1 \\ 0 \\ 0 \end{pmatrix}; \quad \mathbf{j} = \begin{pmatrix} 0 \\ 1 \\ 0 \end{pmatrix}; \quad \mathbf{k} = \begin{pmatrix} 0 \\ 0 \\ 1 \end{pmatrix} \tag{I.40}$$

A linear operator $\hat{A}$ in three-dimensional cartesian space may be represented as a 3×3 matrix $\mathbf{A}$ with elements a_{ij}. The expression $\mathbf{y} = \hat{A}\mathbf{x}$ in matrix notation becomes

$$\mathbf{y} = \begin{pmatrix} y_1 \\ y_2 \\ y_3 \end{pmatrix} = \mathbf{Ax} = \begin{pmatrix} a_{11} & a_{12} & a_{13} \\ a_{21} & a_{22} & a_{23} \\ a_{31} & a_{32} & a_{33} \end{pmatrix} \begin{pmatrix} x_1 \\ x_2 \\ x_3 \end{pmatrix} = \begin{pmatrix} a_{11}x_1 + a_{12}x_2 + a_{13}x_3 \\ a_{21}x_1 + a_{22}x_2 + a_{23}x_3 \\ a_{31}x_1 + a_{32}x_2 + a_{33}x_3 \end{pmatrix}$$

$$\tag{I.41}$$

If $\mathbf{A}$ is non-singular, then in matrix notation the vector $\mathbf{x}$ is related to the vector $\mathbf{y}$ by

$$\mathbf{x} = \mathbf{A}^{-1}\mathbf{y} \tag{I.42}$$

The vector concept may be extended to n-dimensional cartesian space, where we have n mutually orthogonal axes. Each vector $\mathbf{x}$ then has n components $(x_1, x_2, \ldots,$

x_n) and may be represented as a column matrix $\mathbf{x}$ with n rows. The scalar product of the n-dimensional vectors $\mathbf{x}$ and $\mathbf{y}$ in matrix notation is

$$\mathbf{x}^\dagger \mathbf{y} = x_1^* y_1 + x_2^* y_2 + \cdots + x_n^* y_n \tag{I.43}$$

and the magnitude of $\mathbf{x}$ is

$$(\mathbf{x}^\dagger \mathbf{x})^{1/2} = (|x_1|^2 + |x_2|^2 + \cdots + |x_n|^2)^{1/2} \tag{I.44}$$

If we have $\mathbf{x}^\dagger \mathbf{y} = \mathbf{0}$, then the vectors $\mathbf{x}$ and $\mathbf{y}$ are orthogonal. The unit vectors $\mathbf{i}_\alpha$ ($\alpha = 1, 2, \ldots, n$) when expressed in matrix notation are

$$\mathbf{i}_1 = \begin{pmatrix} 1 \\ 0 \\ \vdots \\ 0 \end{pmatrix}; \qquad \mathbf{i}_2 = \begin{pmatrix} 0 \\ 1 \\ \vdots \\ 0 \end{pmatrix}; \qquad \cdots; \qquad \mathbf{i}_n = \begin{pmatrix} 0 \\ 0 \\ \vdots \\ 1 \end{pmatrix} \tag{I.45}$$

If a vector $\mathbf{y}$ is related to a vector $\mathbf{x}$ by the relation $\mathbf{y} = \mathbf{Ax}$ and if the magnitude of $\mathbf{y}$ is to remain the same as the magnitude of $\mathbf{x}$, then we have

$$\mathbf{x}^\dagger \mathbf{x} = \mathbf{y}^\dagger \mathbf{y} = (\mathbf{Ax})^\dagger \mathbf{Ax} = \mathbf{x}^\dagger \mathbf{A}^\dagger \mathbf{Ax} \tag{I.46}$$

where equation (I.10) was used. It follows from equation (I.46) that $\mathbf{A}^\dagger \mathbf{A} = \mathbf{I}$ so that $\mathbf{A}$ must be unitary.

Eigenvalues

The *eigenvalues* λ of a square matrix $\mathbf{A}$ with elements a_{ij} are defined by the equation

$$\mathbf{Ax} = \lambda \mathbf{x} = \lambda \mathbf{Ix} \tag{I.47}$$

where the eigenvector $\mathbf{x}$ is the column matrix corresponding to an n-dimensional vector and λ is a scalar quantity. Equation (I.47) may also be written as

$$(\mathbf{A} - \lambda \mathbf{I})\mathbf{x} = \mathbf{0} \tag{I.48}$$

If the matrix $(\mathbf{A} - \lambda \mathbf{I})$ were to possess an inverse, we could multiply both sides of equation (I.48) by $(\mathbf{A} - \lambda \mathbf{I})^{-1}$ and obtain $\mathbf{x} = \mathbf{0}$. Since $\mathbf{x}$ is not a null matrix, the matrix $(\mathbf{A} - \lambda \mathbf{I})$ is singular and its determinant vanishes

$$\begin{vmatrix} a_{11} - \lambda & a_{12} & \cdots & a_{1n} \\ a_{21} & a_{22} - \lambda & \cdots & a_{2n} \\ \cdots & \cdots & \cdots & \cdots \\ a_{n1} & a_{n2} & \cdots & a_{nn} - \lambda \end{vmatrix} = 0 \tag{I.49}$$

The expansion of this determinant is a polynomial of degree n in λ, giving the *characteristic* or *secular equation*

$$\lambda^n + c_{n-1}\lambda^{n-1} + \cdots + c_1\lambda + c_0 = 0 \tag{I.50}$$

where c_i ($i = 0, 1, \ldots, n-1$) are constants. Equation (I.50) has n roots or eigenvalues λ_α ($\alpha = 1, 2, \ldots, n$). It is possible that some of these eigenvalues are degenerate.

The eigenvalues of a hermitian matrix are real. To prove this statement, we take the adjoint of each side of equation (I.47), apply equation (I.10), and note that $\mathbf{A} = \mathbf{A}^\dagger$

$$(\mathbf{Ax})^\dagger = \mathbf{x}^\dagger \mathbf{A}^\dagger = \mathbf{x}^\dagger \mathbf{A} = \lambda^* \mathbf{x}^\dagger \tag{I.51}$$

Multiplying equation (I.47) from the left by $\mathbf{x}^\dagger$ and equation (I.51) from the right by $\mathbf{x}$, we have

$$\mathbf{x}^\dagger \mathbf{Ax} = \lambda \mathbf{x}^\dagger \mathbf{x}$$

$$\mathbf{x}^\dagger \mathbf{Ax} = \lambda^* \mathbf{x}^\dagger \mathbf{x}$$

Since the magnitude of the vector $\mathbf{x}$ is not zero, we see that $\lambda = \lambda^*$ and λ is real.

The eigenvectors of a hermitian matrix with different eigenvalues are orthogonal. To prove this statement, we consider two distinct eigenvalues λ_1 and λ_2 and their corresponding eigenvectors $\mathbf{x}^{(1)}$ and $\mathbf{x}^{(2)}$, so that

$$\mathbf{A}\mathbf{x}^{(1)} = \lambda_1 \mathbf{x}^{(1)} \tag{I.52a}$$

$$\mathbf{A}\mathbf{x}^{(2)} = \lambda_2 \mathbf{x}^{(2)} \tag{I.52b}$$

If we multiply equation (I.52a) from the left by $\mathbf{x}^{(2)\dagger}$ and the adjoint of (I.52b) from the right by $\mathbf{x}^{(1)}$, we obtain

$$\mathbf{x}^{(2)\dagger}\mathbf{A}\mathbf{x}^{(1)} = \lambda_1 \mathbf{x}^{(2)\dagger}\mathbf{x}^{(1)} \tag{I.53a}$$

$$(\mathbf{A}\mathbf{x}^{(2)})^{\dagger}\mathbf{x}^{(1)} = \mathbf{x}^{(2)\dagger}\mathbf{A}^{\dagger}\mathbf{x}^{(1)} = \mathbf{x}^{(2)\dagger}\mathbf{A}\mathbf{x}^{(1)} = \lambda_2 \mathbf{x}^{(2)\dagger}\mathbf{x}^{(1)} \tag{I.53b}$$

where we have used equation (I.10) and noted that λ_2 is real. Subtracting equation (I.53b) from (I.53a), we find

$$(\lambda_1 - \lambda_2)\mathbf{x}^{(2)\dagger}\mathbf{x}^{(1)} = 0$$

Since λ_1 is not equal to λ_2, we see that $\mathbf{x}^{(1)}$ and $\mathbf{x}^{(2)}$ are orthogonal.

The eigenvector $\mathbf{x}^{(\alpha)}$ corresponding to the eigenvalue λ_α may be determined by substituting the value for λ_α into equation (I.47) and then solving the resulting simultaneous equations for the components $x_2^{(\alpha)}, x_3^{(\alpha)}, \ldots, x_n^{(\alpha)}$ of $\mathbf{x}^{(\alpha)}$ in terms of the first component $x_1^{(\alpha)}$. The value of the first component is arbitrary, but it may be specified by requiring that the vector $\mathbf{x}^{(\alpha)}$ be *normalized*, i.e.,

$$\mathbf{x}^{(\alpha)\dagger}\mathbf{x}^{(\alpha)} = |x_1^{(\alpha)}|^2 + |x_2^{(\alpha)}|^2 + \cdots + |x_n^{(\alpha)}|^2 = 1 \tag{I.54}$$

The determination of the eigenvectors for degenerate eigenvalues is somewhat more complicated and is not discussed here.

We may construct an $n \times n$ matrix $\mathbf{X}$ using the n orthogonal eigenvectors $\mathbf{x}^{(\alpha)}$ as columns

$$\mathbf{X} = \begin{pmatrix} x_1^{(1)} & x_1^{(2)} & \cdots & x_1^{(n)} \\ x_2^{(1)} & x_2^{(2)} & \cdots & x_2^{(n)} \\ \cdots & \cdots & \cdots & \cdots \\ x_n^{(1)} & x_n^{(2)} & \cdots & x_n^{(n)} \end{pmatrix} \tag{I.55}$$

and a diagonal matrix Λ using the n eigenvalues

$$\Lambda = \begin{pmatrix} \lambda_1 & 0 & \cdots & 0 \\ 0 & \lambda_2 & \cdots & 0 \\ \cdots & \cdots & \cdots & \cdots \\ 0 & 0 & \cdots & \lambda_n \end{pmatrix} \tag{I.56}$$

Equation (I.47) may then be written in the form

$$\mathbf{A}\mathbf{X} = \mathbf{X}\Lambda \tag{I.57}$$

The matrix $\mathbf{X}$ is easily seen to be unitary. Since the n eigenvectors are linearly independent, the matrix $\mathbf{X}$ is non-singular and its inverse $\mathbf{X}^{-1}$ exists. If we multiply equation (I.57) from the left by $\mathbf{X}^{-1}$, we obtain

$$\mathbf{X}^{-1}\mathbf{A}\mathbf{X} = \Lambda \tag{I.58}$$

This transformation of the matrix $\mathbf{A}$ to a diagonal matrix is an example of a *similarity transform*.

Trace

The trace $\text{Tr}\,\mathbf{A}$ of a square matrix $\mathbf{A}$ is defined as the sum of the diagonal elements

$$\text{Tr}\,\mathbf{A} = a_{11} + a_{22} + \cdots + a_{nn} \tag{I.59}$$

The operator Tr is a linear operator because

$$\text{Tr}(\mathbf{A} + \mathbf{B}) = (a_{11} + b_{11}) + (a_{22} + b_{22}) + \cdots + (a_{nn} + b_{nn}) = \text{Tr}\,\mathbf{A} + \text{Tr}\,\mathbf{B} \tag{I.60}$$

and

$$\text{Tr}(c\mathbf{A}) = ca_{11} + ca_{22} + \cdots + ca_{nn} = c\,\text{Tr}\,\mathbf{A} \tag{I.61}$$

The trace of a product of two matrices, which may or may not commute, is independent of the order of multiplication

$$\text{Tr}(\mathbf{AB}) = \sum_{i=1}^{n}\sum_{j=1}^{n} a_{ij}b_{ji} = \sum_{j=1}^{n}\sum_{i=1}^{n} b_{ji}a_{ij} = \text{Tr}(\mathbf{BA}) \tag{I.62}$$

Thus, the trace of the commutator $[\mathbf{A}, \mathbf{B}] \equiv \mathbf{AB} - \mathbf{BA}$ is equal to zero. Furthermore, the trace of a continued product of matrices is invariant under a cyclic permutation of the matrices

$$\text{Tr}(\mathbf{ABC}\cdots\mathbf{Q}) = \text{Tr}(\mathbf{BC}\cdots\mathbf{QA}) = \text{Tr}(\mathbf{C}\cdots\mathbf{QAB}) = \cdots \tag{I.63}$$

For a hermitian matrix, the trace is the sum of its eigenvalues

$$\text{Tr}\,\mathbf{A} = \sum_{\alpha=1}^{n} \lambda_{\alpha} \tag{I.64}$$

To demonstrate the validity of equation (I.64), we first take the trace of (I.58) to obtain

$$\text{Tr}(\mathbf{X}^{-1}\mathbf{AX}) = \text{Tr}\,\mathbf{\Lambda} = \sum_{\alpha=1}^{n} \lambda_{\alpha} \tag{I.65}$$

We then note that

$$\text{Tr}(\mathbf{X}^{-1}\mathbf{AX}) = \sum_{\alpha=1}^{n} (\mathbf{X}^{-1}\mathbf{AX})_{\alpha\alpha} = \sum_{\alpha=1}^{n}\sum_{i=1}^{n}\sum_{j=1}^{n} X_{\alpha i}^{-1} a_{ij} X_{j\alpha}$$

$$= \sum_{i=1}^{n}\sum_{j=1}^{n} a_{ij} \sum_{\alpha=1}^{n} X_{j\alpha} X_{\alpha i}^{-1} = \sum_{i=1}^{n}\sum_{j=1}^{n} a_{ij}\delta_{ij} = \sum_{i=1}^{n} a_{ii} = \text{Tr}\,\mathbf{A} \tag{I.66}$$

where $X_{j\alpha}\ (\equiv x_{j}^{(\alpha)})$ are the elements of $\mathbf{X}$. Combining equations (I.65) and (I.66), we obtain equation (I.64).

Appendix J

Evaluation of the two-electron interaction integral

In the application of quantum mechanics to the helium atom, the following integral I arises and needs to be evaluated

$$I = \int\int \frac{e^{-(\rho_1+\rho_2)}}{\rho_{12}}\, d\boldsymbol{\rho}_1\, d\boldsymbol{\rho}_2$$

$$= \int\cdots\int \frac{e^{-(\rho_1+\rho_2)}}{\rho_{12}}\, \rho_1^2\rho_2^2 \sin\theta_1 \sin\theta_2\, d\rho_1\, d\theta_1\, d\varphi_1\, d\rho_2\, d\theta_2\, d\varphi_2 \qquad (J.1)$$

where the position vectors $\boldsymbol{\rho}_i$ ($i = 1, 2$) have components ρ_i, θ_i, φ_i in spherical polar coordinates and where

$$\rho_{12} - |\boldsymbol{\rho}_2 - \boldsymbol{\rho}_1|$$

The distance ρ_{12} is related to ρ_1 and ρ_2 by the law of cosines

$$\rho_{12}^2 = \rho_1^2 + \rho_2^2 - 2\rho_1\rho_2 \cos\gamma \qquad (J.2)$$

where γ is the angle between $\boldsymbol{\rho}_1$ and $\boldsymbol{\rho}_2$ as shown in Figure J.1. The integration is taken over all space for each position vector.

The integral I may be evaluated more easily if we orient the coordinate axes so that the vector $\boldsymbol{\rho}_1$ lies along the positive z-axis as shown in Figure J.2. In that case, the angle γ between $\boldsymbol{\rho}_1$ and $\boldsymbol{\rho}_2$ is equal to the angle θ_2. If we define $\rho_>$ as the larger and $\rho_<$ as the smaller of ρ_1 and ρ_2 and define s by the ratio

$$s \equiv \frac{\rho_<}{\rho_>}$$

so that $s \leq 1$, then equation (J.2) may be expressed in the form

$$\frac{1}{\rho_{12}} = \frac{1}{\rho_>}(1 + s^2 - 2s \cos\theta_2)^{-1/2} \qquad (J.3)$$

At this point, we may proceed in one of two ways, which are mathematically equivalent. In the first procedure, we note that from the generating function (E.1) for Legendre polynomials P_l, equation (J.3) may be written as

$$\frac{1}{\rho_{12}} = \frac{1}{\rho_>}\sum_{l=0}^{\infty} P_l(\cos\theta_2)s^l$$

The integral I in equation (J.1) then becomes

341

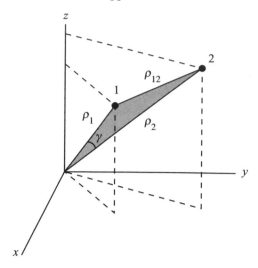

Figure J.1 Distance between two particles 1 and 2 and their respective distances from the origin.

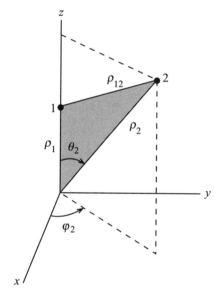

Figure J.2 Rotation of the coordinate axes in Figure J.1 so that the z-axis lies along ρ_1.

$$I = \sum_{l=0}^{\infty} \int \int \frac{e^{-(\rho_1+\rho_2)}}{\rho_>} s^l \rho_1^2 \rho_2^2 \, d\rho_1 \, d\rho_2 \int_0^\pi P_l(\cos\theta_2) \sin\theta_2 \, d\theta_2 \int_0^\pi \sin\theta_1 \, d\theta_1 \int_0^{2\pi} d\varphi_1 \int_0^{2\pi} d\varphi_2$$

The integrals over θ_1, φ_1, and φ_2 are readily evaluated. Since $P_0(\mu) = 1$, we may write the integral over θ_2 as

$$\int_0^\pi P_l(\cos\theta_2)\,\sin\theta_2\,d\theta_2 = \int_{-1}^1 P_l(\mu)P_0(\mu)\,d\mu = 2\delta_{l0}$$

where equations (E.18) and (E.19) have been introduced. Thus, only the term with $l = 0$ in the summation does not vanish and we have

$$I = 16\pi^2\int\int\frac{e^{-(\rho_1+\rho_2)}}{\rho_>}\rho_1^2\rho_2^2\,d\rho_1\,d\rho_2 \qquad (J.4)$$

In the second procedure, we substitute equation (J.3) directly into (J.1) and evaluate the integral over θ_2

$$\int_0^\pi\frac{\sin\theta_2}{(1+s^2-2s\cos\theta_2)^{1/2}}\,d\theta_2 = \frac{1}{s}(1+s^2-2s\cos\theta_2)^{1/2}\Big|_0^\pi$$

$$= \frac{1}{s}[(1+s^2+2s)^{1/2} - (1+s^2-2s)^{1/2}]$$

$$= \frac{1}{s}[(1+s) - (1-s)] = 2$$

The integrals over θ_1, φ_1, and φ_2 are the same as before and equation (J.4) is obtained.

Since $\rho_>$ is the larger of ρ_1 and ρ_2, the integral I in equation (J.4) may be written in the form

$$I = 16\pi^2\int_0^\infty e^{-\rho_1}\rho_1^2\left[\frac{1}{\rho_1}\int_0^{\rho_1}e^{-\rho_2}\rho_2^2\,d\rho_2 + \int_{\rho_1}^\infty e^{-\rho_2}\rho_2\,d\rho_2\right]d\rho_1$$

$$= 16\pi^2\int_0^\infty e^{-\rho_1}\rho_1\{[2 - (\rho_1^2 + 2\rho_1 + 2)e^{-\rho_1}] + \rho_1(\rho_1 + 1)e^{-\rho_1}\}\,d\rho_1 = 16\pi^2(\tfrac{5}{8} + \tfrac{5}{8})$$

Accordingly, the final result is

$$I = \int\int\frac{e^{-(\rho_1+\rho_2)}}{\rho_{12}}\,d\boldsymbol{\rho}_1\,d\boldsymbol{\rho}_2 = 20\pi^2 \qquad (J.5)$$

Selected bibliography

Applied mathematical methods

Three widely used compendia of applied mathematics directed to the needs of chemistry and physics are the following.

G. B. Arfken and H. J. Weber (1995) *Mathematical Methods for Physicists*, 4th edition (Academic Press, San Diego).

M. L. Boas (1983) *Mathematical Methods in the Physical Sciences*, 2nd edition (John Wiley & Sons, New York).

K. F. Riley, M. P. Hobson, and S. J. Bence (1998) *Mathematical Methods for Physics and Engineering* (Cambridge University Press, Cambridge).

Undergraduate physical chemistry

The following undergraduate texts discuss the historical development of quantum concepts and introduce the elements of quantum mechanics.

R. A. Alberty and R. J. Silbey (1996) *Physical Chemistry*, 2nd edition (John Wiley & Sons, New York).

P. W. Atkins (1998) *Physical Chemistry*, 6th edition (Oxford University Press, Oxford; W. H. Freeman, New York).

D. A. McQuarrie and J. D. Simon (1997) *Physical Chemistry: A Molecular Approach* (University Science Books, Sausalito, CA).

History and philosophy of quantum theory

J. Baggott (1992) *The Meaning of Quantum Theory* (Oxford University Press, Oxford).

M. Jammer (1974) *The Philosophy of Quantum Mechanics* (John Wiley & Sons, New York).

M. Jammer (1966)*The Conceptual Development of Quantum Mechanics* (McGraw-Hill, New York).

Some 'classic' quantum mechanics texts

Emphasis on applications to chemistry.

H. Eyring, J. Walter, and G. E. Kimball (1944) *Quantum Chemistry* (John Wiley & Sons, New York).

L. Pauling and E. B. Wilson (1935) *Introduction to Quantum Mechanics: With Applications to Chemistry* (McGraw-Hill, New York; reprinted by Dover, New York, 1985).

Emphasis on applications to physics.
D. Bohm (1951) *Quantum Theory* (Prentice-Hall, New York).
P. A. M. Dirac (1947) *The Principles of Quantum Mechanics*, 3rd edition (Oxford University Press, Oxford) and 4th edition (Oxford University Press, Oxford, 1958). Except for the last chapter, these two editions are virtually identical.
H. A. Kramers (1957) *Quantum Mechanics* (North-Holland, Amsterdam).
L. D. Landau and E. M. Lifshitz (1958) *Quantum Mechanics: Non-Relativistic Theory* (Pergamon, London; Addison-Wesley, Reading, MA).

Some recent quantum mechanics texts

Emphasis on applications to chemistry.
P. W. Atkins and R. S. Friedman (1997) *Molecular Quantum Mechanics*, 3rd edition (Oxford University Press, Oxford).
I. N. Levine (1991) *Quantum Chemistry*, 4th edition (Prentice-Hall, Englewood Cliffs, NJ).
F. L. Pilar (1990) *Elementary Quantum Chemistry*, 2nd edition (McGraw-Hill, New York).
J. Simons and J. Nichols (1997) *Quantum Mechanics in Chemistry* (Oxford University Press, New York).

Emphasis on applications to physics.
B. H. Bransden and C. J. Joachain (1989) *Introduction to Quantum Mechanics* (Addison Wesley Longman, Harlow, Essex).
C. Cohen-Tannoudji, B. Diu, and F. Laloë (1977) *Quantum Mechanics*, volumes I and II (John Wiley & Sons, New York; Hermann, Paris).
D. Park (1992) *Introduction to the Quantum Theory*, 3rd edition (McGraw Hill, New York).
J. J. Sakurai (1994) *Modern Quantum Mechanics*, revised edition (Addison-Wesley, Reading, MA).

Angular momentum

The following books develop the quantum theory of angular momentum in more detail than this text.
A. R. Edmonds (1960) *Angular Momentum in Quantum Mechanics*, 2nd edition (Princeton University Press, Princeton).
M. E. Rose (1957) *Elementary Theory of Angular Momentum* (John Wiley & Sons, New York; reprinted by Dover, New York, 1995).
R. N. Zare (1988) *Angular Momentum: Understanding Spatial Aspects in Chemistry and Physics* (John Wiley & Sons, New York).

Atoms and atomic spectra

H. A. Bethe and E. E. Salpeter (1957) *Quantum Mechanics of One- and Two-Electron Atoms* (Springer, Berlin; Academic Press, New York; reprinted by Plenum, New York, 1977). A comprehensive non-relativistic and relativistic treatment of the hydrogen and helium atoms with and without external fields.

E. U. Condon and G. H. Shortley (1935) *The Theory of Atomic Spectra* (Cambridge University Press, Cambridge). A 'classic' text on the application of quantum theory to atomic spectra.

More advanced applications of quantum mechanics

D. R. Bates (ed) (1961, 1962) *Quantum Theory*, volumes I, II, and III (Academic Press, New York and London). A compendium of articles covering the principles of quantum mechanics and a wide variety of applications.

S. Kim (1998) *Group Theoretical Methods and Applications to Molecules and Crystals* (Cambridge University Press, Cambridge).

I. N. Levine (1975) *Molecular Spectroscopy* (John Wiley & Sons, New York). A survey of the theory of rotational, vibrational, and electronic spectroscopy of diatomic and polyatomic molecules and of nuclear magnetic resonance spectroscopy.

N. F. Mott and H. S. W. Massey (1965) *The Theory of Atomic Collisions*, 3rd edition (Oxford University Press, Oxford). The standard reference for the quantum-mechanical treatment of collisions between atoms.

G. C. Schatz and M. A. Ratner (1993) *Quantum Mechanics in Chemistry* (Prentice-Hall, Englewood Cliffs, NJ). An advanced text emphasizing molecular symmetry and rotations, time-dependent quantum mechanics, collisions and rate processes, correlation functions, and density matrices.

A. J. Stone (1991) *The Theory of Intermolecular Forces* (Oxford University Press, Oxford). An extensive survey of the applications of quantum mechanics to determine the forces between molecules.

Compilations of problems in quantum mechanics

I. I. Gol'dman and V. D. Krivchenkov (1961) *Problems in Quantum Mechanics* (Pergamon, London; Addison-Wesley, Reading, MA; reprinted by Dover, New York, 1993).

C. S. Johnson, Jr. and L. G. Pedersen (1974) *Problems and Solutions in Quantum Chemistry and Physics* (Addison-Wesley, Reading, MA; reprinted by Dover, New York, 1987).

G. L. Squires (1995) *Problems in Quantum Mechanics: with Solutions* (Cambridge University Press, Cambridge).

Index